Observers of the Aurora Borealis in Europe

Series Editor
Denis-Didier Rousseau

Observers of the Aurora Borealis in Europe

Journey into the Learned World of the Enlightenment

Eric Chassefière

WILEY

First published 2023 in Great Britain and the United States by ISTE Ltd and John Wiley & Sons, Inc.

Apart from any fair dealing for the purposes of research or private study, or criticism or review, as permitted under the Copyright, Designs and Patents Act 1988, this publication may only be reproduced, stored or transmitted, in any form or by any means, with the prior permission in writing of the publishers, or in the case of reprographic reproduction in accordance with the terms and licenses issued by the CLA. Enquiries concerning reproduction outside these terms should be sent to the publishers at the undermentioned address:

ISTE Ltd
27-37 St George's Road
London SW19 4EU
UK

www.iste.co.uk

John Wiley & Sons, Inc.
111 River Street
Hoboken, NJ 07030
USA

www.wiley.com

© ISTE Ltd 2023

The rights of Eric Chassefière to be identified as the author of this work have been asserted by him in accordance with the Copyright, Designs and Patents Act 1988.

Any opinions, findings, and conclusions or recommendations expressed in this material are those of the author(s), contributor(s) or editor(s) and do not necessarily reflect the views of ISTE Group.

Library of Congress Control Number: 2022947541

British Library Cataloguing-in-Publication Data
A CIP record for this book is available from the British Library
ISBN 978-1-78630-792-7

Contents

Introduction ix

Chapter 1. The Aurora Borealis Issue of the Affirmation of the Cartesian Mechanism and the Dispute Between Paris and Montpellier: The French Choice 1

  1.1. Introduction 1
  1.2. The two main systems of the aurora borealis 2
    1.2.1. Halley's system 2
    1.2.2. Mairan's system 5
  1.3. History of the aurora borealis in the volumes of the Académie Royale
  des Sciences between 1716 and 1733 8
    1.3.1. The silence on Halley's system in *Mémoires* and *Histoire* 8
    1.3.2. The memoir refused by the Parisian Academy of François de Plantade . . 13
  1.4. The Montpellier actors: François de Plantade and the Société Royale
  des Sciences 20
    1.4.1. François de Plantade, founder of the Société Royale de Montpellier 20
    1.4.2. The Société Royale des Sciences de Montpellier 21
  1.5. The Parisian actors: Bernard le Bovier de Fontenelle and
  Jean-Jacques Dortous de Mairan, the Académie Royale des Sciences 26
    1.5.1. The Académie Royale des Sciences 26
    1.5.2. The permanent secretary Bernard le Bovier de Fontenelle 30
    1.5.3. Jean-Jacques Dortous de Mairan 37
  1.6. The London actors: Hans Sloane and Edmond Halley, the Royal Society . . . 43
    1.6.1. Hans Sloane 43
    1.6.2. Edmond Halley 45
    1.6.3. The Royal Society and its relations with the Académie Royale
    des Sciences 49
  1.7. Discussion of the reasons for rejecting Plantade's submission 51

Chapter 2. Joseph-Nicolas Delisle: Grandeur and Vicissitudes of a Newtonian Scientist with Thwarted Ambitions 55

2.1. Introduction 55
2.2. Delisle in the period before his departure for Russia (1710–1725) 61
    2.2.1. Delisle's beginnings in astronomy and optics, a Newtonian 61
    2.2.2. Delisle's setbacks at the Académie Royale des Sciences 71
    2.2.3. Delisle's great project: Histoire Céleste. 83
    2.2.4. Epilogue concerning the Parisian period 89
2.3. The invitation to St. Petersburg and Delisle's Russian period
(1726–1747). 90
    2.3.1. The cartographic objective of Delisle's mission 90
    2.3.2. Delisle's means at the St. Petersburg Observatory 97
2.4. Brief synthesis of Delisle's scientific trajectory. 109
2.5. Conclusion 112

Chapter 3. The Creation Ex-nihilo and the Beginnings of the Imperial Russian Academy of Sciences: The Influence of Christian Wolff 115

3.1. Introduction 115
3.2. The foundation of the Imperial Academy of Sciences in St. Petersburg 117
    3.2.1. Historical context 117
    3.2.2. Peter the Great's Imperial Academy of Sciences project 120
    3.2.3. The birth of astronomy in Russia 122
3.3. Christian Wolff, the aurora borealis and their first observers at the
Academy of Sciences in St. Petersburg. 125
    3.3.1. Historical context 125
    3.3.2. Christian Wolff's conference 126
    3.3.3. The quartet of aurora observers at the Academy of Sciences of
St. Petersburg. 131
    3.3.4. The rejection of aurora observations by Mayer 135
    3.3.5. Euler's physical–mathematical explanation 143
    3.3.6. Mayer's philosophical position and possible reasons for his
abandonment of aurora observation 146
3.4. The Imperial Academy of Sciences of St. Petersburg 149
    3.4.1. The setting up of the Academy 149
    3.4.2. The clerical and noble opposition 151
    3.4.3. Wolffians versus Newtonians 155
    3.4.4. The problems of the functioning of the Academy in the decades
1730–1740 161
    3.4.5. The regulation of 1748 refounding the Academy 164
3.5. Conclusion 167

Chapter 4. Anders Celsius and the European Observation Networks, Setting Up a Science Society and an Astronomical Observatory in Uppsala 171

4.1. Introduction 171
4.2. The life of Celsius 173
   4.2.1. The first years 173
   4.2.2. The European journey 176
   4.2.3. Maupertuis' expedition in Lapland 179
   4.2.4. The last few years. 181
4.3. Three European networks for the observation of natural phenomena 184
   4.3.1. The observations of the aurora borealis around de Mairan 185
   4.3.2. Monitoring the variations of the magnetic needle according to Anders Celsius 190
   4.3.3. Thermometry and meteorological records around Joseph-Nicolas Delisle. 199
4.4. The Royal Society of Uppsala and Celsius' legacy 211
   4.4.1. Historical context of the Enlightenment in Sweden 211
   4.4.2. Birth and development of the Royal Society of Sciences in Uppsala. . . . 214
   4.4.3. Relations between the Royal Society and the University 219
   4.4.4. Celsius' legacy 222
4.5. Conclusion 228

Chapter 5. Genesis of the Academies of Bologna and Berlin, the Involvement of Women in Astronomy and the Gender Issue. 231

5.1. Introduction 231
5.2. Three examples of "astronomical households" 236
   5.2.1. The Kirchs: an artisanal-type household inspired by the guild tradition . . 238
   5.2.2. The Manfredis: a household with a humanistic coloration inherited from the Renaissance 247
   5.2.3. The Delisle family: an artisanal household where women took care of the family scientific heritage. 255
5.3. Two examples of astronomical institutions: the academies of Bologna and Berlin and their observatories. 259
   5.3.1. The Academy and the Bologna Observatory 262
   5.3.2. The Academy and the Observatory of Berlin. 270
5.4. Astronomical households, institutions and gender in Bologna and Berlin . . . 280
5.5. Conclusion 287

Conclusion	289
Appendix	301
References	313
Index	331

Introduction

The subject of this book is not the observation and the scientific interpretation of the aurora borealis, which we have dealt with in a previous book (Chassefière 2021a), but the human, institutional and philosophical context in which the principal scientists involved in the observation of the phenomenon evolved in the first half of the 18th century. The aurora borealis by itself, as a physical phenomenon, was only approached when the understanding of the scientific fact enabled clarifying the human or philosophical context in which it is inscribed. This book is dedicated to the observers of the Northern Lights, for the most part astronomers whose main interests were found in astronomy and its applications to cartography, and to their individual and collective trajectories, in an institutional and philosophical space undergoing profound change. The choice of the aurora borealis as the main theme of the narrative is due to the unifying character of the phenomenon, which, because of its very high altitude, can only be correctly characterized by measurements made simultaneously in places thousands of kilometers away, a scale that is that of Europe, and not of each of its states. But before starting our European journey in the academic world of the Enlightenment, it is useful to review the human and scientific landscape that constitutes its framework.

The aurora borealis was a phenomenon still relatively unknown at the beginning of the 18th century. The abbot Pierre-Nicolas Bertholon provides a very long historical and scientific analysis of the phenomenon in the article "Aurore Boréale" of the *Dictionnaire de Physique*[1] (1793). He quotes Aristotle, witness of an aurora in Macedonia, who compared the appearance of the aurora borealis, when

---

1 The *Dictionnaire de Physique*, published in 1793, is the first volume of a set of four encyclopedias of which the second, third and fourth volumes are called *Encyclopédie méthodique*, published, respectively, in 1816, 1819 and 1822.

it was extended, to "the blaze of a campaign whose bale is burned". Aristotle gave pictorial names to the jets of light observed on this occasion: "lighted firebrands, torches, lamps, burning beams". According to him, the most common colors of the aurora borealis were "purple, bright red and the color of blood". Bertholon reports numerous testimonies from the past, from Antiquity to the very beginning of the 18th century, mentioning scenes of panic among the populations, and frequent associations of the aurora borealis with a fatal omen, announcing wars and desolation. In particular, the fear of the Apocalypse, which seized Europe at the end of the 15th century, and lasted until the middle of the 17th century, conferred to the aurora borealis and to other unexplained phenomena, such as rainbows, solar or lunar halos, or flying fires (meteoroids entering the atmosphere) a particularly disturbing character (Schröder 2005). Superstitions in this area were still alive at the turn of the 18th century. The astronomer Maria Winkelmann-Kirch, who observed the aurora borealis from Berlin in March 1707, wondered in a letter to Gottfried Wilhelm Leibniz in November of the same year: "I am not sure what nature was trying to tell us" (Schiebinger 1987, p. 183). The sustained resumption of the aurora borealis in 1716, after a long period of interruption (attributable to low solar activity, identifiable by the smaller number of spots visible on the disk: the so-called "Maunder Minimum", which lasted from 1645 to 1715), with the exception of the timid resumption of 1686 and 1707, marked an important turning point in the popular understanding of unexplained atmospheric phenomena. In Halle, Germany, the philosopher and physicist Christian Wolff, one week after the event, gave a conference on the subject at the request of the citizens of his city, to provide his interpretation of the aurora borealis, a natural phenomenon according to him, completely explicable without calling upon any divine intervention (Schröder 2005).

The irruption of the aurora borealis phenomenon aroused a considerable craze among the scientists of the time, and this all the more so as it was still poorly documented, and the only rational explanations of the phenomenon were still strongly inspired by Aristotelian meteorology. This one attributed the origin of the aurora borealis and other atmospheric luminous phenomena to dry vapors raised in the atmosphere by the heat of the sun. As opposed to wet vapor, remaining close to the Earth's surface and giving rise to rain, dry vapor rises and extends to the boundary between the Earth's sphere and the rotating celestial sphere, contact with the moving ether in movement provoking its ignition: "We must think of what we just called fire as being spread round the terrestrial sphere on the outside like a kind of fuel, so that a little motion often makes it burst into flame just as smoke does: for flame is the ebullition of a dry exhalation" (Aristotle 2009, Book I, Part 4). Galileo, who probably coined the term aurora borealis to designate the phenomenon, wrote in a 1619 text signed by his pupil Mario Guiducci that the phenomenon "has no other origin than that a part of the vapor-laden air surrounding the Earth is for some reason unusually rarefied, and being extraordinarily sublimated has risen above the cone of the Earth's shadow so that its upper parts are struck by the sun and made able to reflect its splendor to us, thus forming for us (this northern dawn – questa boreale aurora)" (Siscoe (1986), quote taken from Drake and O'Malley (1960, p. 53)). Thus, while he considered, like Aristotle, that the material cause was due to vapors emanating from the Earth, he attributed the light of the aurora borealis, not to an inflammation, but to the reflection by these vapors of the sunlight. A similar explanation was provided by Pierre Gassendi, witness of the aurora borealis in 1621, last insisting on the planetary character of the phenomenon, related, according to him, to a particular interior arrangement of the terrestrial globe expelling vapors through a significant part of its surface (Bernier 1684, p. 266). René Descartes in his essay *Les Météores* published in 1637 advanced three possible explanations of the aurora borealis:

> The first is that there are many clouds in the air, sufficiently small to be taken for so many soldiers; and falling onto one another, these enclose enough exhalations to cause a quantity of small flashes, and to throw small fires, and perhaps to cause small noises to be heard, by which means these soldiers seem to do battle. The second cause is also that there are such clouds in the air; but instead of falling on one another, they receive their light from the fires and lightning flashes of some large storm, which occurs so far away that it cannot be perceived at that location. And the third cause is that these clouds, or some other more southern ones, from which they receive their light, are so high that the rays of the sun reach right to them (Descartes 1965, p. 331).

The first explanation, which assimilates the aurora borealis to a small storm, is closely inspired by the Aristotelian vision, although differing on the mechanism of the inflammation, associated in Descartes' work to a movement occurring within the atmosphere. The third explanation is directly inspired by the idea of Galileo and Gassendi, the second being an intermediate hypothesis, the vapors transmitting a light which is not that of the sun, but comes from a distant storm.

Bertholon detailed no less than 12 possible explanations for the aurora borealis, proposed during the 18th century. Edmond Halley, the first to take a scientific stand after the marked resumption of the aurora in 1716, published in 1717 an explanation based on the idea that the aurora borealis was the result of magnetic matter circulating in the great Earth magnet, inspired by Descartes' magnet theory (Halley 1717). After having exposed, in the same article, the theory of Aristotelian inspiration of exhalations igniting in the atmosphere, which he thought could not explain the considerable geographical extension of the aurora, he advanced his own explanation according to which it was the magnetic subtle matter, which he supposed to leave by the boreal pole of the Earth and circulate in the ether towards its southern pole, which generates the luminous phenomena of the aurora at high latitude. Jean-Jacques Dortous de Mairan, a few years later, conceived a system of the aurora based on the idea that the subtle matter at the origin of the Northern Light came from the sun, whose atmosphere, to which he attributed with Jean-Dominique Cassini the origin of the zodiacal light and which could extend beyond the orbit of the Earth, precipitated in the subtle air constituting the upper atmosphere of the Earth. By mixing with the atmosphere, the solar matter stratified, and produced the luminous structures of the aurora. Mairan's treatise on the aurora borealis, completed in 1731, was published two years later (de Mairan 1733). It is around these two great systems, as well as that of the ignition of the dry exhalations inherited from the meteorology of Aristotle which remained defended by many scientists until the middle of the century, that the scientific thought of the time developed as regards sciences of the atmosphere and meteorology. A fourth system was proposed by Leonhard Euler about 15 years later (Euler 1746), using an effect of the thrust of the sun's rays on the particles of the upper atmosphere, these rays being supposed to expel particles towards space in the same way that they push back, according to Euler, the envelopes of the comets to form their tails. Bertholon mentioned other theories developed at the same time, inspired by the ideas of Galileo or Gassendi who saw in the sun the source of auroral radiation. Thus, Abbot Hell of the Observatory of Vienna supposed an effect of the particles of ice in suspension in the atmosphere reflecting and refracting the rays of the sun or of the moon as it occurred in the parhelias. Others appealed to the snow and ice cover of the polar regions, which would reflect towards the upper layers of the atmosphere the low-angled light of the sun (placed just below the horizon), these layers reflecting in their turn this light towards the observer located at the surface of the Earth. But, in the second half of the 18th century, all these systems were supplanted by the hypothesis of an electrical origin for the auroras, this on the basis of the similarities of texture and color of the auroras with the lights generated within previously electrified enclosures. John Canton, in 1753, published the first electrical model of the aurora borealis, defending the idea that the aurora was produced by a discharge between positively and negatively electrified clouds passing through the upper atmosphere, whose electrical resistance was lower (Canton 1753). Bertholon also cited the work of Benjamin Franklin and his own, on the hypothesis of an effect of electricity. It was not until the end of the 19th century, with the discovery of the precipitation of solar particles, that the true mechanism, involving the solar wind particles, the Earth's magnetic field and the circulation of solar particles along the field lines, came to synthesize the explanations proposed by Halley, Mairan and Canton.

In his treatise of 1733, Mairan argued that, if his system was correct, namely if the aurora borealis were the result of the precipitation of matter from the solar atmosphere in the terrestrial atmosphere, the aurora must be more frequent when the Earth is closer to the sun, that is, in the vicinity of its perihelion on its orbit around the sun, around the winter solstice, and when the northern hemisphere of the Earth points in the direction of the movement of the Earth on its orbit, in the period between the summer solstice and the winter solstice, catching the solar matter head on. He believed that future observations of the aurora borealis would test the validity of his hypothesis. In the second edition of the treatise, published in 1754, applying the method traced 20 years earlier, he reviewed the existing observations of aurora borealis, in particular those which intervened since the publication of his first treaty. His goal was to statistically analyze the frequencies of appearance of auroras in the vicinity of the perihelion (December–January) and aphelion (June–July) of the Earth on its orbit around the sun within the framework of his system, a system from which, according to him, "results the constant connection of the Aurora Borealis & of its appearances, with this luminous fluid or illuminated by the Sun, which extending sometimes until the Earth & beyond, must by the laws of the gravitation, fall in the terrestrial Atmosphere, & produce there this Phenomenon". He began by analyzing the auroras listed by a professor of philosophy in Helmstadt, Germany, Jean-Nicolas Frobès, who published in 1739 a catalog of 796 auroras observed between 500 and 1739, of which three quarters were later than 1716, thus describing the advantage of the statistical approach which, by the multiplication of observations, as well as their results as their interpretations, smoothed and finally canceled the biases related to the measurements and individual appreciations:

> It is that the principle of frequency which we are talking about being true, all these differences disappear on the great masses of time and numbers; everything finally compensates itself according to the Doctrine of Chance, and the sought-after relationship manifests itself. If the way of seeing or judging, of an Observer, of a Historian, his attentions, his prejudices, his superstition even, making him multiply or omit certain Phenomena, contrary dispositions in another will make him reject what this one has admitted, & retain what he had rejected […] If the long twilights of Summer, & longer in one climate than in the other, cause us to lose some small Aurora Borealis, the dark nights of Winter, & whose length is relative to these climates in inverse proportion to the days & twilights, rob us of others of the same kind. […]
>
> It is thus from this very diversity, of times, of countries & of writers, & from these great masses of years, of observations & of numbers, that our inductions on the correspondence of which it is a question, will draw their greatest forces (de Mairan 1754, pp. 486–487).

He thus analyzed the series of observations which were delivered to him by several observers of aurora borealis from various European countries, whose references he provides. Two hundred and thirty-three observations came from Joseph-Nicolas Delisle, director of the Imperial Observatory of St. Petersburg, that in the following we will call for simplicity St. Petersburg Observatory, where they were realized by himself and his colleagues in the Observatory, Friedrich Christoph Mayer and Georg Wolfgang Krafft. The 57 aurora borealis observed by the younger brother of Delisle, Louis de La Croyère, also posted in St. Petersburg, during his trip to Siberia in 1727–1730, were not used by Mairan in his analysis because they were taken, according to him, at too high a latitude, in extreme conditions of the diurnal cycle of sunshine likely to bias the statistics of the observed auroras. Two hundred and twenty-four observations of auroras were provided by Anders Celsius, director of the Astronomical Observatory of Uppsala, some observed by him, others by his Swedish colleagues. One hundred and six were made by Christfried Kirch, director of the Royal Observatory of Berlin and 91 by Johann Friedrich Weidler from Wittenberg, a city located about 100 km southwest of Berlin. Eighty-eight observations were made by Eustachio Zanotti, successor of Eustachio Manfredi, another aurora observer, at the direction of the Astronomical Observatory of Bologna and Jacopo Bartolomeo Beccari of the same city. Other observations were published in London in 1749 by Thomas Short, listing 148 aurora borealis. Among all these observations, Mairan eliminated those which corresponded to the same aurora seen by various observers, which occurred frequently taking into account the very great height of the phenomenon, on average 175 leagues, that is, 700 km, which meant that it could be seen from very far, the aurorae taking place most often between 400 km and 1,200 km of altitude (de Mairan 1754, pp. 433–434). Examining thus the cases of more than 2,000 aurorae, in great majority posterior to 1716, Mairan deduced that the aurorae were more frequent during the winter months, when the Earth is closer to the sun, while noting the bias which could come from the fact that summer is also the period during which the nights are the shortest, the phenomenon being then likely to be masked by the greater ambient luminosity.

The question of the capacity of a system to be tested by the collection of observations over time, an essential asset of Mairan's aurora borealis theory, was important, and was part of a debate that shook the entire first half of the 18th century around the question of the spirit of the system, in this period of reversal of the deductive mechanistic approach inherited from Descartes to the benefit of the inductive approach based on the results of observation and experience alone, as advocated by the proponents of English empiricism led by Isaac Newton. As soon as he took office as Permanent Secretary of the Académie Royale des Sciences in 1699, Bernard le Bovier de Fontenelle in his foreword to the *Histoire de l'Académie*

Royale des Sciences[2] (Fontenelle 1699), which we will refer to in the following as simply the *Histoire*, advocated the patient accumulation of observations as a prerequisite to the elaboration of any system, "because systematic physics must wait to build buildings until experimental physics is in a position to provide it with the necessary materials". For Fontenelle, as for Mairan, who expresses it perfectly in his foreword to the dissertation on ice, *Dissertation sur la Glace* (de Mairan 1749), the system is indeed the accomplished form of knowledge (Mazauric 2007), but it can in no way be posited a priori, since it must on the contrary result from the observation of phenomena. To the "spirit of system" inherited from Descartes, which he challenged, Jean le Rond D'Alembert opposes in his preliminary speech of the *Encyclopédie* of 1751 a "systematic spirit" consisting of "reducing, as far as possible, a large number of phenomena to a single one that can be considered as the principle" (D'Alembert 1893, p. 23), in which the system is not situated upstream of the phenomena that we seek to explain, but at the articulation between these phenomena, which we must compare and study in a reflective manner. This question of systems forms an important background of the scientific life of the time, as we relate it in this work.

Almost all of the aurora observers quoted by Mairan in his memoir of 1754, for the most part also involved in daily measurements of meteorological parameters (temperature, pressure, wind speed), were in regular contact with each other, as we will describe in the pages of this book. Thus, Joseph-Nicolas Delisle maintained a close correspondence with Kirch and Celsius, and also exchanged regularly with Zanotti and Weidler, during the long years that he spent in St. Petersburg. Celsius, during his European journey of 1732–1737, which he concluded by his participation in the expedition led by Pierre Louis Moreau de Maupertuis in Lapland to measure the shape of the Earth, made observations with Kirch in Berlin, then with Manfredi in Bologna, went to Wittenberg to meet Weidler, then spent time in Paris, where he met Delisle's sister (the latter was then in St. Petersburg), who occasionally interacted with the Académie Royale des Sciences and the Collège Royal of which her brother remained a member despite his absence. In Padua, Celsius met Giovanni Poleni and transmitted to him his observations of aurora borealis. Poleni, a partisan

---

2 Each volume, usually one per year, of the proceedings of the Académie Royale des Sciences (French Royal Academy of Sciences), consists of a series of memoirs (the *Mémoires de Mathématiques* and the *Mémoires de Physique* of the Académie Royale des Sciences de Paris (Royal Academy of Sciences of Paris) preceded by a section composed of articles by Fontenelle introducing and popularizing the science presented in the memoirs, which will be called the *Histoire* without further precision in the rest of this book), and followed by a section devoted to the Eulogies of the deceased academicians.

of Mairan's system, aware of the latter's expectations, tabled the observations of his Swedish colleague and transmitted them to Mairan, who then met directly with Celsius during his time in Paris. Delisle realized thermometers that he sent throughout Europe, in order to allow temperature measurements that could be compared from one country to another. In the early 1740s, Celsius, with his associate Olof Hiorter, observed magnetic needle agitation during certain aurora borealis, and the needle's declination measurements, hitherto mostly dedicated to mapping declination in sight to improve navigation, also taking an interest in the interpretation of the aurora borealis.

These observers, on the whole, did not take explicit sides on the system of the aurora which they favored, and even, for some, like Celsius, refrained from doing it, limiting themselves to the systematic and patient recording of the data with a view to future scientific work, work of which the statistical analysis carried out by Mairan in 1754 is the perfect illustration. We know that Weidler, as well as François de Plantade, who observed auroras in Montpellier, were supporters of Halley's system. Mayer, in St. Petersburg, favored the hypothesis of the inflammation of sulfurous materials. Poleni, who observed some auroras from Padua, and many other scientists in Europe were seduced by Mairan's cosmo-atmospheric system. They were all astronomers and, as such, dedicated most of their time to the observation of planets and stars, especially their conjunctions (eclipses of the moon and the sun, occultations of stars by the moon or of its satellites by Jupiter, transits of Mercury...), as well as comets in a period where the verification of the conformity of the trajectories of comets to the laws of Newton was a major subject for Newtonians as Delisle or Celsius. These astronomers, in their observatories, were also in charge of meteorological observation. This included the aurora borealis, which until then had been considered a meteorological phenomenon in the Aristotelian tradition. Because of the long nights spent observing, they were led to witness the aurora borealis, which were particularly frequent at the latitudes of Uppsala or St. Petersburg. Their observation instruments allowed them to angularly characterize the structures of the aurora borealis (arcs, jets, etc.) relative to the surface of the Earth and to the direction of the geographical north, leading in particular to estimates of the height of the phenomenon, thanks to parallax measurements made from distant observations of the same structure, or to idealized models, such as Mayer's, allowing the height of the aurora to be estimated from a measurement in a single point. The estimated heights were considerable (several hundreds, even more than a thousand kilometers), suggesting an atmosphere much higher than it was believed in the previous century (typically less than 100 kilometers; see Chassefière 2021a, Chapter 7), and leading some scholars like Euler to attribute to it a purely cosmic origin.

These aurora observers all had important responsibilities in the then emerging institutional system of scientific academies and observatories in European countries and in Russia. Celsius was the one who, after his European trip, took over the Royal Society of Sciences of Uppsala and its Observatory project. Manfredi was the founder, at the end of the previous century, of the Accademia degli Inquieti on which the Academy of Sciences of the Institute of Bologna was built and its Observatory, whose directors were Manfredi and later Zanotti. The Kirch family, Gottfried Kirch and his wife Maria Winkelmann as well as their son Christfried Kirch and his sisters, played an important role in the scientific production of the Royal Observatory of Berlin, the centerpiece of the Academy founded by Gottfried Wilhelm Leibniz in the same city at the turn of the century, of which Gottfried Kirch and later Christfried were directors. As for Joseph-Nicolas Delisle, he was the major architect of the creation ex nihilo of an astronomical observatory in St. Petersburg, within the framework of the Imperial Academy of Sciences wanted by Peter the Great, and set up with the help of Christian Wolff, a follower of Leibniz, on a scheme identical to that of the Royal Academy of Sciences of Prussia that we will often call for simplicity the Academy of Berlin. To these actors of the observation, it was necessary to join Halley and Mairan, the creators of the two great systems in dispute with the Cartesian vision of the Meteors, the first having played an essential role in the conduct of the Royal Society of London and its Observatory (as secretary of the society from 1713, then, from 1720, as director of the Royal Observatory of Greenwich), the second having taken in 1741, for three years, the succession of Fontenelle to the perpetual secretariat of the Académie Royale des Sciences de Paris. This group of scholars alone covered the main academies of the time (London, Paris, Uppsala, Berlin, Bologna, St. Petersburg), and many of its members played prominent roles in the creation, or the direction, of these academies and their observatories. These observatories constituted, until the middle of the 18th century, the essential of the establishments of importance in astronomy of Europe.

The institutionalization of astronomy on the European continent was initiated in the second half of the 17th century through the creation of two great observatories: the Observatoire Royal de Paris in 1667 within the framework of the new Académie Royale des Sciences de Paris (Maury 1864) founded the year before; and the Royal Observatory of Greenwich in 1675 within the framework of the Royal Society of London (Mailly 1867) founded in 1660. In the following century, many other academies were founded: the Royal Academy of Sciences of Prussia in Berlin in 1700, the Royal Society of Sciences of Uppsala in 1710, the Academy of Sciences of the Bologna Institute in 1714, the Imperial Academy of Sciences of St. Petersburg in 1725, and about 15 others in the course of the 18th century (Sigrist 2013). The Academies of Paris, London, Berlin, Uppsala, Bologna and St. Petersburg, which we will study in particular in this work, all of which had astronomical observatories as an integral part of their structure, gathered between 30 and 40% of the European astronomers of the time, whose number approached half a thousand in the second half of the 18th century (ibid.). These astronomers did not observe only in the observatories attached to these academies, far from it. In Paris alone, about 10 private observatories were used by astronomical academicians (Passeron 2013). At the European level, more than 60 mobile quadrants were delivered in the 17th and 18th centuries in about 50 different observatories (Turner 2002), translating the existence of several tens of valuable observatories at that time. The observation of the meteorological parameters, as well as the aurora borealis, constituted in the public observatories attached to the academies activities of service, recognized and to which were devoted means in terms of men and material.

The aurora borealis was an emblematic example of an object, by nature observable simultaneously by observers from different countries, the observation of which by witnesses several hundreds or thousands of kilometers away was necessary to precisely characterize its morphology and dimensions, and as such calls for the establishment of coordinated observation networks. Contrary to the astronomical phenomena, whose observation required coordinated campaigns on a global scale, for example, to measure the parallaxes of the moon, the planets or the sun providing their distances, the aurora borealis are unpredictable events and thus required constant monitoring, and the accumulation of observations in large numbers, especially for statistical purposes as we have seen. The cooperative aspect of the observation of the aurora borealis constitutes an important dimension, even if it represented only a relatively secondary aspect of the activity of the astronomers of the time, whose principal field of expertise was that of astronomy and of the prediction of the trajectories of celestial bodies, in particular in this period of penetration of Newtonian mechanics, of which some, like Delisle or Celsius, set themselves the goal of providing observational validation. The aurora borealis was (almost) never mentioned in Delisle's correspondence with his European interlocutors, the reason being that their observation could not give rise to any calculation intended to predict their course in an unequivocal manner, as is the case with the trajectories of celestial bodies. The system proposed by Mairan, using the sun and the Earth's orbit as partial determinants, nevertheless contrasted with this state of affairs by giving the aurora borealis an astronomical dimension, providing relatively clear validation criteria. The present work, based on the circle of about 15 astronomers providing aurora observations previously described, is not mainly concerned, as we have said, with the science of the aurora borealis, which has already been the subject of lengthy developments in a previous book (Chassefière 2021a), but rather with the human, institutional and philosophical context in which these aurora observing astronomers evolved.

A particularity of the aurora borealis phenomenon, in this first half of the 18th century which saw the penetration of Newtonianism on the continent, was that the systems proposed to describe it did not fit into a clear duality between Cartesian and Newtonian influences. The system proposed by Halley, a close associate of Newton, took up Descartes' magnet theory, while that of Mairan, defender of Cartesian vortices, claimed the inclusion of the mathematical law of gravitation proposed by Newton to define the size of the accretion zone of solar matter by the Earth on its orbit, even if this particularity did not constitute a central aspect. Pieter van Musschenbroek, a Dutch physicist professing the ideas of Newton, defended the Cartesian system of inflammation of exhalations. Many aurora observers, as mentioned, did not take sides in favor of a particular system. The aurora borealis is not by itself, if we stick to the question of its observation, an object cleaving in terms of belonging to such or such philosophical school, the cooperation required to observe and characterize it largely prevailing, and welding communities of various obediences. The development of the Mairan system, in the second half of the 1720s, was nevertheless part of a particular period of marked opposition between Cartesians and Newtonians. Several authors, such as Pierre Brunet (1970) and later John Bennett Shank (2008), described the period that began in 1727 and lasted until the end of the 1730s as a time of strong Cartesian reaction to the increasing penetration of Newtonianism. For the former, the Cartesian reaction began at the beginning of the 18th century, with opposition becoming increasingly intense from 1727 onwards, whereas for the latter, the process did not really begin until 1715 due to a publicly displayed rejection of Descartes by the Newtonians, Fontenelle's praise of Newton in 1727 amplifying the awareness of the Newtonian threat by the Cartesians of the French Academy (a synthesis of the periodizations proposed in the literature can be found in Crépel and Schmit (2017, pp. 21–40)). In this particular context, Mairan's system, which appealed to subtle matters, was, as we shall clearly see, established by Fontenelle, secretary of the French Academy and uncompromising defender of the theory of vortices, as a monument of Cartesian thought against the growing influence of Newton's ideas.

In Chapter 1, we question the silence maintained over several decades in the volumes of the Académie Royale des Sciences on Halley's aurora borealis system. This silence was reinforced at the end of the 1720s by the rejection by the Academy of a memoir of the Société Royale des Sciences de Montpellier, written after the great aurora borealis of 1726 by François de Plantade, proposing a system similar to that of Halley's. The arguments put forward against Plantade to justify the rejection of his memoir by various Parisian interlocutors did not seem to be able to explain the decision taken, firstly because Plantade, a recognized scientist and founder of the Société Royale des Sciences de Montpellier, explained that his system differed sufficiently in detail from that of Halley's to deserve to be exposed in the volumes of the Parisian memoirs. Secondly, the crisis triggered by the rejection of the memoir threatened the organic link established in 1706 between the two institutions, a consequence far more serious than the publication of a memoir whose descriptive part at least had been judged of good quality. Based on an analysis of the facts, and of the positions of the main actors involved, namely the perpetual secretary Fontenelle, and Mairan himself, who was very close to Fontenelle, and by placing the episode in the context of the intensification of scientific relations between the Académie Royale des Sciences and the Royal Society of London, we suggest that the arguments communicated to Plantade by various Parisian interlocutors to justify the rejection of his memoir masked a "political" motivation, in the sense of the defense of the Cartesian sphere of influence, in a period of hardening conflict between Cartesians and Newtonians, the affirmation of Mairan's aurora borealis system constituting a piece of the Cartesian counter-offensive.

Chapter 2 presents the life and study trajectory of Joseph-Nicolas Delisle, an astronomer and academician invited to St. Petersburg, the new capital of Russia, by Peter the Great to take the direction of an Astronomical Observatory within the framework of the new Imperial Academy of Sciences. He observed more than 200 aurora borealis in the decade following his arrival in 1726. The context of the publication of these observations in 1738 within the framework of the Russian Academy of Sciences is surprising in more than one way. The isolated character of this publication, in which were announced others which needed to succeed it in order to carry out a Celestial History of all the European astronomical observations, and which, however, never appeared, the late character of the publication of the experiments of diffraction of light carried out 20 years earlier in Paris by Delisle, published in the same work, in a collection where they were not meant to appear, the questioning of the reasons which pushed Delisle to accept the tsar's offer to come and settle in Russia, are so many questions which deserve to be examined. The first interests of Delisle, an essential actor of the diffusion of the Newtonian theses on the old continent in the first half of the 18th century, were in celestial mechanics and the astronomical surveys, essentially of eclipses, intended to measure the position of geographical locations for cartographic purposes. This is the main reason why, even before the mission of setting up an observatory, he was invited to Russia. We show to what extent, in the years 1715–1725, Delisle's scientific projects in France were thwarted because of his Newtonian convictions, probably influencing his decision to leave France. This chapter, although moving away from the theme of the aurora borealis, announces the following one devoted to the Imperial Academy of St. Petersburg, which brings us back to it in part.

Chapter 3 focuses on the creation of the Imperial Academy of Sciences in St. Petersburg and the difficulties it encountered in its early days. We start in this study using a particular fact, which is the conference given by Christian Wolff in his city of Halle one week after the great aurora borealis of March 1716, addressing the citizens of his city in order to give them a rational explanation of the phenomenon. Many of the scientists invited to join the St. Petersburg Academy were recommended by Wolff, and we illustrate Wolff's influence on the work achieved at the Academy by taking the example of his follower Friedrich Christoph Mayer, an observer of the aurora borealis, partner of Joseph-Nicolas Delisle at the Observatory, who gave up the observation of the aurora borealis because he did not believe in his observations, which indicated an altitude of the auroral matter that was one hundred times higher than the one predicted by the system he defended, inspired by that of Wolff, and which attributed a stormy origin to the phenomenon. We show how Leonhard Euler, in contrast to Mayer, whom he worked with for several years at the St. Petersburg Observatory, solved the paradox by proposing 20 years later a new system of the aurora borealis, placing it very above the atmosphere. We discuss the scientific attitudes of the two men with regard to their respective schools of thought. We then examine, on a broader level, the evolution of the young academy under the sometimes contradictory effects of Wolffianism and Newtonianism and the obstacles encountered, in particular religious censorship, as well as the serious dysfunctions in the administrative management of the Academy, which, combined with the general political instability which reigned then in Russia, provoked the premature departure of several scientists of great stature, like Daniel Bernoulli, Euler and finally Delisle. We conclude by explaining the dynamics of the evolution of the Imperial Academy of St. Petersburg throughout the 18th century which, in spite of the numerous difficulties encountered, finally led to a Russian scientific community, notably astronomical, emerging towards the end of the century.

More than 200 observations of auroras among those analyzed by Mairan in 1754 were made by Anders Celsius, who published a compilation of them during his European trip, while he was in Germany, and it is to Celsius, and to his asserted international dimension, that Chapter 4 is devoted. In addition to his observations of auroras, Celsius discovered the relation between aurora borealis and magnetism by highlighting the irregular variation of the magnetized needle during the aurora, documented very precisely by his assistant Olof Hiorter in a series of thousands of measurements carried out at the beginning of the 1740s, some of which corresponded to those of the clockmaker George Graham in London. Celsius traveled for five years in Europe in the mid-1730s and established a network of correspondents among the best astronomers of the time. In addition to the astronomical observation itself, notably on the problem determining longitudes, or around the observation of comets and the calculation of their orbits according to the laws of Newtonian mechanics, then defended by a few scientists, including Joseph-Nicolas Delisle, networks of scientists were formed around the observation of the aurora borealis, variations in the declination of the magnetic needle or meteorological parameters. Celsius, like Delisle, was a very active member of these different networks, which were strongly interconnected. In this chapter, we are interested in the life of Celsius and the observational campaigns in which he was involved in different capacities, as well as in the genesis of the Royal Society of Uppsala and its Observatory, as well as their counterparts in Stockholm, of which Celsius was the main architect, in the context of a chronic lack of means in terms of infrastructure and equipment. It was only in the middle of the 18th century, only two years before the death of Celsius, that the Astronomical Observatory of Uppsala was officially established, following three decades of persistent efforts by Swedish astronomers to observe, at all costs, and to try to obtain from the political authorities, with more or less success, means commensurate with their ambitions.

During his European trip, Anders Celsius noticed the astronomical competence of the sisters of the astronomers he visited: Christfried Kirch first in Berlin, then Eustachio Manfredi in Bologna. In Paris, he stayed with the sister and mother of Joseph-Nicolas Delisle, then in St. Petersburg. Thus, a large number of the astronomers of the 18th century involved in the observation of the aurora borealis, and more generally in astronomical observation, were members of astronomical households, informal units in which men and women shared on an equal footing, at least within the household, the work of observation, the men being the only ones entitled to publicly exercise the profession of astronomer because of the gender conventions of the time. Chapter 5 is devoted to the description of these households. This was a period of establishment of academies of great fame, as in Berlin or Bologna, and the private circles constituted by the astronomical households, and even the still relatively informal embryos of future academies at work in the Italian salons, coexisted for some time with the new institutions, in ways that varied from one country to another. In this chapter, we analyze the constitution and the modalities of operation of the Kirch, Manfredi and Delisle households, for the first two were in close connection with the academies and astronomical observatories born in Berlin and Bologna from the Enlightenment approach, aiming at developing experimental science, whose birth and rise to power phase we describe. We analyze the similarities and differences between these different households, and the different institutions to which they were linked. We extend, in the case of Italy, the question of gender to other women scientists than the Manfredis, women whose careers were emblematic of the Italian specificity in this field.

The aurora borealis in itself is not the main subject of the book, which uses it rather as a device for the narrative, a particular thread linking the different actors, this one being able to take more density in certain passages, as in Chapter 1, about the quarrel which opposed Paris to Montpellier about the refused memory of Plantade, or in Chapter 3, involving the renunciation of Mayer to the observation of the aurora borealis. But it is to the personalities or to the life paths of some of the outstanding scientists who observed and studied the phenomenon, and to the institutions (academies, observatories) within which these scientists evolved, that the book is above all devoted. These scholars, for the most part, were primarily interested in astronomical observation, and even in meteorology activities that also required the networking of instruments and methods, the observation of the aurora borealis, and even the elaboration of systems supposed to explain them, constituting only a limited part of their activity. Speaking of men, of the ideas they defended, of the institutions they modeled, it is thus not possible, nor even desirable, to restrict the scientific field of our narration to the sole question of the aurora borealis. The question, for example, of the role of women in the production of knowledge, is much wider, concerning all the sectors of astronomy and beyond science in general. The reading of the present work must be undertaken as a retrospective journey through the Europe of the first aurora observers, during which we will often cross paths with the same scientists, but each time in different contexts, or following different angles, thus trying to account for the swarming of ideas and encounters that constituted the development of experimental science in the pivotal era of the Enlightenment.

1

The Aurora Borealis Issue of the Affirmation of the Cartesian Mechanism and the Dispute Between Paris and Montpellier: The French Choice

1.1. Introduction

We previously evoked the two great systems of the aurora borealis seen at the beginning of the 18th century, in dispute with the Aristotelian vision of the inflammation of dry vapors taken over by René Descartes; one by Edmond Halley published in 1717, invoking an effect of the magnetic matter circulating in the great Earth magnet, the other by Jean-Jacques Dortous de Mairan, published in 1733, appealing to the matter of the solar atmosphere precipitating in the high layers of terrestrial subtle air. However, between publication dates of the two systems, in a period where the aurora borealis occurred regularly giving place, almost annually, to articles in the *Mémoires de l'Académie Royale des Sciences* that we will call hereafter simply the *Mémoires*, systematically taken up and commented by its permanent secretary Bernard le Bovier de Fontenelle in the columns of the *Histoire de l'Académie Royale des Sciences*, Halley's system is never quoted, except, in passing, in Mairan's treatise of 1733, for a brief criticism in half-tone. The fact that Halley's system is not mentioned in the French Academy's volumes, notably in the *Histoire*, whose role was to contextualize and synthesize the science presented in the *Mémoires*, was not only by omission. In 1727, a memorandum of François de Plantade communicated by the Société Royale des Sciences de Montpellier to the Parisian Académie des Sciences for publication in its *Mémoires*, in accordance with the agreements which bound the two institutions, in which Plantade related the aurora of 1726 and proposed a system close to that of Halley's, was refused by the Academy, causing a crisis between the two institutions. In a first step, we will briefly describe the two major systems in question of the aurora borealis. In a second step, we trace the history of the aurora borealis in the volumes of the *Histoire* and *Mémoires* between 1717 and 1733. Then, we devote three specific sections to the actors of the considered facts: the first to the Montpellier inhabitants (Plantade at the Société Royale des Sciences de Montpellier), the second to the Parisians (Fontenelle and Mairan at the Académie Royale des Sciences), the third to the Londoners (Hans Sloane and Halley at the Royal Society). Finally, we examine the question of the silence of Halley's hypothesis in the French community, and the rejection of Plantade's memoir for publication in the French Academy's memoirs, which we propose constitute elements of an affirmation strategy of Cartesian thought in a period of tension generated by the progressive penetration of Newtonianism on the continent.

1.2. The two main systems of the aurora borealis

The first synthetic description of these two systems, completed by that of the system proposed some 20 years later by Leonhard Euler, was published by Morton Briggs (1967). More recently, Stéphane Le Gars has focused on Mairan's theory, discussed in relation to Halley's (Le Gars 2015), and these systems, as well as others developed in the 18th century, have been described in a previous book (Chassefière 2021a). A common feature of the memoirs of the period reporting on the aurora borealis was the detailed description, judged "Baconian" by Briggs, of the forms and colors of the aurora, as well as their evolutionary dynamics in the course of the same aurora. This luxury of detail, present in Halley and Mairan's work, and found in many others, was considered necessary in order not to miss a possible new fact, which would help provide an explanation of the phenomenon. Briggs quotes a Harvard professor, Isaac Greenwood, who in 1731 gave a detailed description of the smallest details of the aurora observed at a temporal resolution of 5–10 minutes over a period of several hours. The systems proposed by Halley and Mairan both break with previous theories attributing the cause of the aurora to exhalations igniting in the atmosphere.

1.2.1. *Halley's system*

Halley, who witnessed the great aurora of March 1716, proposes in *Philosophical Transactions* of the following year (1717) the first scientific interpretation, answering the mandate which was entrusted to him by the Royal Society to provide a rational explanation of the phenomenon (Briggs 1967). No event of this type appears in the English annals since 1574. Halley put forward in his article a first explanation inspired by Aristotle in terms of sulfurous vapors, quite similar to the one he gave two years later for a flying fire (a meteoroid entering the atmosphere) that flew over England. He noted the sporadic and unpredictable character of the auroras, which brought them closer to earthquakes, suggesting at first that the two phenomena could be due to the heating of the water vapor inside the Earth: this escaping towards the atmosphere by carrying with it sulfurous vapors. This conception was inherited from the ideas of Greek antiquity, and more particularly from Thales, which supposes that the Earth, and all the universe, are carried by water (Taton 1957, p. 220), the earthquakes resulting from eruptions of hot water occurring inside the Earth. According to whether this escape of water under pressure was violent, and succeeded in fracturing the Earth's crust, or moderate, being satisfied to filter through the surface, the phenomenon would give place, either to an earthquake, or to an aurora borealis. In support of this thesis, he cited the experiments of heating gunpowder under vacuum (or at low pressure) carried out by the Reverend Whiteside in Oxford, which showed that heated matter ignited spontaneously. In the first half of the 18th century, the supporters of a chemical cause linked to the ignition of exhalations to explain different types of meteors, such as thunderbolts, flying fires, shooting stars or the aurora borealis, are numerous, such as Pieter van Musschenbroek and Pierre-Charles Lemonnier. For the supporters of this explanation of the aurora borealis, inherited from Aristotle and taken up by Descartes, the flammable matter could rise from a northern region of the Earth, and the burial of this matter by an earthquake produces long periods of cessation of the auroral phenomenon, explaining the interruption observed during the greater part of the 17th century.

But Halley then expressed his doubts about the fact that an aurora borealis, seen from Ireland, Russia and Poland, extending over 30° of longitude and 50° of latitude over northern Europe, could be linked to an earthquake, the spatial scale of which is generally not so extensive. He then proposed another hypothesis, radically different. Taking up the ideas of William Gilbert and then René Descartes who compared the Earth to a big magnet, he had the intuition of the role of magnetism in the appearance of auroras, which was suggested to him by the location at high latitude of auroral phenomena. He explicitly referred to Descartes' theory of the circulation of subtle magnetic matter through the Earth, but, contrary to Descartes, who supposed that this matter circulated in both directions (which he did not mention), only admitted, with Christian Huygens (1680) and Nicolas Hartsoeker (1707), only one direction of circulation. Like Hartsoeker, he considered that the magnetic matter came out of the northern pole of the Earth and circulated in the ether towards its southern pole. Halley argued that the auroras were most often seen in the northwest (Iceland, Greenland), precisely in the vicinity of the North Magnetic Pole at the time. The light jets were identified with columns of magnetic vapors rising vertically from the surface of the Earth, and arranged like iron filings around a magnet. The crowns seen near the zenith resulted from beams rising all around the observer and converging at a great distance above him. He evoked the fact that these beams, rising above the Earth's shadow, could be illuminated by the sun, like a twilight sky. He did not exclude that magnetic vapors could also radiate their own light. For Halley, the magnetic vapors rose in the ether, without reference in the article to an atmospheric character of the aurora borealis. The co-rotation observed between auroral structures and the Earth's atmosphere, noted by Halley, did not mean for him, strictly speaking, that the aurora occurred in the atmosphere, contrary to the opinion expressed by other scientists of the time, like Jacques Philippe Maraldi (1716). Halley did not comment on the nature of the matter composing the magnetic effluence, noting that it may be luminous by itself, or it may carry from the depths of the Earth atoms capable of producing light in the ether, such as those in water vapor. Halley ended his paper by briefly revisiting his idea of the existence of an inner Earth, rotating out of step with the shell on which we live, to explain the magnetic pole drift (Halley 1692). He suggested that the space between the concave roof of the upper crust and the inner Earth could shelter the luminous matter responsible for the auroras, which would thus maintain a perpetual daylight on the inner Earth. This matter would escape in the vicinity of the poles because the Earth's crust must be less thick there, the Earth being flattened at the poles (according to the Newtonian theory of gravitation).

Figure 1.1. *Drawing of an aurora borealis published in* Philosophical Transactions *in 1728 (the description of an Aurora Borealis 1728)*

1.2.2. *Mairan's system*

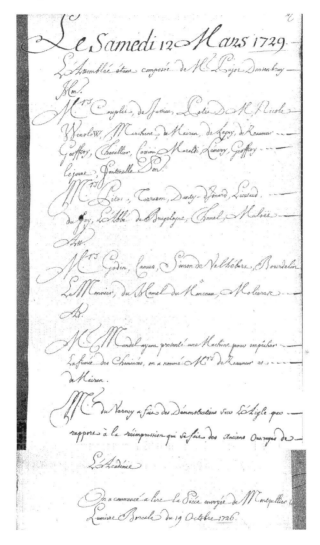

Figure 1.2. *Extract from the minutes of the Académie Royale des Sciences for the year 1729 (Bibliothèque numérique en ligne Gallica)*

Mairan did not witness the aurora of 1716, but he began to take an interest in the phenomenon in the years that followed, his colleague at the Academy Jacques Philippe Maraldi regularly publishing accounts of aurora borealis in *Mémoires*, as we will return to. It seems that Mairan began to foresee the system that he would publish in 1733 as early as the middle of the 1720s, probably on the occasion of the great aurora borealis of November 1726, which he witnessed. In a letter to Jean Bouillet from May 1730 (Unpublished letters from Mairan à Bouillet 1860), he declared to have "taken date" of his system "at the Academy more than one year ago", on the occasion of the reading of a treatise on the aurora borealis sent from Montpellier by François de Plantade. We find trace of the reading of "the piece sent from Montpellier on the Northern Light of October 19, 1726" in the minutes of the year 1729 (Bibliothèque numérique en ligne Gallica), this one having extended over three meetings: March 12, 16 and 23.

Thus, in March 1729, Mairan's system was revealed to the members of the French Academy, and this precisely on the occasion of the reading of Plantade's memoir, in which we will see that he took sides with the magnetic hypothesis. Mairan's treatise was not finalized until 1731, and published two years later (1733).

The idea behind this system was expressed 40 years ago by Jean-Dominique Cassini after his discovery of zodiacal light, that a "very large sphere of atoms concentric to the Earth" can stop and bring together in abundance "the sphere of atoms of the sun" (Cassini 1693, p. 11). It is this concept that Mairan took back by taking it further, and by introducing the Newtonian attraction as a driving force of the incorporation of solar matter to the Earth's atmosphere. Based on the fact that the top of the zodiacal light frequently deviates from the sun of an angle higher than 90°, which shows that it can reach the terrestrial orbit at certain times, Mairan argued that, when the Earth crosses the solar atmosphere, it must be attracted by it within a certain distance that he evaluated by balancing the gravitational attraction forces of the Earth and the Sun calculated according to Newton's law. This calculation, physically incorrect because it did not take into account the driving forces related to the rotation of the Earth and the Sun around their common center of gravity, provided a capture distance of 240,000 km, less than the true distance of equilibrium (that of the point of Lagrange L1 to the Earth, that is, 1.5 million kilometers), but it allowed Mairan to quantify the distance of interaction necessary for the accretion of solar matter to take place. Mairan's theory, which borrowed much from Cartesianism by its use of subtle matter, also integrated the idea of Newtonian universal gravity, even if it did not bring any quantitative added value to the theory. Mairan claimed in his treatise his refusal to place himself in a pre-existing current of ideas, "because we will try, as much as it will be possible for us, to preserve to our research the advantage to be supported with all the systems by admitting only Observations and facts that can be admitted on both sides" (de Mairan 1733, p. 30). In the Cartesian conception, the atmosphere of the sun is constituted of a great number of parts of different sizes, which according to Mairan had to be translated by a vertical structuring of the solar matter in the high terrestrial atmosphere, the various solar parts stabilizing at the level of the terrestrial parts of the same density. Mairan, from measurements of parallax of the auroral structures, estimated the height to several hundreds of kilometers, "from where it follows, or that the Aurora Borealis consists of a rarer matter, & lighter than the higher parts of our air, some rare & some light & loose that it must be at these great distances, according to the common opinion, or that the Atmosphere is much higher than one believed it until now; which is, according to us, much more probable, & that we hope to prove" (de Mairan 1733, pp. 6–7). The second hypothesis, which he favored, implied that the atmospheric matter within which the solar matter precipitates is itself subtle, extending far above the coarse matter of the atmosphere which is supposed to reach only about 70 km of altitude on the basis of the estimation method deduced from the duration of twilights (de La Hire 1713). The concept of subtle air, less subtle than ether but more subtle than coarse air, was introduced in the previous century, or at the very beginning of the current century, to explain various phenomena: inequality in the level of mercury in different barometers, suspension of mercury at a great height in inverted tubes, "mercurial phosphorus" (light barometers), considerable degree of adhesion between joined polished planes, these phenomena, with the exception of light barometers (Bernoulli 1701), being described in the treaty (de Mairan 1733, pp. 43–51).

Mairan, like Halley, did not discuss the nature of emitted light, this being able to come from the solar matter entering into contact with the subtle air or from the solar light reflected by it. A significant difficulty in his theory was explaining the polar location of the aurora borealis, inherent to Halley's competing theory. Mairan attributed the concentration of solar matter in the polar regions, where the effect of dispersion by the shocks is minimal, the speed of the atmospheric particles being weak or nil (on the pole), to the scattering of the solar matter, due to the shocks with the particles of the atmosphere endowed with a high linear speed in the equatorial regions. Mairan established in his treatise a complete catalog of the auroras observed for more than a thousand years, and was interested in the fact that the auroras appeared less at certain times than at others. He observed that between 1621 and 1686, no marked aurora was observed, that the resumption in 1686 lasted only a few years and that the phenomenon started to reappear only in 1707, to resume with intensity only from 1716. On a seasonal scale, he tried to demonstrate that the aurora borealis was more frequent when the Earth was close to its perihelion on its orbit around the sun towards the end of December, or when its northern hemisphere pointed in the direction of its speed on its orbit, between the northern summer solstice and the northern winter solstice, the meeting with the solar matter occurring there in a frontal way. He found indeed the expected imbalance, while noting the bias related to the greater duration of twilight towards the summer solstice, which decreased the duration of the dark periods during which, a priori, the aurora was easier to see. Mairan also pointed out a possible link between sunspots and the aurora borealis, an important idea that went unnoticed at the time, and that would not be concretized until a century later by a systematic observation of the link between the two phenomena.

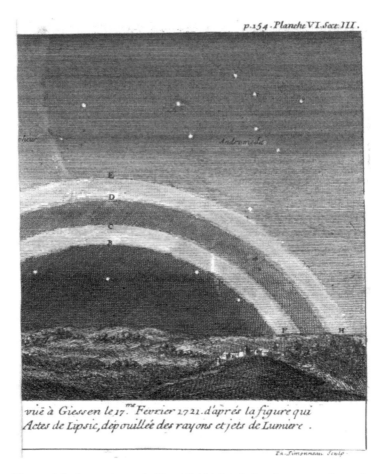

Figure 1.3. *Auroral arcs (without light jets) during the aurora borealis seen in Giessen on February 17, 1721 (de Mairan 1733, p. 154)*

1.3. History of the aurora borealis in the volumes of the Académie Royale des Sciences between 1716 and 1733

1.3.1. *The silence on Halley's system in* Mémoires *and* Histoire

Jacques Philippe Maraldi was the only French scientist to publish a report of the aurora of 1716 (Maraldi 1716). He noted that its light "was not celestial, but rather attached to our atmosphere & […] is different from the light that was discovered on the Zodiac by Mr. Cassini" (Maraldi 1716, p. 96). Contrary to Halley, he deduced from the observed co-rotation of the Earth and the structures of the aurora borealis the atmospheric character of the aurora. He mentioned that this aurora was also seen in Brest, Dieppe, Rouen, on the coasts of Languedoc, then cited comparable phenomena observed in Copenhagen and Berlin in 1707, and mentioned the observations of 993, and of 1621 (made by Pierre Gassendi). He wrote:

> From the observations we have just reported, it seems that all these phenomena are more or less of the same nature, although there have been greater, clearer and better finished ones than the others; that there are times of the year more suitable for these kinds of apparitions, which are February, March, April & September, although according to the testimony of the Saxon analyst [10th century, identity not specified], they still appeared in December towards the Winter Solstice; that these phenomena appeared in serene weather & after one or more days of hot weather. This is what Gassendi & Kirchius [Gottfried Kirch, the first director of the Berlin observatory] testify, and this is also what happened in the two apparitions of this year (Maraldi 1716, p. 107).

Fontenelle, in *Histoire* of the same year, related the aurora observed by Maraldi. Neither Maraldi, nor he, mentioned Halley's observation of the same aurora, even less the explanation given by the latter. In the *Mémoires* of 1717 appeared new observations by Maraldi, introduced in the *Histoire* by Fontenelle, who advanced an explanation inspired by the ideas of Descartes:

> From all the similar Observations spread in both ancient and modern Authors, & that Mr. Maraldi has carefully collected, it seems that this phenomenon, quite independent of the season of the year, is due to a weather that is warm for the season & dry. The phenomenon must rather follow this weather and be the effect of it than precede it and announce it, because it seems to be formed only by sulphurous exhalations which will have risen higher than usual, which during a long enough drought will not have been soaked by aqueous vapors & will have accumulated in great quantities, & finally will have caught fire (Fontenelle 1717, p. 4).

Similar testimonies by Maraldi, as well as by Jacques d'Allonville de Louville, appear in the *Mémoires* of 1718, 1719, 1720 and 1721, preceded by comments by Fontenelle in *Histoire* of the same years. After six years of observation, Maraldi noted in 1721 a certain correspondence between the occurrence of the auroras and the dryness of the climate which was not the case, however, in 1720. The five years which followed, the question was not approached in *Histoire*, nor in the *Mémoires*, for the reason that the aurorae seen between 1721 and 1725 were only of weak extent. The aurora of October 19, 1726 was very intense and spectacular. Jean-Jacques Dortous de Mairan and Louis Godin gave a very detailed report in the *Mémoires*. Fontenelle, after having commented on the observations of Mairan, and noticing that the height of the phenomenon estimated by Mairan was more than 20 leagues (80 km), "which would increase by more than twice the height of the atmosphere determined by the barometer", reiterated his explanation of 1717, but by specifying it:

> The air being certainly denser and heavier under the pole, it must by its weight make the light matters rise higher, which make the exhalations of the sulfurous, nitrous, ferruginous earth, finally all those which are suitable to be ignited. They can form a fairly large mass before fermentation occurs to ignite them. [...] If the cluster of exhalations catches fire & if the flames come out of the lower part as well as the upper part of the zone, the inhabitants of the Pole will see on their heads during the night a light, & lightning similar to those of our thunder. But if these phenomena remain the same, the spectator moves away from the Pole, he will see the top of the zone always lowering towards the horizon, & the spherical zone which he saw as a whole will only appear to him as an arc, or rather as a circular zone which will have a higher middle point, & its two extremities leaning on the horizon (Fontenelle 1726, p. 5).

The fact that we do not see auroras during the summer is due to the summer sun which heats the atmosphere and prevents the particles from "coming closer, joining and mixing more intimately".

Godin, on the other hand, in his relation of the aurora, quoted Halley's observation of 1716, but without mentioning his explanation of the phenomenon, as well as those of Maraldi carried out since, but did not say more, not wanting to "copy" their research. He quoted the idea of Arnelius of sheets of ice in suspension reflecting the sunlight, but concluded that the sun is too low below the horizon for that to be possible. Then, he put forward an explanation inspired by chemistry experiments carried out by Nicolas Lémery on a mixture of iron filings and pulverized sulfur reduced to a paste, then buried, which shows that after fermentation, after a few hours, the mixture spontaneously pierces the Earth and produces flames. A second experiment by the same scientist shows, according to Godin, the production of red vapors from a mixture of iron filings and niter spirit (nitric acid). Now, according to Godin, the vapors that rise from the Earth necessarily contain niter, iron and sulfur. In the torrid zone, fermentation will take place at a low level, because the very rarefied air will not be able to support them very high. Fermentation will make the exhalations catch fire and will produce the meteors of summer: thunder, lightning. In the polar zone, on the contrary, the air is very dense, and the Earth gives proportionally less exhalations which rise very high, and have less tendency to dry out. They will close less strongly and ignite less quickly, making also less noise. In a temperate zone, we will have one or the other situation according to whether the sun is closer or further from its zenith. Auroras were still seen in 1729 by Jacques Cassini. Fontenelle, in the *Histoire* of that year, wrote: "Perhaps it will please the public to announce that Mr. de Mairan has undertaken to reduce the whole thing to a regulated system which will appear shortly" (Fontenelle 1730, p. 9).

In 1732, Fontenelle mentions in the *Histoire* the Mairan treaty which was finalized in 1731 and published the following year. He said of him: "He undertook this work with all the more ardor, because having glimpsed for quite some time a singular and bold explanation of the Phenomenon, he found that the facts seemed to conform more and more to this idea." And here is how Fontenelle introduced Mairan's system:

> The ordinary or average Aurora Borealis, little & badly observed, could seem a Meteor formed in the terrestrial Atmosphere, like the Lightning, the Falling Stars, the Flying Fires. But when we came to reflect on the great frequency of this Phenomenon, on its appearance attached to certain Seasons of the year almost exclusively to the others, on its place always marked in the North, even on its magnificence, when it is what we call complete, it was difficult to believe that it was a simple fortuitous Meteor, which was not essentially due to the general constitution of the World, or of our whole Solar System, in a word, which was not cosmic. But how can a Meteor be cosmic? These are two ideas that seem to exclude each other, and that Mr de Mairan has found the secret of combining. Thus the Aurora Borealis will hold a middle ground between pure Meteors & pure cosmic phenomena, such as all those of Astronomy, & this disposition seems to be enough of Nature's genius (Fontenelle 1732, p. 3).

Thus, Mairan's hypothesis distanced itself from Cartesian ideas, recreating a new paradigm that placed the Earth in a world system involving the entire solar system. In none of the numerous articles published in the *Histoire* and in the *Mémoires* since 1716 was Halley's system cited, Halley himself, as an observer of the aurora borealis of 1716, having been quoted only once by Louis Godin. Mairan was the first, in his treatise published in 1733, to describe and criticize Halley's system, but he devoted only one page of his work to it. In the second edition of his treatise, published in 1754, Mairan said that he had tried to obtain criticism of his system from Halley through Godin who traveled to London in 1734 to obtain from Halley instruments for the Peruvian expedition, but without success, and he seems to have conceived a certain bitterness:

> I did not delay long in acting accordingly. Mr. Godin having then a trip to make to London, I begged him to obtain from Mr. Halley some remarks on my hypothesis, or rather some objections against it; for I did not flatter myself to have brought this famous Astronomer to my feeling, on a matter where we had taken such directly opposite roads. Mr. Halley makes the Aurora Borealis come from the luminous Atmosphere of the small magnetic Earth which he supposes to be at the center of our Globe imagined as a hollow Sphere. From there, according to him, escape from time to time vapors from the Poles of the upper crust that we inhabit, or at least from its Boreal Pole; whereas, according to me, the origin of the Phenomenon is none other than the Sun or the Solar Atmosphere. Nothing was more capable of providing me with strong & learned objections. I had reiterated the request by an ostensible letter sent to Mr. Godin; but all these requests were only worth to me from Mr. Halley some politeness on the way I had treated my subject, without consequence for the hypothesis (de Mairan 1754, pp. 301–302).

An obvious interest of Mairan's system, which was at no time given as sure, neither by him, nor by Fontenelle, was that it could be tested more and more precisely with the accumulation of observation data over the years. He thought indeed to be able to explain the temporal variability of the aurora borealis with his hypothesis, the auroras being more frequent when the Earth was closest to the sun in its annual rotation around the sun, or when its northern hemisphere was pointing in the direction of its speed, which he believed to be the case. Fontenelle noted on this subject:

> All these consequences on the times of the appearance of the Aurora Borealis are so necessarily & so particularly drawn from the System of Mr de Mairan, that if it is not true, they will be infallibly contradicted by the facts. This kind of touchstone could be applied to the whole sequence of the Aurora Borealis, of which one will have observations that will mark the times of the year. To research it in all the Books was a work of erudition, which belonged to another Academy, but Mr de Mairan showed that he could have been one of its most worthy members (Fontenelle 1732, p. 13).

The aurora continued in 1732 and 1733, giving rise to a veritable scientific saga that Fontenelle exalts in his *Histoire*, a saga that precisely constituted the symbol of the adventure of the mind that Fontenelle sought to constitute in his great work of *Histoire*:

> He could not, therefore, gather too many observations to justify his System on this subject, & however numerous were those which had already served to establish its foundations, he was obliged to continue them. There is a great likelihood that he will not soon dispense with this care and attention, in favor of an idea so new, so bold, and up to now so plausible. It is a kind of commitment that he has contracted with the Public, & which he begins to fulfil here by giving us the Observations of the Aurora Borealis, of Zodiacal Light, & of all that has some relation to Phenomena seen in Paris, or in the vicinity, during the course of the years 1732 & 1733. One will find there new examples, & new evidence of what he advanced in his treatise on this subject. Nothing is more in keeping with the spirit of the Academy than to collect materials in this way, the assembly of which alone may one day reveal the secret of Nature (Fontenelle 1733a, p. 24).

1.3.2. *The memoir refused by the Parisian Academy of François de Plantade*

A significant event intervened during this period, which was the refusal by the Académie Royale des Sciences of the memoir of François de Plantade on the aurora borealis that the Société Royale des Sciences de Montpellier sent to him under the title of the memoir that the Academy was required to publish annually in its volumes, following the agreements made at the foundation of the society in 1706. Plantade observed in detail the aurora of October 1726, and recorded his observations in a memoir, the second part of which consists of a physical explanation of the phenomenon, which seems to be close, at least in its first principle, to that imagined by Halley. Plantade was an attentive observer of the aurora borealis, as evidenced by his observation of an aurora borealis in December 1737, of which a figure was transmitted to Mairan, "carefully drawn & colored"[1] (de Mairan 1754, p. 429).

---

[1] The image has recently been restored to its original colors thanks to a team that includes the LIRMM, the CNRS, the Bibliothèque Universitaire des Sciences and the IUT de Béziers, url: https://www.umontpellier.fr/articles/laurore-boreale-de-1737-devoilee.

14 Observers of the Aurora Borealis in Europe

Figure 1.4. *Engraving representing the aurora borealis of December 16, 1737 observed in Montpellier by François de Plantade (de Mairan 1754, p. 430)*

The memorandum relating to the aurora borealis of 1726 was sent to Paris in spring 1727. We learn in the *Histoire de la Société Royale de Montpellier* that "the Academy offered to print the first part containing the details of the observation, but it refused to do the same honor to the second part, which seemed to him too much in conformity with what Mr. Halley had given on the cause of the phenomenon" (*Histoire de la Société Royale de Montpellier* 1778, p. 20). Stung to the core, Plantade withdrew the entirety of his memoir, which was never published, and of which trace has been lost. Here is what is reported on this subject in the *Histoire de la Société*:

> M. de Plantade has always maintained that his system on the Aurora Borealis had nothing in common with that of M. Halley's, but the general attribution of the Phenomenon to magnetic matter that, in addition, the two hypotheses differed infinitely by the details. He at least flattered himself that he had extended and greatly embellished the idea of the English Philosopher. We do not wish to discuss here

this claim of an Author who could be suspect in his own cause. He said several times that unfortunately for his physical explanation, another system quite different had been born a short time before in the heart of the Académie des Sciences, where it had made a rather great fortune, so that the minds had found themselves in general little disposed to taste other conjectures & other views (*Histoire de la Société Royale de Montpellier* 1778, pp. 20–21).

The other system of which Plantade speaks here is the one imagined by Mairan, of which we have seen that the latter presented it to the French Academy in March 1729, precisely on the occasion of the reading of Plantade's memoir. A letter written two months earlier, on January 13, 1729, by Abbé Jean-Paul Bignon, the architect of the renovation of the Académie Royale des Sciences in 1699, to François de Plantade, referring to two distinct works of Plantade, the one, refused, on the aurora borealis, the other on a meteorological phenomenon, explicitly reproached him for wanting to create systems too hastily:

> I have always been convinced that nature was not yet sufficiently known to us, and that it would take many years for men to be in a position to form well-founded systems on the causes of the various events which strike our eyes and excite our research. I therefore still believe that the most important object, as of now, would be to gather the facts, and not to omit the slightest circumstances, leaving to those who will come after us the care of drawing from these different facts notions which could establish more certain systems. If I am in this thought, it is not only by persuasion of the difficulty, not to say of the impossibility to succeed at present in making systems, it is still by another more delicate reason. For I must not conceal from you that I have sometimes noticed too much that our system-makers, far from simply gathering the facts and explaining in detail all the circumstances and all the parts of them, let themselves be led, by the love of the system, not to report with complete accuracy those circumstances which would not be sufficiently favorable to it. It is this fear that has always made me preach to our physicists to be on guard against systematic ideas and to stick rather to the sole indication of what they have found, without omitting anything in the world. Forgive, Sir, this reflection on the zeal I have always had for the improvement of science […].

> A second reason that confirms me in these ideas is that the time given to the composition of a system is as much taken from that which should be given to observations, for which life is already too short. Moreover, it is to be feared that a philosopher filled with his system

will not notice more singularly than the circumstances which are favorable to it. We have had a sad experience of this in the late Mr. Perrault, who, in the dissection of animals, having too often begun by imagining the reasonings he could make about causes and events, has also too often forgotten many things which could have destroyed these same reasonings (Castelnau 1858, p. 124).

Thus, the reason invoked by Bignon is not a too great conformity to Halley's system, but the very fact that Plantade proposed a system. As we mentioned in the introduction, Fontenelle in his foreword to the *Histoire* of 1699 advocates the necessary precedence of observation and experimentation over the elaboration of systems. The latter nevertheless constituted for him the ultimate objective of physics. Mairan, in his foreword to his *Dissertation sur la Glace*, precisely criticizes the refusal a priori of any system. He writes, about the system he defends, that of subtle matters inherent to the doctrine of Cartesian vortices:

> I am afraid that it will take away from me some suffrages which would be precious to me: because system or Chimera seem to be synonymous terms today in the mouths of many people, moreover skilful, & who distinguish themselves by their works. It is a system, often makes the whole criticism of a book; to declare oneself against systems, & to assure that what one is going to give to the public is not one, has become a commonplace of forewords.
>
> After that, I dare to use part of it to show that prejudice has been carried beyond its proper limits, and all the more so, since it has an air of solidity and maturity of mind, which imposes itself; in a word, that it is false to assert that systems, sometimes even the most hazardous, are as contrary to the advancement of Science, and as unfruitful as is commonly believed (de Mairan 1749, pp. v–vj).

In this foreword, he protests against the common opinion that the secrets of nature are inaccessible to us. For is the boundary between the knowledge that is accessible to us and that which is forbidden to us, between effects and causes, so well-marked? "Have those who condemn us to an eternal ignorance of first principles seen so perfectly the depths of things that there is no more exception, nor revision to propose after them? What is certain is that one must know a great deal to decide in this way on the scope of the human mind, present and future" (de Mairan 1749, p. vij). We can abuse systems, and we have often done so, but "what do we have that is not susceptible to abuse?" Is it not necessary to have a system in mind to carry out good experiments?

> In vain will it be said that the systematic spirit has always caused the Philosophers to fall into the greatest errors. This spirit is no less valuable, and more necessary, for arriving at the most sublime knowledge, as well as for performing the greatest things. For in what does this spirit consist, if it is not in a natural disposition turned into a habit of making a reasoned plan of our object, a whole of what composes it, according to what we know of it, to ascend from there by degrees to what we do not know of it, and what it is important for us to know? We abuse the terms when we understand it otherwise, & of this very spirit, when we use it to forge systems & hypotheses without necessity & without examination (ibid., pp. ix–x).

As Ellen McNiven Hine (1995) notes, Mairan advocated the constant interplay between experiment and reason as the only way to decipher how nature works. Those who reject the search for first causes were for Mairan as much in the wrong as those who had the permanent tendency to create systems. Both extremes were, according to him, as harmful as each other. To create systems, it is also to give ourselves a guide for the observation, not risking missing observations by the absence of preparation, as testifies his system of the aurora borealis which calls for a precise long-term temporal follow-up of the frequency of the aurora's occurrence. About the systems, he specifies at the end of the warning introducing the treatise: "One will observe on a new plan, & with new views, which is always useful, there being an infinity of objects in Nature, which escape us, for lack of suspecting their existence, & that we will never see that after having been warned that we must see them" (de Mairan 1733, p. 2). Bignon, in his letter, seems to side with the intransigent opponents of the systems.

Bignon's letter was sent at the beginning of 1729, that is, almost two years after the Parisian Academy received Plantade's memoir, which he wanted to be published in the 1727 volume of *Mémoires*. The memoir was not read at the Academy until March 1729, during a session that saw Mairan also take note of his own system, probably under study since the observation of the aurora of October 1726, the same one that had inspired Plantade's system. It was only on December 9, 1730, as can be read in the minutes of that year (Bibliothèque numérique en ligne Gallica), that the Academy explicitly decided against the publication of Plantade's work in the *Mémoires* of 1729, preferring another memoir on a medical subject. Junius Castelnau, a French magistrate and historian from Montpellier in the first half of the 19th century, attributed to a crisis between the two institutions (which formed a single body since, by the Letters Patent of 1706, the Société de Montpellier had been totally integrated into the Académie des Sciences), provoked by the refusal of the Parisian Academy to publish Plantade's memoir, the "gap of four years" (in reality three years: 1726, 1727, 1728) in the annual insertions to the memoirs of the Parisian Academy. A letter by Bignon dated July 22, 1730 confirms the extent of the crisis, which almost resulted in the division of the constituted body in 1706:

I am not surprised to find in the letter that you did me the grace of writing to me on the 14th of this month, less vivacity than in the previous one, with regard to the complaints of Mr de Plantade. I even regret having thought myself obliged to communicate this previous letter to the Academy. It is a great misfortune in all societies, but even greater in those of scholars, to give in to harassment. Thus, I will be careful not to communicate your last letter in the same way. I will only try to make use of it with Mr de Maurepas if it should happen that Mr de Maisons, at the request of some academics, should propose changes to what has been done up to now. I think that things could not be better than by remaining on the same footing. But my thoughts do not decide on those of others, and all that there would be to fear for our Society is that I would not be consulted, if it were a question of some new regulation. I will take care of it with attention, and will forget nothing to continue to give proof of my zeal to our Society (Castelnau 1858, p. 125).

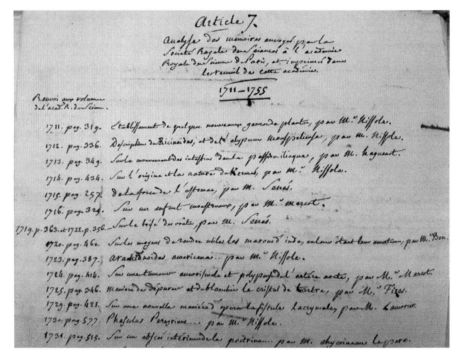

Figure 1.5. *Excerpt from the list of memoirs sent by the Société Royale des Sciences to the Académie Royale des Sciences de Paris, and printed in the collection of this academy (1711–1755) (Archives départementales de l'Hérault – Pierresvives)*

Thus, it was only after the crisis had subsided in the middle of 1730, three years after the memoir was received, that the Academy formally ratified its refusal to publish Plantade's memoir, a decision clearly taken upstream by the Academy's governing bodies, and which seems to be related to the choice of Mairan's system against that of Halley's. Castelnau specified that the matter was subsequently settled, at the price of "some sacrifices of self-respect":

> These difficulties on the choice or the insertion of the memoir sent by the Société royale do not seem to have been renewed. The Académie des Sciences, as I will say elsewhere in speaking of the works of the Society inserted in its Collection, generally exercised only a very discreet and barely perceptible control over the works sent to it; and the deference of the Société royale, if it had to show any, was all the more natural since it could not conceal from itself how much, since the beginning of the century, the respective positions had changed. The progress of centralization and the great and continuous services rendered to the sciences by the first of these bodies, tended constantly to assure it a preponderance, as it were, outside the line, which the Société royale had to recognize without difficulty; and it had to judge that the honor of belonging to it could well be paid for by some sacrifices of self-respect, if they were required of it (ibid., pp. 125–126).

While the *Encyclopédie de Diderot et d'Alembert* (1751) says nothing about Halley's system in its article "Aurore Boréale", the *Dictionnaire de Physique* (1793) mentions it in the same article, and quotes Plantade:

> The illustrious Mr. Halley, to whom science and especially astronomy owe so much, believed that the aurora borealis owed its origin to magnetic matter [...] According to this author, the terrestrial globe is like a hollow sphere in the center of which is a small magnetic earth. It is from this central magnet that the magnetic fluid emanates, which, escaping through the northern pole of the upper crust that we inhabit, circulates around the surface of the earth. This bright light, these sparkling fires that we see shining in the atmosphere during the aurora borealis are, according to him, an effect of the magnetic fluid which ignites like iron filings. Several physicists, & among others Mr. Plantade, of the Société royale des sciences de Montpellier, were of this sentiment.

In good scientific logic, Plantade's dissertation, whose observational part was considered valid, and whose theoretical part, according to the scientist, was far from Halley's system in detail, should have been published. Plantade was a recognized scientist and founder of the Société Royale des Sciences de Montpellier. The refusal of the dissertation jeopardized the cohesion of an institution dear to its founders, who were still in office at the time, and this refusal could only be justified by a reason of a higher order, the same one that made the board of the Academy choose to support the system proposed by Mairan. But before going further in our analysis, we need to better define the human and philosophical contexts in which the actors of this sequence evolved.

1.4. The Montpellier actors: François de Plantade and the Société Royale des Sciences

1.4.1. *François de Plantade, founder of the Société Royale de Montpellier*

The eulogy of François de Plantade written by de Ratte (1743, pp. 5–31), an astronomer and mathematician who was made permanent secretary of the Société Royale de Montpellier in 1743, provides biographical information on the life of Plantade. He was born in Montpellier in 1670. After rather literary studies, during which he acquired the mastery of Greek, Latin and Hebrew, he turned to astronomy and went up in 1700 to Paris, where he came into contact with Jean-Dominique Cassini with whom the Plantade family had, it seems, although it is not certain, family ties. Cassini introduced him to the techniques of astronomical observation. He traveled to England and the Netherlands in 1698 and 1699, where he learned the languages of these two countries, and returned to Montpellier in May 1700, to undertake the construction of an astronomical observatory. Cassini came down shortly after to Montpellier on the occasion of the extension of the meridian of Paris, and Plantade attended the operations. He then conceived the project of the establishment of a society of sciences in this city. With his friends François-Xavier Bon, adviser at the Court of Accounts, Aids and Finances of Montpellier, and the astronomer Jean de Clapiès, they gathered a whole group of scientists from astronomy, physics and natural sciences, and obtained in 1706 from the king the letters-patents allowing the creation of the Société Royale des Sciences de Montpellier. Bon dealt with physics and natural history, and built up a cabinet of instruments and a collection of minerals (Castelnau 1858). He wrote a memoir on spider silk, a substance that he was the first to spin and weave. He also made meteorological observations. Initially an antiquarian, he later became an honorary member of the Académie des Inscriptions et Belles-Lettres de Paris. Jean de Clapiès, a native of Béziers, taught mathematics in Montpellier. Like Plantade, he became a correspondent of the Académie Royale des Sciences in 1702. 1706 was the year of a total eclipse of the sun that Plantade and Clapiès observed with great precision. Plantade realized then numerous observations of sunspots. He became a lawyer-general in 1711, a position from which he resigned in 1730 to devote himself more exclusively to science. In 1729, Plantade made numerous maps of the province of Languedoc in the field. In 1732, he carried out measurements of pressure at the top of the highest peaks of the Pyrenees, suggesting that the density of the air did not follow the Boyle–Mariotte law at great heights, which led Jacques Cassini to revise upwards the height of the atmosphere (Cassini 1733) (we can consult on this question; Chassefière 2021b). In 1730, following his observations of the aurora borealis, he established a system, although, said de Ratte, "he was always little sensitive to this kind of pleasure". He attributed the aurora to the magnetic matter circulating from one pole to the other of the Earth, a theory that Halley had already proposed in 1717, which he seemed to have ignored when developing his own system. But he would have been flattered to have had the same idea, knowing that his explanation proposed an improvement. de Ratte did not mention in his eulogy the quarrel with the Académie Royale des Sciences that we have mentioned. He spoke about the difficult observation of the transit of Mercury of November 1736 realized successfully by Plantade. He mentioned the death in 1737 of Antoine Gauteron, secretary of the Société Royale de Montpellier, who was succeeded by Plantade. Wishing to repeat pressure measurements, this time on the Pic du Midi, Plantade died at the age of 71, on August 25, 1741, during the ascent of this mountain.

1.4.2. *The Société Royale des Sciences de Montpellier*

The *Histoire de la Société Royale des Sciences de Montpellier* (1766) and the historical and biographical memoir of Junius Castelnau on this society (1858) relate the genesis of its creation. Montpellier was the site of a renowned medical university founded in the 12th century by the followers of Avicenna and Averroës. In 1671, Sylvain Régis arrived in Montpellier and, according to the author of the article in the *Histoire*:

> Zealous for the glory of Descartes, M. Régis saw with pleasure Cartesianism, or, if you like, modern philosophy, acquire in Montpellier by his care a great number of new followers. True physicists are geometers: people were convinced of this truth, to which they even paid a kind of homage; but the difficulty of studying mathematics, much greater than it is today, often put off those who tried to apply themselves to it. Most of them had only a superficial knowledge of it, for lack of the necessary help take it further (Histoire de la Société Royale de Montpellier, Volume One 1766, p. 4).

Régis stayed in Montpellier for seven to eight years, spreading his ideas through teaching which many people from wealthy backgrounds came to attend. Descartes' ideas proved to be stimulating and contributed to the development of a taste for science in this city. At the end of the 17th century, a core group of educated men, animated by a common taste for natural sciences, and meeting periodically to discuss them, was formed:

> There, says Gauteron [future secretary of the Society], were agitated questions of astronomy, physics, anatomy and natural history that the progress of science put on the agenda. Each one brought to these conferences his share of lights, and provided to the maintenance by the account of his own observations and the result of his work. It was already, in a word, the substance of an Academy, minus the title which signals it to the outside world and the forms which regulate its action (Castelnau 1858, p. 18).

Father Jean Picard was sent in May 1674 to Montpellier, where the weather was generally more favorable than in Paris, to observe the passage of Mercury on the sun. The overcast weather that day prevented the observation, but Picard took advantage of it to determine the height of the pole and to make numerous observations of the refraction of starlight. In 1676, Saporta and Rheyle observed in Montpellier an eclipse of the sun. At the beginning of the 1700s, anatomists, physicists and natural history specialists met frequently, under the leadership of Plantade, Bon and Clapiès, and conceived the desire to make these meetings official with respect to their political authorities, in order to improve the dissemination of their studies to the public. Famous Parisian scholars, like Jean-Dominique Cassini and his nephew Jacques Philippe Maraldi are hosted in 1701 at the Plantade's, on the occasion of the surveys intended to extend the meridian of the Observatoire de Paris. François de Plantade spoke to Cassini about his project. The quote below shows how receptive he was:

> Mr Cassini agreed that the beauty of the climate, the serenity of the air, the fertile soil of Montpellier in plants of all kinds, the vicinity of the sea, the University of Medicine, & the great number of foreigners who come to this city favored infinitely the establishment that was being meditated, which could only be very advantageous to the progress of Arts & Sciences.

> M. Cassini did not remain there: he promised to use all his power for the success of the project which one wanted to entrust to him, and even let hope that the negotiation would not be fruitless; it was not a question any more from then on to find the occasion to ask the King for his approval (Histoire de la Société Royale de Montpellier, Volume One 1766, p. 7).

The war related to the succession of Spain, which opposed England to France in this period, delaying the project, but the three friends, supported by the bishop of

Montpellier, Charles-Joachim Colbert, who had opened to the members of the Society his abundant library, used as a meeting place, continued to exchange with Cassini and the Parisian Academy. In 1705, Plantade addressed an official request to Cassini, who immediately spoke about it to Abbé Bignon. The Father approved the project and asked the Chancellor Louis Phélypeaux de Pontchartrain to speak about it to the king. Louis XIV's decision was positive:

> It was only a matter of giving this body a suitable form, and Statutes by which it could support itself and conduct itself in perpetuity. Abboy Bignon did not regulate himself in all things by the Memoirs which had been sent to him from Montpellier on the request which he himself had made, & it was a pleasant surprise for all the individuals who were accustomed to meet, when in the month of February 1706, they saw their Society created by the Letters Patent into an academy body, under the name of Société Royale des Sciences, having the King as its Protector, illustrated by six honorary Academicians, equally distinguished by their rank & by their birth, united with the Académie Royale des Sciences de Paris & forming with it a single body, regulated by statutes which are the same, except for a few changes, which the difference of places, and other considerations of this nature made indispensable (ibid., pp. 9–10).

The Letters Patent establishing the Société Royale des Sciences de Montpellier are dated February 1706 (ibid.). They summarize the general motivations for the establishment of academic societies (national and international influence, scientific and technical progress in the service of the State, dissemination of knowledge to the public) and those specific to Montpellier (pre-existence of a high-level scientific community, quality of the climate for astronomical observation):

> The particular attention we have always given, and the care we have taken in the most important occupations of our State, to make the Arts and Sciences flourish in our Kingdom, having made us aware that the most suitable means to contribute to their perfection, have been the various Regulations we have given at different times, to favor the Scholars, either by establishing new assemblies of People of the Humanities, or by giving a new form to the old ones. We usefully employed this last means, by re-establishing in January 1699, under a new order, our Académie Royale des Sciences of our good city of Paris, whose Academicians endeavoring since then, by their application and their discoveries, to answer worthily to the honor which we made to them, seem to have spread the spirit and the taste of sciences in all Europe, to have animated quantity of scientists of our Kingdom to seek the ways of making their studies & their knowledge

> useful to our service, & profitable to the Public. In this view several subjects of our city of Montpellier, who for a long time have maintained between them a close bond of friendship & study, have very humbly pointed out to us, that applying themselves for several years to the various parts of Mathematics & Physics, & besides living in a City where the temperature & the serenity of the air put them in a position to make more easily, than in any other place, important & curious observations & research, they could thus contribute a lot to the desire that we have to see the perfection of all these sciences, if we wanted to allow that, to work on it with more fruits, they could assemble under our Royal protection (ibid., pp. 11–12).

The Société de Montpellier was the first of the scientific societies to be instituted by the Letters Patent, the same treatment having been previously applied only to literary societies (Académie Française, Académies de Nîmes affiliated to the Académie Française, Académies de Toulouse and de Caen). Moreover, contrary to the 10 or so scientific societies in the provinces that were formed in the course of the 18th century (Bordeaux in 1712; Lyon in 1724; Marseille in 1726; Dijon in 1740, etc.), the Société de Montpellier was totally integrated into the Académie Royale des Sciences, "of which the said society would be considered only as an extension and a part". Bignon, who had just restructured the Académie des Sciences, thought "that it might be useful to give this great academy a counterpart at another end of the kingdom, where the sciences were successfully cultivated". The first article of the statutes of the Society was explicit on this point: "The Société Royale des Sciences, established in Montpellier, will always remain under the protection of the King, in the same way as the Académie Royale des Sciences established in Paris, with which it will maintain the most intimate union, as if they were one body" (ibid., p. 16). The Montpellier society had six honorary members, with intelligence in mathematics and physics, from among whom the president was chosen, and 15 associates who had to be established in Montpellier, and from among whom the secretary was chosen, as well as 15 students. The associates included three mathematicians, three anatomists, three chemists, three botanists and three other physicists who worked on other parts of the natural sciences, one of these 15 associates playing the role of secretary. The role assigned to the secretary was the same as for the Parisian Academy:

> The Secretary will be exact to collect in substance all that has been proposed, shaken, examined & resolved in the Society, to write it on his register, in relation to each day of assembly, & to insert the treaty of which will have been read; he will sign all the acts which will be delivered, either to those of the Society, or to others who will have interest to have it, & will give to the public an extract of his registers, or a reasoned history of what will be done of most remarkable in the Society, each time that he will have enough material (ibid., p. 22).

Articles XXIX to XLII of the statutes concern the links between the Parisian Académie des Sciences and the Société de Montpellier, and as such were of capital importance for the successful functioning of the whole. The two societies were obliged to send each other all the memoirs that they printed. The Société de Montpellier sent every year, just before Easter, one of its memoirs to the Académie des Sciences, which printed it with its own memoirs of the same year. The Académie des Sciences could ask the Société de Montpellier to examine a particular scientific question, and vice versa. Finally, the members of the Académie des Sciences present in Montpellier could attend the meetings of the Société de Montpellier, and vice versa. The Société de Montpellier held its first public meeting on December 10, 1706 in the presence of the States of Languedoc. Plantade, as director, gave a speech detailing all the skills and scientific works of the members of the Society, presided over by the Bishop of Montpellier. Antoine Gauteron elected Permanent Secretary of the Society, received in 1707 letters from Fontenelle, his counterpart at the Académie des Sciences, detailing for him as an example the way in which the latter exercised his function. During the same period, the Society completed its constitution and chose correspondents, among whom the Italian naturalist and astronomer Louis-Ferdinand Marsigli, one of the founding fathers of the Academy of Sciences of the Institute of Bologna, and Jean-Jacques Dortous de Mairan then still in Béziers.

Figure 1.6. *Cover page of the book containing the speech given by François de Plantade at the first assembly of the* Société Royale des Sciences de Montpellier *(de Plantade 1706)*

1.5. The Parisian actors: Bernard le Bovier de Fontenelle and Jean-Jacques Dortous de Mairan, the Académie Royale des Sciences

1.5.1. *The Académie Royale des Sciences*

The Académie Royale des Sciences de Paris was founded in 1666 (Maury 1864). Like its English and Italian counterparts, it grew out of pre-existing informal circles of scholars and amateurs who met regularly to discuss their studies and share their discoveries. In France, these meetings took place first at the home of Henri Louis Haber de Montmor, a poet and collector of instruments, who gathered in his home a circle of scholars including Descartes, Gilles Personne de Roberval, Marin Mersenne, Pierre Gassendi, Blaise Pascal and his father. Thomas Hobbes participated in these meetings during his stay in Paris in 1640. Colbert, eager to hasten the progress of knowledge, formed the project to make this circle a royal institution. He perfectly realized the practical importance of the research in progress, until then purely speculative. The core of the academic society, which took place next to the already institutionalized Académie Française, consisted of seven mathematicians, among them Christian Huygens, Roberval and Adrien Auzout. The group was then enlarged to include observational sciences: anatomy, physics and chemistry:

> This gathering of hard-working men, of elite minds, of genius intelligences, was to devote itself in common and in an even more assiduous way than in the past, to the study of questions whose observation or calculation promised the solution. The king ensured the existence of the academicians by means of pensions and placed at their disposal a fund intended to provide for the expenses of their experiments and their instruments (ibid., p. 13).

On December 22, 1666, the Assemblée opened its meetings in one of the rooms of the King's Library. It was decided that the Society would meet twice a week: the mathematicians on Wednesday, the physicists (naturalists and physiologists) on Saturday. Works and discussions had to be kept secret, to avoid any risk of plagiarism. The king chose as Secretary of the Academy the oratorian Jean-Baptiste Duhamel, who had a perfect command of Latin, the language in which the minutes were written until 1699. Duhamel taught philosophy and was a mathematician. The treasurer, second officer of the Academy in the order of protocol, had only a restricted role, because he did not manage the finances. The meeting room was also at that time a real laboratory, where experiments and observations were set up and their results discussed. Claude Perrault defined in 1667 a work plan for physics (all natural history). Geometers and physicists did not remain in their disciplines, but communicated between them. Among the geometers, Edme Mariotte distinguished himself by his ability to derive mathematical principles from the phenomena he studied. The debates were less lively among the astronomers, who built together tables of ephemerides, more and more complete.

Under the influence of Colbert, the king tried at that time to attract foreign scientists to increase the influence of France, competing in particular with the Royal Society and its Greenwich Observatory. Huygens was integrated from the beginning, in 1666. Ole Chistensen Rœmer was elected in 1672, Jean-Dominique Cassini, director of the Observatoire de Paris since its foundation in 1669, in 1673. Others, like Ehrenfried Walther von Tschirnhaus, Domenico Guglielmini, Nicolas Hartsoeker, the Jean brothers and Jacques Bernoulli and Isaac Newton were elected as foreign associates in the last decade of the 17th century, but did not take French nationality. Alfred Maury, a French scholar of the 18th century, attributed the refusal of many foreign scholars to settle in France to the direct tutelage that the king exercised over the academicians, whom he in fact paid, which did not allow the exercise of an independent science. This was one of the essential differences between the Académie Royale des Sciences and the Royal Society which was financed (with great difficulty) by the contributions of its members, paying at the same time its independence with the price of poverty. In 1679, Colbert asked Philippe de La Hire and Jean Picard to supervise the establishment of a map of France, and it was the occasion for the astronomers to make numerous observations, benefiting from the royal financing granted for the enterprise. The death of Colbert replaced by Louvois in 1683 marked a break with a period of great activity. As much as Colbert respected and understood the importance of letting scientists do their research, diverting them as little as possible from their objective, Louvois considered them as "people paid by the king to satisfy his curiosity, to answer him about the rain and shine and to help his masons, his officers and his architects" (ibid., p. 37). Louvois mobilized the energy of the Academy on the development of the gardens of the Palace of Versailles, or on the improvement of weapons of war. At the very end of the 18th century, Louis Phélypeaux de Pontchartrain, Secretary of State in charge of the King's Mission and the Academies, wished to develop the Academy. He placed at its head his nephew, Father Jean-Paul Bignon, whose culture, literary qualities and eloquence made him an ideal director for a scientific society. Pontchartrain and Bignon, with the agreement of Louis XIV, proposed a new regulation of the Academy, signed in Versailles in January 1699, and read at the Academy the following February 4. The proposed organization, which was used almost word for word for the Société Royale des Sciences de Montpellier was as follows:

> Not only had the framework of the Society been considerably enlarged, but a hierarchy was introduced which allowed the entry of professional scholars, young men promising to become such, and great lords wishing they were not; in other words, according to the new constitution, there were honorary members, members, associates,

and students. The honorary members had to be, according to the rules, recommendable by their intelligence in mathematics and physics. These places, ten in number, were reserved for high-ranking persons, for the courtiers began to aspire to a title which had seemed at first too modest to their ambition, but to which they later realized that some fame could be attached. The members were the real academicians; they included three geometers, an equal number of astronomers, mechanics, anatomists, chemists, botanists, plus a secretary and a treasurer. To each of the sections of three members were aggregated two associates. There were also eight foreign associates and four free associates. Finally, the students had to be attached to the persons of the members, who each had one; these students had to be at least twenty years old (ibid., pp. 41–42).

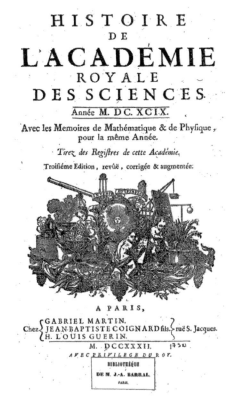

Figure 1.7. *Cover page of the first volume of the* Histoire de l'Académie Royale des Sciences *(1699)*

The solemn opening of the new Academy took place on June 29, 1699, in premises in the old Louvre made available by the king. Bernard le Bovier de Fontenelle was chosen to be the Permanent Secretary. The king appointed the members. The secretary and the treasurer were permanent also. The president, the vice-president, the director and the deputy director had an annual mandate. Bignon was president the first two years, then every other year, then every third year, on average until 1734. The presidents who alternated with him changed from one time to the next. They were rarely scholars, more often men of influence serving as intermediaries with the king. Bignon was to control the evolution of the Academy, in good agreement with his secretary Fontenelle, and to put the academicians in regular contact with scholars from the provinces or from abroad. Fifteen foreigners were associated between 1699 and 1730, coming from Germany, Switzerland, England, Italy and Spain. Bignon had to manage the stormy introduction in France of the infinitesimal calculus of Leibniz and Newton, which provoked divisions within the Academy, by appointing a commission on the subject, which included Thomas Gouye, Jean-Dominique Cassini and Philippe de La Hire. Pierre Varignon and Bernard Joseph Saurin were very offensive in favor of the new method. After five years of debate, the case was decided in favor of the infinitesimal calculus. Newtonian physics gave rise to much longer disputes, which went far beyond the framework of the Académie des Sciences alone, and were only resolved in the middle of the 18th century with the triumph of Newtonianism following the Swedish and Peruvian expeditions to measure the shape of the Earth. The Academy led a very active policy of opening to foreign countries, in particular by the institution of scientific competitions. Here is what Maury said about it:

> What contributed most to establish relations between the Académie des Sciences and foreign scholars were the annual competitions hosted by the Society; the winning works were published in a special collection following the Mémoires and which began to appear in 1721. The idea of these prizes belonged to a Councillor of the Parliament, M. Rouillé de Meslay, who, that year, had donated a sum of money to the Academy for this purpose. The most eminent scientists from abroad did not believe that they were in derogation, by coming to compete for the approval of an illustrious assembly, and one finds, in the Collection of prizes, memoirs composed by some of the men of genius whose names I have quoted above. Euler was four times crowned for questions of physics and mathematics. The marquis Poleni competed four times and won the prize three times. Daniel Bernoulli won ten times. L'Académie des sciences in Paris was not the only one to hold such competitions. For a century, various cities of France had received from the king the confirmation of the academies which had been founded there and where physics, geometry and humanities were cultivated at the same time. Caen, Arles, Soissons,

Villefranche-en-Beaujolais, Nîmes, Angers, Montpellier, Bordeaux, Lyon, etc., had successively established their academies, whose competitions also called for and sometimes obtained the memoirs of distinguished men (ibid., p. 171).

We will now discover Bernard le Bovier de Fontenelle, key figure of the French scientific community at that time, and Jean-Jacques Dortous de Mairan, creator of the aurora borealis system defended by the Académie des Sciences and who briefly succeeded Fontenelle as secretary of the Academy between 1740 and 1743.

1.5.2. *The permanent secretary Bernard le Bovier de Fontenelle*

The secretary of the Académie des Sciences, who worked for almost the entire first half of the 18th century, was an important figure, whose personality and idea of his function, as well as his philosophical position, had to be clearly defined in order to understand what followed. The main part of the portrait of Fontenelle that we draw here is based on the elements of his correspondence published in 1766 in the 11th volume of his works. They do not so much paint an objective portrait of the man as the image he sought to project of himself, with the obvious aim of establishing his authority. He presents himself in these letters as a man of even temperament and peace, who did not enjoy quarrels, as he said to Jean Le Clerc in a letter from 1707: "Fat any rate, I am not at all in a polemical mood, & all quarrels displease me" (Fontenelle 1766a, p. 5), or to Father Louis Bertrand Castel in 1729:

> But, my reverend Father, I think it is better to leave all these little facts there, which those who report them almost always report very unfaithfully. One would never cease to complain, to accuse, to suspect; and peace of mind is preferable to all the possible powers and roots of all the numbers. Besides, we must not allow all these minutiae to distract us from serious studies (ibid., p. 154).

Fontenelle presented himself to Father Castel as open-minded, a quality that he considered rare, differentiating himself from his fellow students: "All the advantage I can have, and which is nevertheless quite rare, is that I am not prejudiced towards any system, and that I will not reject any opinion that is contrary to my own." The portrait of Fontenelle made in 1726 by Madame de Forgeville insists a lot on his kindness, "his uniform conduct, and everywhere his principles". Fontenelle was described as an honest man, "modest in his speech, simple in his actions. The superiority of his merit shows itself, but he never makes it felt" (ibid., p. 223). This portrait, as conventional and superficial as it is, shows to what extent Fontenelle adhered to the image he forged for himself through his correspondence.

He also appeared as a man exercising an important power in the academic world of the Enlightenment, as shown by the letter sent to him in June 1741 by his friend David Renaud Boullier who had just learned that he was resigning from the position of secretary of the Académie des Sciences: "Please allow me, Sir, to mourn the irreparable loss that the Sciences are making in this respect, & that they are making rather than one would have expected. You were for them a kind of Prime Minister" (ibid., pp. 34–35). Fontenelle's success at the helm of French science was fully appreciated and recognized by royal power, as attested by the answer of the Count of Maurepas, Secretary of State and Minister of the Navy, to Fontenelle's request in 1730 to be dismissed from his position: "I agree, Sir, that the learned Moulin [the Academy] must always go on, & that it must go on well; but for that very reason it is necessary that the one who has led it so well for a long time, does not get sick of it, & continues to give it his care, at least until he has put a successor in a position to make people regret his loss less" (ibid., p. 104). The epistolary relations between Fontenelle and Cardinal de Fleury, Louis XV's first minister, shows, despite the polite words used by Fontenelle, a certain familiarity between the two men, as shown by the exchange of May 1740, in which Fontenelle asked the royal power to be relieved of his duties:

MY LORD,

Ten years ago I obtained from Your Eminence written permission to abdicate my sole dignity as Secretary of the Académie des Sciences. The reasons I had then have been strengthened, and I ask you very humbly and sincerely for confirmation of the same grace. Count de Maurepas is informed of everything.

RESPONSE

You are only an idle man & a libertine; but one must be indulgent to these kinds of characters. We shall see (ibid., pp. 186–187).

Fontenelle's aura placed him in a position to influence the creation and development of provincial academies, such as that of Rouen, for which he was widely consulted between 1740 and 1743 by Claude Nicolas Le Cat and Bettencourt, a lawyer at the Parliament of Rouen, who proposed him to be the patron of the new academy and asked him to write the draft of the Letters Patent, which in the end was strongly inspired by the Parisian organization. In 1750 and 1751, Fontenelle was named a member of the Academies of Prussia and Nancy, respectively, for which he thanked Samuel Formey, Professor of Philosophy at the French College in Berlin and member of this Academy, who had invited him to join, and the King of Poland, Duke of Lorraine and Bar.

Fontenelle also had a long-standing professional relationship with Father Jean-Claude Bignon, a man of considerable influence who was the architect of the reorganization of the Académie Royale des Sciences in 1699. Bignon was president of the Academy every other year during most of Fontenelle's term as secretary. In a letter from January 1716, which followed the national inquiry launched by the Duke of Orleans, the new regent, on the natural resources of France, a survey placed under the responsibility of René Antoine Ferchault de Réaumur, Bignon expressed his gratitude and unwavering attachment to Fontenelle. He said about this survey: "It is properly your work, and I have no intention of forgetting all the virtues and all the friendship for me, that you have made appear in the last place. Also you must be persuaded that, if it were possible, it would be enough to redouble the esteem & the attachment with which I will be all my life" (ibid., p. 103). The duo of Fontenelle and Bignon was soon joined by the young Réaumur (Shank 2008, pp. 76–86), whose meteoric rise at the Academy, from the rank of student in 1708 to that of member and director on numerous occasions from 1714 onwards, was due, even more than to his scientific talents and exceptional pedagogical qualities, to an ever more marked orientation towards the applications of science. From 1708, the year in which Réaumur, only 25 years old, turned away from mathematics to focus on applications, a strong link was created between Réaumur and Bignon that would not be denied thereafter.

Fontenelle made abundant use of the power of his position to defend his own interests. In his correspondence, we notice the frequent mention of his works, which he recommended to the attention of other scholars. In his letter from 1728 to Johann Christoph Gottsched, professor in Leipzig, he mentions that he sent his book *Éléments de la géométrie de l'infini* (*Éléments* in the following text) to Mr. Hausen, also from Leipzig, asking him to say something about it in the *Acta Eruditorum* of Leipzig, and at the same time he asked Gotssched "to obtain from him the grace that he reads me". But he sometimes went further, as in an exchange with Willem Jacob's Gravesande in the early 1730s, in which he asked his interlocutor to review his criticism. Gravesande, who held Newtonian views, had in fact published a review of *Éléments* in the journal published in The Hague, which had displeased Fontenelle, and the latter did not hesitate to write to him: "I have reason to fear that your difficulties, which come from such good hands, will make too much of an impression," and to ask him explicitly to review his judgment publicly: "I already have as guarantors a great number of suffrages of the greatest weight; I would infinitely wish that yours could be one, & that at least you give at the end of your extracts a general judgment, which would perhaps be more favorable to me than the detailed judgments: but I have no intention of asking you anything against your conscience" (Fontenelle 1766a, p. 41). Gravesande published a correction which did not satisfy Fontenelle, with him reacting and attacking the editorial staff of the newspaper, leaving himself "mortified". In a similar register, a dispute opposed in 1728 and 1729 Fontenelle and Father Castel. Castel was a brilliant Jesuit, admitted in 1721, on Fontenelle's recommendation, to the team of editors of the *Mémoires de Trévoux* and taught mathematics. In 1727, he published a short work of popularization, which was immediately subjected to harsh criticism by some of Fontenelle's academic colleagues. Castel's draft reading note on the *Éléments*, probably bearing the mark of this dispute, was not to Fontenelle's liking, and he enjoined Castel to revise the first version and to "kindly finish [his] third extract with a general [favorable] judgment, as would be quite natural to do. This is the impression, or at least the strongest one, that remains with most readers" (ibid., p. 153). Castel complied, but the situation became increasingly serious, leading to a quarrel and the interruption of the correspondence between the two men. These examples show, on the one hand, to what extent a positive judgment by Fontenelle constituted for an author, a guarantee of the scientific value of his work, and, on the other hand, how Fontenelle used the power that his fame conferred on him to influence the criticisms that were made of his own books. We discover a determined and offensive Fontenelle, concerned with his scientific interests and those of the Academy, over whose destiny he presided.

In a letter dated January 30, 1706 to Antoine Gauteron, secretary of the Société Royale des Sciences de Montpellier, Fontenelle describes his vision of his role as permanent secretary, clearly distinguishing his role from that of Abbé Bignon:

> We do not give oral speeches at public assemblies. Eloquence is not accepted here, […] at the most the President says a few words to announce to the public that everything is going to happen in the usual way […] Up to now it has always been Abbé Bignon who has presided over the public assemblies. After the reading of each one, he summarizes what has been said, gives it to the public in abbreviated form, and usually in clearer terms, adds such reflections as he wishes, and that in a way that everyone is charmed.
>
> […]
>
> 3. I do not profess any science like all the others, and I am the Ignorant of the Society. I have been taken on for no other reason. […] My work consists of writing these Histoires which are given every year, and in which what is properly called Histoire is mine, to the exclusion of the Mémoires. […] I am not a born orator by virtue of my position, and I am even less so by virtue of my character.
>
> 4. When someone from the outside wants to introduce himself to the Society, he addresses the President and not me. I do not interfere in introducing him. However, anyone can suggest to the President or to

the Assembly that someone is asking to enter and speak. All this is done without any regulated ceremony [...].

6. I speak about myself in my Registers in the first person, and in the Histoire in the third;

[...]

9. We have no formal attire. Each one comes to the Assemblies in his ordinary dress (Castelnau 1858, pp. 119–120).

In this text, Fontenelle displays his neutrality, not failing, as usual in his letters, to show an excessive modesty, corresponding to the image of honest scholar of the Republic of Letters that he sought to give. It is clearly the writing of *Histoire* that Fontenelle put forward. He clearly differentiated between his registers, where "everything that is read in the Assemblies, whatever it may be, is copied word for word", and his *Histoire*, which "is not written in the Registers; neither do I keep the original, because I am naturally an enemy of useless papers, and I am relieved of them as much as I can". In the same letter from January 1706, he specifies that he did not make "any profit from the printing of the *Histoire*, nor from the Society either".

The drafting of a "reasoned history" appeared in the 1699 regulations. The objectives were "the dissemination of the work of scholars, the communication to the public of the activities conducted by the academicians and, in the background, the exaltation of the king's scientific policy". In a break with the writings of the previous secretary, Jean-Baptiste Duhamel, written in Latin, *Histoire* was written in French:

> On the one hand, the Histoire does not merely summarize or comment on the memoirs published in the second part of the volume; it has its own content, which the secretary of the Academy considers just as useful for the knowledge of the institution as the work of the academicians. On the other hand, this part is not aimed at an academic audience, but rather at curious readers who, without having much scientific knowledge, have enough to appreciate the theories discussed at the Academy (Seguin 2012, p. 370).

Fontenelle was not content to simply popularize; he also let his own vision shine through his thematic choices and comments, as expressed by D'Alembert in his eulogy of the scholar: "Fontenelle, without ever being obscure, except for those who do not even deserve to be clear, spares himself both the pleasure of implying, and that of hoping that he will be fully heard by those who are worthy" (D'Alembert

1821, p. 138). He mixed sources, rewrote, added comments of his own. He analyzed the conditions in which discoveries were made, discussed the relevance of methods, explained difficulties and drew conclusions, "not only on the validity of the knowledge thus obtained, but also on the legitimacy of the discourse elaborated by scholars, even on the very possibility of producing a scientific discourse, notably in certain fields such as the life sciences" (Seguin 2012, p. 373). The articles from *Histoire* are daily testimonies of the progress made in the different fields (physics: general physics, anatomy, chemistry, botany; mathematics: algebra, geometry, astronomy, optics, acoustics and mechanics), and were considered as a knowledge base for a synthesis that would be realized later. These articles could only be written by members of the Academy. We find, for example, in the correspondence between Isaac Newton and Roger Cotes (Edleston 1850), a reference to a letter from Edmond Halley written to Fontenelle in 1716 asking him to insert in the memoirs of the Académie Royale des Sciences an excerpt from a text by John Keill, popularizer in England of Newton's theory, written in reaction to a recent article in the *Mémoires*. Fontenelle replied that "we do not yield here to the English even in esteem and veneration for Mr Newton. And the Academy would very much like it to be possible" (ibid., p. 187) to insert Keill's article in its *Mémoires*, but that it was the invariable rule to admit only articles written by the members of the Institution.

Fontenelle seemed to communicate little, at least directly, with foreign scholars. In a letter from August 1731 to Hans Sloane, first secretary between 1693 and 1713, then president of the Royal Society from 1727 onwards, he acknowledged that he did not yet know him, and it was only in a letter from 1733 that he claimed to have entered into a closer relationship with him. This may seem all the more astonishing since Sloane had been a foreign member of the Académie Royale des Sciences since 1708, a privilege he shared at the time of the correspondence with Fontenelle with only three other Englishmen: Newton, admitted in 1699, Thomas Herbert of Pembroke received in 1710, and Halley, made a member only two years earlier. In fact, it was mainly through Bignon that the exchanges between the French and English communities of the time took place (Bonno 1948). Sloane, who had many French scholars elected to the Royal Society, about 15 between 1698 and the date of the correspondence with Fontenelle, the latter not entering the Royal Society until 1733, was in charge of transmitting to Bignon, and to certain scholars, such as Cassini or Mairan, copies of the *Philosophical Transactions* and of scientific books of famous Englishmen such as Roger Bacon or Newton. Fontenelle regularly displayed his admiration for English science, as in his letter to Lockmann in 1744: "I would have taken great pleasure in comparing the genius of the two nations. I am already familiar with the genius of yours [England], on works of strength, so to speak, on geometry, physics, metaphysics, and I know that it goes as far as it is possible to go," and a little further on: "I am proud to be a little English, since the

Royal Society of London has kindly received me into its illustrious body" (Fontenelle 1766a, pp. 59–60). Fontenelle was regularly kept informed of English scientific life by his correspondents in London, such as Father de la Pillonière who, in June 1730, informed him of the publication of an abridged explanation of Newton's *Principles*, which Sloane judged "certainly [...] very capable of shedding light on a philosophy as little developed as it is worthy of being heard". As a foreign member of the Academy, Newton was the subject, after his death in March 1727, of a eulogy written by Fontenelle. In June of the same year, Fontenelle approached Jacques Serces, a French Protestant who had taken refuge in London, to collect from John Conduitt, who had succeeded Newton as Master of the Mint, the biographical information necessary for the writing of the eulogy. This eulogy, which gave pride of place to Descartes, was not well received by the English, as Father Castel wrote to Fontenelle at the beginning of 1729: "the English find that, in the eulogy of Newton, you have exalted Descartes too much to the detriment of their Hero" (ibid., p. 168).

Fontenelle appears in his correspondence as a level-headed man, gifted with an obvious sense of diplomacy, and at the same time quick to defend his own interests and those of the institution he represented, which fully embraced his role as secretary of one of the leading European scientific institutions of the time. Without being himself a scholar of great stature, he provided the scientific community with the animation and coordination necessary for its structuring, and thus played a decisive role. The period of his secretariat, which covered almost the entire first half of the 18th century, is of considerable historical importance, since it corresponds to the phase of penetration of Newtonianism on the continent, which did not take place without resistance. Fontenelle was Cartesian in the sense that he rejected attraction at a distance and remained a supporter of vortices and of action by contact. But he differed from Descartes by his historical approach, not claiming for scientific truths any kind of eternity (because they can always be made to evolve), which he expressed perfectly in his foreword to the *Histoire* of 1699:

> Let us always gather truths of Mathematics and Physics at random, it is not to risk much. It is certain that they will be drawn from a fund from which a great number have already been found useful. We can presume with reason that from this same fund we will draw several, brilliant from their birth, of a sensitive and incontestable utility. There will be others which will wait for some time until a fine meditation or a happy chance discovers their use. There will be some which, taken separately, will be sterile, and will cease to be so only when one decides to bring them together. Finally, at worst, there will be some that will be eternally useless (Fontenelle 1699, p. xj).

The system was indeed for Fontenelle the completed form of knowledge, but we need not precipitate its elaboration, because premature systems were dangerous. This is precisely the criticism that Bignon addressed to Plantade, taking up to the letter Fontenelle's recommendations, which were also widely shared at the time in the Republic of Letters. "The Académie des Sciences thus advocates an empiricist and skeptical approach, which invites to keep as close as possible to the facts, and to be satisfied, in the absence of a system, with simple hypotheses" (Mazauric 2007, p. 351). The spirit of the Enlightenment is thus marked at the base by an inductive empiricist approach, embodied by Newton, in dispute with the Cartesian deductive approach. But the Cartesians of the Academy, such as Fontenelle or Mairan, reproached Newton for the idea of attraction at a distance irreconcilable with their mechanistic vision of the world, that they judge likely to make science regress towards obscurantism. They remained partisans until the end of the vortex theory and of the subtle matter proposed by Descartes. More deeply:

> The only function of the reference to Descartes is to mark a moment, the moment of access to reason, an access to reason that founds the right to make use of it as he did, and thus possibly to make use of it in a different way, even against him. It legitimizes from then on this great enterprise of historical construction of the truth that the history of science has the function of telling, a history of science that, in so doing, can thus fulfil the philosophical function that Fontenelle had assigned to it from the start: to attest the reality of human progress (ibid., p. 354).

Far from being regressive, Fontenelle's Cartesianism marked an essential stage in the gradual mutation of the scientific approach towards a process of constant back and forth between observation/experiment and reason, which is the foundation of modern scientific practice.

1.5.3. *Jean-Jacques Dortous de Mairan*

The principal features of Mairan's character, as emerging from his correspondence with his old friend from Béziers Jean Bouillet (Unpublished letters from Mairan à Bouillet 1860), a doctor, physicist and astronomer, remind by some aspects (the reflective character, the detestation of frontal conflicts) those of Fontenelle. Bouillet and Mairan, both natives of Béziers, had known each other for a long time, and were co-founders of the Academy of that city in 1723. The correspondence between the two men, which spanned almost 50 years, allows us to better understand Mairan's personality. It reveals in particular the repeated and insistent steps of Bouillet with his friend so that he supported him in obtaining the Letters Patent which would make official the Académie de Béziers and obtain the means to make it function. As a wise man, Mairan did not cease to temper the ardor of his friend, trying to convince him not to want to go too fast, and to develop the scientific activities, and the quality of the productions, of the society before thinking of asking for the official recognition by the king. Mairan nevertheless began to take cautious steps towards Cardinal de Fleury, the first minister of Louis XV, whom he had the opportunity to meet at Versailles. In 1729, the patents were refused on the grounds that another important academy was present in the region (the Société Royale des Sciences de Montpellier, strongly supported by Jean-Dominique Cassini) and that Béziers is a small city, meaning it did not seem likely to provide enough scholars to compose an academy. But the door was not definitively closed by the royal power (the patents were finally granted in 1766). Bouillet, on the other hand, did not cease to consider going to Paris, where he wanted to be appointed to the Académie Royale des Sciences, an undertaking of which Mairan did not cease in their exchanges to point out the difficulty. The correspondence between the two men focused much on the sending of parcels of books and volumes of the memoirs of the two academies, as well as of supply to Bouillet of material of good invoice for astronomy (quadrant) and meteorology (thermometers with mercury and spirit of wine). Mairan did not cease to ask Bouillet for temperature readings, of which he estimated that regular taking was essential, just as it was for him the cataloging of the aurora borealis, in the spirit defined by Fontenelle in his foreword of 1699, promising to have them inserted by Fontenelle in the *Histoire*. Throughout these years, Mairan showed a great constancy to recall to Fontenelle's memory the meteorological observations of his friend.

In addition to Mairan's long-term loyalty to Bouillet, and the efforts he made to help him in his endeavors, these exchanges also reveal a refusal to engage in quarrels between schools of thought, combined with a keen sense of diplomacy designed to protect him from them, very reminiscent of Fontenelle's position. He wrote, for example, in his letter from May 20, 1734:

> It is a deplorable thing that the harassments, the deaf maneuvers, the envies and the injustices which are mixed in the processes of the scholars and the people of letters. But finally, they are men, and consequently exposed to all the miseries of humanity. This should not prevent those among them who think and act more roundly and more nobly, from going their way and being useful to the public by their vigils, as much as their faculties and their health can allow it (ibid., p. 12).

He said he avoided any direct exchange with scholars whose views he did not share, on subjects likely to oppose them. For example, on the question of the active forces of which Jean Bernoulli was a defender, and about a letter he wrote to him to present him a work of Bouillet's, here is how Mairan expressed himself: "it would have been imprudent of me to hit a nerve, neither directly nor indirectly; because I will always do my best to live well with persons of such distinguished merit. Our letters were full of the active Forces before 1723; but since this memoir we do not speak of them anymore" (ibid., p. 177). Although he criticized Newton's optics and Newton's gravitation, Mairan wrote: "For although I always speak of Mr Newton with all the respect due to him and that I truly have, as much for his person as for his great knowledge and his works, I do not want anything to leave my pen that smacks of criticism of him" (ibid., p. 31). He went so far as to advise his friend to adopt the same attitude as him, for example, towards René-Antoine Ferchault de Réaumur. As we have seen, he was a man of great influence at the Academy, with whom he admitted to be in a difficult relationship, and seemed not to interact much anymore: "As for the little discussion you have with Mr de Réaumur about your beetles, you will drop it, if you believe me. It is certainly the best you can do, in the circumstances you are in and the views you have, and it is only a trifle. You will benefit in this from an advice whose solidity I learned at my expense" (ibid., p. 147). Mairan was a man of consensus, as shown by his affirmation of the Cartesian and Newtonian character of his system of the aurora borealis and more generally his tendency to put on the same footing Newton and Descartes without taking sides, for example, when he wrote in his letter of March 1732 about sunspots that "if they were planets they should, both by the laws of Newtonian central forces, and by those of Cartesian vortices move all in large circles whose planes would pass by the center of the globe of the sun" (ibid., p. 124).

In an August 1730 letter to his friend, Mairan confided his relief that Fontenelle, who wanted to leave his post as Secretary, which the king wished Mairan to take over, had finally decided to stay: "I will enjoy the glory of having been chosen, without running the risks, for we have engaged Mr de Fontenelle to stay; although older than I am, he is still full of health, and he can very well give my eulogy. Be that as it may, I feel very little inclination for this function which takes me away from my ordinary and most cherished occupations" (ibid., p. 241). Mairan was not a man of power, nor of money (he wrote it often), wishing to devote the maximum of his time to the scientific studies which interested him. Anything that could keep him away from his studies, especially the risk of illness, which was a constant preoccupation of his letters, both concerning him and his interlocutor, was perceived as a threat. The need to take it easy, to live and work, permeated all his correspondence. The significant success of his aurora borealis system constituted for him an undeniable pride, as he expressed it in a letter to Bouillet in March 1734, a few months after the publication of the treaty:

> There are few people who have all the necessary parts to judge it as you do Sir. All that I receive from this small number, that is to say, from the greatest geometers and astronomers of Europe, shower me with so much praise for this book that the most unbridled vanity of authorship would be satisfied. I have even been honored, on this occasion, to ask that my name be inscribed in the catalog of the Academicians of some of the societies to whom I have sent it, and there are such great geometers as M. Bernoulli, who seem to me to be even more convinced of the truth of my explanation than I am, although I am not badly so, and who have gone on from there to imagine very ingenious things about the structure of the world (ibid., p. 247)

But he said this only to Bouillet, and would not display it in the public square, so as not to risk appearing to want to put himself forward:

> I must say all this to you as to my friend and compatriot; otherwise it would be misery on my part. It is not that we should not be sensitive to an honest glory, and that we should not have it as an object in our work; but to be too busy with it, and especially to daze others with it, is not forgivable. An honest leisure with much health would undoubtedly be better than this little smoke (ibid.).

Mairan was clearly very attached to his image of a tenacious and modest man of science, keeping out of any polemic, and refusing the exercise of power. His system of the aurora borealis that Fontenelle fully endorsed and inscribed in *Histoire* after having displayed at first his support to the Aristotelian vision of the inflammation of exhalations defended by Jacques Philippe Maraldi, became one of the proudest of the Académie Royale des Sciences, of which Fontenelle said, as we have already mentioned: "Thus the Aurora Borealis will hold a middle ground between the pure Meteors & the pure cosmic phenomena, such as all those of Astronomy, & this disposition seems to be enough of the genius of Nature" (Fontenelle 1732, p. 3).

Like Réaumur, Mairan experienced a rapid rise in the Académie Royale des Sciences (Shank 2008, pp. 94–107; Bruneau and Passeron 2015). Originally from Béziers, and initially intending to enter the Orders, he won three times, in 1716, 1717 and 1718, the prize of the Académie de Bordeaux on works applying the Cartesian mechanistic approach to questions of physics, notably the formation of ice. Like Réaumur, he was interested in analytical calculus at the very beginning of his scientific life, but quickly lost interest in abstract mathematics and devoted himself to the elucidation of physical problems through clear mechanistic explanations based on simple geometrical demonstrations. Like Fontenelle, and Réaumur, he was a convinced follower of experimental philosophy for whom a solid science begins with the collection and interpretation of empirical data. This did not prevent him, as we have seen, from defending the elaboration of systems as frameworks of thought within which to build relevant strategies of observation and experimentation. Renouncing his ecclesiastical vocation, Mairan settled in Paris in 1717, and entered directly as an Associate of the Académie Royale des Sciences at the very beginning of 1719, benefiting from the highest level of support. He was quickly made a member, and in 1721 he was entrusted with the board of the Academy:

> Dortous de Mairan's power in the academy after 1720 was thus great, and this alone placed him alongside Réaumur and Bignon. His scientific outlook, moreover, was also exceedingly compatible with theirs. Like Réaumur, Dortous de Mairan was a devoted empiricist who believed that sound science began with the collection and interpretation of empirical data. He was perhaps more attached to mechanistic system building than Réaumur, for as Kleinbaum states, "He could not single out any problem about the cosmos without presenting an entire cosmology." Nevertheless, despite his particular inclination toward grand structures, Dortous de Mairan's science was compatible with Réaumur's, especially in its concrete, empirical approach to nature (Shank 2008, p. 102).

Unlike Réaumur, Mairan frequented literary exhibitions, as did his colleague and friend Fontenelle. The three men, sharing a close relationship with Jean-Paul Bignon, playing a decisive role in the evolution of the institution:

> Overall, this triumvirate exerted an enormous influence over the actual work done inside the institution. Just as important, it also shaped the wider public perception of the academy and its work. Abstract mathematical mechanics did not fare well in this new climate, despite Fontenelle's continuing centrality and advocacy of it. The geodesic and astronomical work of the Cassinis and their astronomical colleagues; Dortous de Mairan's elegant mechanistic models of ice and the aurora borealis; the latest natural curiosity unearthed by Réaumur; the advances being made in metallurgy, clock making and the other publicly useful utilitarian arts: these were the topics that dominated the public discourse of the academy after 1715. The shift to these new priorities reflected the new prominence of Bignon's agenda's within the institution as well as the ascent of his supreme

triumvirate – Fontenelle, Réaumur, and Dortous de Mairan – to the center of French public science. They also paved the way for the new understanding of Newtonian science that began to take hold in France after 1715 (Shank 2008, p. 104).

Mairan's admiration of Newton was real, and was not simply the result of the recognition of Newton's objective genius, as it may have been the case for Fontenelle. It was a fact that Mairan constantly claimed in his work, as in his treatise on the aurora borealis, the Newtonian influence as well as the Cartesian filiation, refusing the very principle of siding with one of the two schools of thought, and taking from each of them, according to him, the best of their ideas. Ellen McNiven Hine suggests, concerning Mairan, a particular proximity to Newtonian empiricism:

> In conclusion, an examination of Mairan's published works and his unpublished correspondence reveals the complexity of his understanding of scientific methodology. The picture that emerges indicates a fascination with the Newtonian method, a wariness where speculative systems are concerned and an appreciation of the importance of the scientific imagination – all of which serve to throw light on the argument that he puts forward in the preface [to the *Dissertation sur la Glace*] (Hine 1995, p. 64).

Underlining Mairan's distrust of speculative systems (which were not sufficiently supported by observation), in fact rejected by the Newtonians, she seems to place Mairan, at least partially, in the Newtonian camp, coining for him the neologism "cartonian" (for Cartesian–Newtonian). Hine's vision was criticized by Schmit, who considered it too simplistic, especially because the term Cartesian covered a diversity of approaches at the beginning of the 18th century that could not be reduced to the refusal of the vacuum and the defense of vortices which were the basis of Descartes' cosmology, and invoked the influence of Malebranche's philosophy on Mairan (Schmit 2015). Be that as it may, Mairan was unquestionably apart from the Cartesian family of the Academy, if only by his constant claim to Newtonian input to his thought.

We will now focus on two outstanding personalities, albeit in different capacities, of the English scientific community of the time, both of whom held important responsibilities at the Royal Society: Hans Sloane, promoter with Abbé Bignon of the rapprochement between the English and French scientific communities; and Edmond Halley, an astronomer with a very broad spectrum of interests, close to Newton, whose work, notably on comets, made him an essential protagonist in the debate between Newtonians and Cartesians at the beginning of the 18th century.

1.6. The London actors: Hans Sloane and Edmond Halley, the Royal Society

1.6.1. *Hans Sloane*

Hans Sloane, born in 1660 in Ireland and of Scottish origin, studied medicine and cultivated a taste for the natural sciences (Fouchy 1753). He studied for four years in London, where he became friends with famous physicists, among them Robert Boyle. Then, he traveled to France in 1683, staying in Paris and Montpellier, before returning to London at the end of 1684. He was admitted to the Royal Society in January 1685, and also joined the Royal College of Physicians two years later. Around this time, he spent 15 months in Jamaica, from which he brought back a significant collection of plants (800 species), which contributed to his fame, and which he would show to his numerous foreign guests throughout his life. On November 30, 1693, he became Secretary of the Royal Society, and restarted the publication, interrupted a few years earlier, of the *Philosophical Transactions*, the journal of the Society. The Society had been founded in 1663 and had about a hundred members, including eight foreign members. It brought together sciences and humanities, unlike the French Académie Royale des Sciences, which was distinct from the Académie Française. Henry Oldenburg, one of its two secretaries at the beginning, published the first issue of the *Philosophical Transactions* in 1665. The Greenwich Observatory was founded in 1675, and John Flamsteed, who became a Fellow of the Royal Society in 1676, took over its direction. By this time, Oldenburg was already in regular correspondence with more than 70 scientists abroad, where the Royal Society had quickly gained a great reputation. From 1691, the year of Robert Boyle's death, the publication of the *Philosophical Transactions* resumed under the impulse of Halley after an interruption of several years. The number of new members elected each year increased from about 10 before 1691 to about 20 in the final years of the 17th century, which made it possible to increase the number of subscriptions, after a period of marked poverty, with many members being in arrears. Hans Sloane became Secretary of the Royal Society in 1693, a position he held for 20 years. During this period, but especially after his nomination as president of the Society in 1727, he was the main architect of a considerable development of scientific relations between France and England. As soon as he took office, he gave a strong impulse to the scientific correspondence of the Society, as he wrote in a letter to its members:

> The Royall [sic] Society are resolved to prosecute vigorously the whole designs of their institution, and accordingly they desire you will be pleased to give them an account of what you meet with or hear of, that is curious in nature, or any ways tending to the advancement of natural [sic] knowledge, or usefull [sic] arts. They in return will always be glad to serve you in any thing in their power (Weld 1848, pp. 356–357).

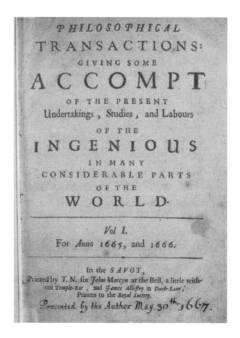

Figure 1.8. *Cover page of the first volume of* Philosophical Transactions *(1665)*

Sloane published a catalog of plants brought back from Jamaica around this time. Jean-Paul Grandjean de Fouchy, Mairan's successor as permanent secretary of the Académie des Sciences in 1744, in his eulogy after his death in 1750, noted his interest in physics:

> The following year [1702] the first volume of the voyage to Jamaica appeared in folio: Mr. Sloane's occupations delayed the printing of the second until 1725.

> In a detailed foreword which is at the head of the first volume, he establishes the pleasures & the necessity of the study of Physics; he emphasizes the advantage that this Science has, to be almost everywhere based on facts, & He points out the advantage of this science, to be almost everywhere based on facts, and therefore less subject to error, to rise through the contemplation of created things to the knowledge of the Creator, and finally to teach men the use of the countless treasures that they have from the divine liberality, and whose price their ignorance hides from them (Fouchy 1753, p. 311).

The year 1708 was marked by the election of Sloane to the French Académie Royale des Sciences. Newton was at that time president of the Royal Society and Sloane became its vice-president. In 1713, more and more taken by his occupations as doctor (notably for the queen), he abandoned the task of secretary, and was replaced in this post by Edmond Halley, another great promoter of scientific relations with the continent. He was knighted by the king in 1716, and became the physician of his armies. In 1716, he was appointed President of the Royal College of Physicians of London. Following the death of Isaac Newton, in 1727, he became president of the Royal Society, a position he kept until 1740. At the end of his eulogy, Fouchy insisted on his exceptional national and international influence:

> Mr. Sloane was a member of almost all the academies of Europe, of Berlin, of Petersburg, etc. He was a Doctor of Oxford University and a member of the College of Physicians of Edinburgh […] He was also in exchange with the late Abbé Bignon; the King himself deigned to send him as a gift the collection of engravings from his Cabinet, a gift which is usually only given to the most distinguished people, & which proves both, & the great reputation of the English philosopher, & the value that the French Monarch knows how to give to merit (ibid., p. 319).

1.6.2. *Edmond Halley*

Halley, today known especially for the comet which bears his name, is a scientific personality of first importance, distinguished by Pierre Simon de Laplace among the greatest English astronomers of the time, with John Flamsteed and James Bradley (Mailly 1867). His field of study was wide, ranging from terrestrial magnetism to meteorology, and from the study of comets to the observation of stars, passing by the observation and the interpretation of phenomena such as the flying fires (meteoroids entering the atmosphere) or the aurora borealis. Born in 1656, he entered Oxford University at the age of 17. Mairan said of him in his eulogy of 1747, the year of his death:

> He was barely 19 years old when he gave his Direct & Geometrical Method for finding the Aphelia & the eccentricities of the Planets, a work that the most consummate Astronomers of that time could envy, and which ended a famous dispute that existed between them on this subject. Descartes began his Geometry with a Problem where the Ancients had stopped; the first road that Mr Halley opened led him to all that is most hidden and subtle in Astronomy (de Mairan 1747, p. 112).

Halley began with a two-year journey, between the end of 1676 and the end of 1678, to St. Helena Island, the southernmost land that the English had under their domination, to complete with new stars the catalog of fixed stars established by Ptolemy and Tycho Brahe and thus to help in this task Flamsteed and Johannes Hevelius. On his return, he published a catalog of southern stars. He returned with an observation of a transit of Mercury in front of the sun and a method of determination from the transits of the solar parallax that he proposed to apply to the transit of Venus planned for 1761, almost a century later. In 1679, he visited Hevelius, an astronomer of high reputation in Europe, in his city of Danzig, and the two men became friends. He went to Paris in early 1681, and visited Jean-Dominique Cassini at the Observatoire de Paris on May 10. Between April 1681 and January 1682, the stages of Halley's journey in France and Italy were recorded in the register of observations of Jean-Dominique Cassini: Saumur, Toulouse, Narbonne, Montpellier, Marseille, Toulon, Fréjus, then Genoa, Livorno, Florence, Siena and finally Rome, then Paris at the beginning of 1682, where he met Cassini again (Debarbat 1986). In June 1686, the Royal Society decided to finance the measurement of a degree of latitude, that Halley wanted to realize, as Jean Picard did before him probably to confirm the dimension of the Earth deduced by the latter, used by Newton in his *Principles*. The archives of the Observatoire de Paris contain a letter from Halley to Cassini dated 1686. In it, Halley said he regrets that Cassini had interrupted his relations with the Royal Society which he hoped was not due to "some aversion" that Cassini would have conceived against it. He told him that he was in charge of the correspondence of the Royal Society, and encouraged him to renew the links: "What you will be pleased to communicate to us, will be always dear and pleasant to us, and we will not fail to send you all that will happen here of curious which will seem to us worthy of you" (ibid., p. 154). In 1687, Halley offered Cassini a copy of the 1687 edition of Newton's *Principles*, which Cassini read and annotated. Mairan, in his eulogy, mentions the compendium of comet astronomy, in which Halley gathered in a table the parameters of the orbits of 24 comets. This table, assimilated to parabolas having the sun as focus, was published in 1705 in *Philosophical Transactions* (Halley 1705). Then, he described the works and discoveries of Halley in various fields, in particular in that of the terrestrial magnetism and the aurora borealis. He paid tribute, to finish his eulogy, to Halley's intellectual boldness:

> His genius led him to bold systems. This Globe, this small Earth that we said he imagined at the center of the hollow Globe of the large one, to give reason for the magnetic variations, he still employs it to the explanation of the Aurora Borealis; because he supposes that the interval included between the concave surface of the one & the convex surface of the other, is filled with a light & luminous vapor,

which coming to escape in certain times by the Poles of the terrestrial Globe, produces all the appearances of this phenomenon.

[…]

He admitted the real and boundless space, the mutual attraction of bodies, and consequently he believed the stars to be infinite in number, because if they were not balanced on all sides and infinitely by reciprocal tendencies, they would all incessantly gather around a common center.

[…]

In a word, M. Halley was not afraid to clash with common opinions, and did not scruple to imagine, to propose hypotheses, & to conjecture according to his observations & his particular ideas. It is to this boldness, often fortunate, because it was always enlightened, that we owe the admirable theory of the variations of the Compass, & most of the other discoveries with which he enriched the learned World & Society (de Mairan 1747, pp. 147–150).

He praised his human qualities and his broadmindedness in these terms:

The glory of others did not bother him, an anxious & jealous emulation had never had access to his heart; he was also unaware of those outrageous prejudices in favor of a nation, offensive to the rest of the human race. Friend, compatriot & follower of Newton, he spoke of Descartes with respect (de Mairan 1747, p. 155).

Despite the review of the first edition of Newton's *Principles* of 1687, which appeared in the *Journal des Savants* of August 2, 1688 (Bonno 1948), in which the journal's author did not seem to grasp all the implications of the new theory, the work, which Halley had offered to Cassini the previous year, was not significantly disseminated in France, where Newton remained essentially an unknown for about 20 years. In 1700, Le Sage de la Colombière published in Geneva a book exposing the theory of universal attraction, followed the next year by John Keill in England. Keill's book, republished in 1705 and 1715, soon spread throughout Europe, marking the beginning of the long penetration of Newtonianism on the continent. Reviews of the book were published in Dutch newspapers in French, and in the *Mémoires de Trévoux*. On the French side, Father Philippe Villemot was the first to counter-attack in 1707. Academicians such as Joseph Saurin, then Nicolas Malebranche, also began to react. Roger Cotes published the second edition of the *Principles* in 1713, and it was reported in the *Journal des Savants*. Gravesande took up the Newtonian theory to explain various phenomena, such as the tides or the orbits of the planets and comets. Father Castel opposed Gravesande by denigrating "a system as ancient and as decried as that of the vacuum, of attractions and gravities independent of all mechanical forces and structures" (ibid., p. 129). For Fontenelle or Jean Bernoulli this system "is founded on principles of which one can form no idea", and here is what Fontenelle declared in his eulogy of Newton:

> He [Newton] clearly declares that he gives this attraction only for a cause which he does not know, and of which only he considers, compares, and calculates the effects, and to save himself from the reproach of calling it the *occult Qualities* of the Scholastics, he says that he establishes only *manifest* and very sensible qualities by phenomena, but that in his truth the causes of these qualities are *occult*, and that he leaves the search for them to other Philosophers. But what the Scholastics called Occult Qualities, were they not Causes! they also saw the Effects well. Moreover, these occult causes that Mr. Newton did not find, did he believe that others would find them! will we set out with great hope to seek them out? (Fontenelle 1766b, pp. 302–303).

The Cartesian opponents to Newton wanted "to maintain in physics the character of intelligibility by which Descartes had definitively broken with the errors of Scholasticism" (Bonno 1948, p. 130).

Edmond Halley was clearly a Newtonian. In an article published in 1687 in *Philosophical Transactions*, he said that he did not understand the vortex theory and its application to the explanation of gravity by the supposed effect of a subtle matter:

> Des Cartes his Notion, I must needs confess to be to me Incomprehensible, while he will have the Particles of his *Celestial matter*, by being reflected on the Surface of the Earth, and so ascending therefrom, to drive down into their places those *Terrestrial Bodies* they find above them [...] neither he, nor any of his Followers can shew how a Body suspended in *libero œthere*, shall be carried downwards by a continual Impulse tending upwards, and acting upon all its parts equally (Halley 1687, pp. 3–4).

In 1705, Halley published in *Philosophical Transactions* the elements of the parabolas in their orbit near the sun of 24 comets established according to the laws of Newtonian mechanics (Halley 1705). Three years later, Jean-Dominique Cassini continued to affirm in the *Mémoires* that the comets are in orbit around the Earth (Cassini 1708). Comets posed a complicated problem for the supporters of the Cartesian system of vortices, not allowing the trajectories of the planets to cross each other. The retrograde motion of some of them, in particular, posed an insoluble problem, and Jacques Cassini in a series of memoirs published between 1723 and 1729 tried to show that this retrograde motion was only an appearance. It took 30 years for Cassini to finally recognize the circumsolar character of cometary orbits, pointed out by Halley in 1705: "Although we could, by the reasons we have alleged, relate its movement immediately to the Earth we did not believe we had to depart from the most commonly received feeling of astronomers, that these are planets that make their revolutions around the Sun, with respect to which they describe very eccentric orbs" (Cassini 1737, p. 177). On the broader question of the explanation of gravity, for which the precise measurement of cometary orbits was precisely a means of testing the law of universal attraction, concentrating as we shall see the efforts of many astronomers such as Joseph-Nicolas Delisle or Anders Celsius, for example, several publications were made by Cartesians in the sense of defenders of the vortex theory, in the 1720s such as Jean Bouillet (the secretary of the Académie de Béziers and a friend of Mairan's), Father Castel or Georg Bernhard Bilfinger of the Imperial Academy of Sciences of Saint-Petersburg (1728, Academy prize winner), for whom "it was necessary to try everything before abandoning the vortices" (Bilfinger, quoted in Bonno (1948, p. 131)). Newton noted the incompatibility between Kepler's laws and the vortex theory. But Nicolas Saulmon from hydrodynamic experiments conducted between 1712 and 1716, and then, in 1728, Joseph Privat de Molières, professor of philosophy at the Collège de France, responded to Newton's objections. Jean Bernoulli made complex assumptions to reconcile Kepler's laws with the vortex theory. Thus, Halley, whose calculations of celestial mechanics provided the key to the explanation of gravity, was a central figure on the Newtonian scene, and we cannot doubt that he crystallized the opposition of the predominantly Cartesian leadership of the Académie Royale des Sciences. The interruption of the relations between Jean-Dominique Cassini and the Royal Society mentioned above, and that Halley told him he regretted during his visit to Paris in 1686, was most probably the beginning of this tense relationship.

1.6.3. *The Royal Society and its relations with the Académie Royale des Sciences*

Sloane and Halley welcomed in the first third of the 18th century many French scholars in the English capital, especially after the Peace of Utrecht of 1713, which followed 13 years of war between the two nations (Bonno 1948). In 1715, Jacques-Eugène d'Allonville de Louville came to London to observe a total eclipse of the sun. He met Sloane and Halley. The latter helped him to obtain the necessary equipment to observe the eclipse. Louville was elected member of the Royal Society immediately after, at the same time as the French mathematician Pierre Rémond de Montmort and the botanist Claude Geoffroy, brother of Etienne, already a member for 15 years. The decade 1720 was the one of the first explicit rallies of recognized French scientists to the Newtonian theory. Joseph-Nicolas Delisle, who left to set up an observatory in Saint Petersburg at the invitation of Peter the Great, as well as Pierre-Louis Moreau de Maupertuis, who organized a few years later the expedition to Lapland to measure the shape of the Earth, made the trip to London. To Delisle, who stayed there in 1724, Halley offered a copy of his astronomical tables not yet published, asking him to verify them. Delisle took several years to realize this work with the help of his colleagues in Saint Petersburg. Maupertuis spent six months in London in 1728, and met several of Newton's followers: Jean-Théophile Desaguliers, Henry Pemberton and John Keill. Both of them were appointed, on the occasion of their trip, members of the Royal Society. Louis Godin, a former student of Delisle, went to London in 1734 to prepare the expedition to measure the meridian arc in Peru, complementary to the one of Maupertuis in Lapland, that he was about to undertake with Charles Marie de La Condamine and Pierre Bouguer in particular. He was welcomed there, like his predecessors, by Sloane, and met Halley who provided him with the necessary optical instruments. He was also named a member of the Royal Society. But many Cartesian-inspired scientists also entered the Society at the same time, such as Joseph Privat de Molières in 1729 and Father Louis Bertrand Castel in 1730, followed in the next five years by some of the pillars of the Académie des Sciences, such as Fontenelle, Bignon (who was never a member of the Académie Royale des Sciences) or Mairan. Symmetrically, although to a lesser extent, because of its much smaller membership, the Académie Royale des Sciences opened up to foreign scholars, especially English ones. From a total of 16 members at its creation in 1666 (15 French members and Huygens), it grew to 50 members when it was restructured in 1699, 12 of whom were associates, including eight foreigners. Of the 20 or so foreign associates appointed to the Academy between 1699 and 1740 included the German Nicolas Hartsoeker, the Swiss brothers Jean and Jacques Bernoulli, the Dutchman Herman Boerhaave, and four English scholars (Newton, Sloane, Pembroke and Halley). A dozen British foreign correspondents, including Flamsteed, director of the Greenwich Observatory, were moreover designated. At the same time, the *Journal des Savants* and the *Mémoires de Trévoux* opened their columns to the British scholars, as well as the *Mercure de France* from 1724. We also find extracts of *Philosophical Transactions* in the *Mémoires littéraires de la Grande Bretagne* and in the British Library. The obedience of the scholars, Cartesian and Newtonian, clearly had nothing to do with the choice of foreign members that were made on both sides by the academies, whose essential goal was to be open to all influences, in a context of rapid development of cooperation, especially in astronomy.

As soon as he became President of the Royal Society, Sloane initiated major changes to formalize the Society's role at the highest level and to improve its financial situation, as well as to increase its international influence. One of his first acts after his appointment was to have the Council present an address to King George II, asking him to grant his patronage to the Society. The King accepted his request and entered his name in the Society's charter book. Sloane then took care to make it more difficult to admit members and to ensure the collection of dues. He first had the principle accepted that all applications for membership must be submitted to the Council for prior approval, and the following statute was adopted in 1730:

> Any person seeking election to membership in the Royal Society shall be nominated and recommended at a meeting of the Society by not less than three members, who shall deliver to one of the secretaries a certificate signed by them, containing the name, qualifications, occupation and principal titles of the candidate for election, and giving his ordinary residence. A legible copy of this certificate, with the date it was presented, shall remain posted in the meeting room for ten regular sittings, before the candidate may be balloted upon (Mailly 1867, p. 38).

This statute did not apply to peers or sons of peers, members of the King's Privy Council, and any prince or foreign ambassador who, upon the nomination of a single member, was elected on the day of their nomination. Another statute authorized the Council to sue members in arrears. Foreign members, whose number in 1730 was 79, including about 20 French, were at the same time exempted from all contributions. The period of Sloane's tenure as Secretary of the Royal Society was a flourishing one, marked by the increasing internationalization of scientific relations. Exchanges within the Society took place in Latin, English, French and even Italian (Lamoine 1993). In 1732, Sloane wrote exclusively in Latin. Cromwell Mortimer, the secretary of the time, used three languages for his letters: English, Latin or French. The exchanges involved not only the members, but also the numerous correspondents throughout Europe and America. Sloane's declining health forced him to resign in 1741. He left the Society in a satisfactory financial state. Many efforts had been necessary to balance the accounts, directed in particular towards increasing the membership of the Society. In 1740, the number of members had risen to almost 300, a threefold increase compared to November 1666. At the same time, the Académie Royale des Sciences had only 70 members, to which were added 80 correspondents. By this date, half of the members of the Royal Society had paid their dues, and most of the others had signed the obligation to pay them. On November 30, 1741, the Royal Society elected a man of letters, Martin Folkes, as president. Halley died the following year, with Sloane surviving him by eight years.

1.7. Discussion of the reasons for rejecting Plantade's submission

The reason of a too great conformity to Halley's system, which would have been officially advanced by the Academy to reject the theoretical part of the memoir, is not convincing. Plantade mentioned the numerous differences that existed between his explanation and that of Halley's, whose only common point was to call upon terrestrial magnetism to explain the aurora borealis. The first part of his memoir, describing the aurora of 1726, was considered by the Academy as publishable in the *Mémoires*, and justified a publication on its own, this one being introduced by a critical analysis of Fontenelle in *Histoire* putting Halley–Plantade's hypothesis in its historical context. The motive developed by Bignon in his letter to Plantade, whom he reproached for wanting to create a system, at a stage when observations were still too early, is theoretically admissible, given the centrality of this question in the scientific thought of the time. But Plantade was a recognized scientist, founder of the Société Royale des Sciences de Montpellier. He was well aware of the empiricist approach that led to the rejection of any premature system. Bignon's argument, not based on a scientific analysis of the memoir's content, with Bignon not being a scientist, seems more like a pretext elaborated a posteriori to support a rejection based on other reasons, knowing that Plantade's reaction was strong, and caused a crisis that lasted more than three years. This argument lends itself to the criticism of the "double standard", what was granted to Mairan being denied to Plantade. It is true, however, that Halley, and most likely Plantade, did not engage in the considerable archival excavation work that Mairan did in order to best support his system through observation.

The refusal of Plantade's *Mémoire* was not the result of an ego quarrel either, given Mairan's reputation and eminent position in the Academy, a hypothesis which, moreover, did not fit well with his personality as it appeared in his correspondence. The three criticisms addressed by Mairan to Halley's system were serious (de Mairan 1754, pp. 77–78). Firstly, the fact that the aurora borealis declines most often towards the North-West, of 14 or 15 degrees, a value which is approximately that of the declination of the magnetic needle, does not convince him of the magnetic origin of the aurora, as the declination of the aurora presented brutal variations over short periods of time, whereas the magnetic declination was only very slowly variable. Secondly, he did not see how the magnetic matter, even more subtle than that of light, could reflect to us "the light [of the sun] being carried to two or three hundred leagues of height, that is, infinitely above the region of the twilight", or, in the case of a radiation by terrestrial exhalations themselves, how it could raise these exhalations to such considerable altitudes. Finally, the circulation of magnetic matter, in Descartes' vision, was permanent, and could not "agree with the cessations & recoveries of the Aurora Borealis", a point on which his system was unquestionably superior to that of Halley's. These various objections were well founded, and were sufficient to place Mairan in a strong position in any contradictory scientific debate organized around the two systems. We cannot doubt that this debate took place in the assembly of the Academy during the reading of the Montpellier memoir in March 1729, a reading which, according to Mairan, was the occasion for him to take date of his own system.

As we have seen, it took a little more than three years for the decision not to publish the memoir to be officially taken by the Academy, in December 1730. However, a letter written by Bignon on July 22, 1730 shows the beginning of a normalization of the situation between Paris and Montpellier (Castelnau 1858, p. 125), suggesting that Bignon waited for the tension to subside before scheduling an official vote at the assembly of the Academy to ratify the rejection of the memoir. During this three-year period, no memoir by the Société de Montpellier was published in the *Mémoires parisiens*, and the resumption of publications was precisely by a memoir chosen against that of Plantade's, visibly sealing the end of the crisis between the two institutions. Significantly, this period (1727–1730) coincided with the gestation phase of Mairan's system. This suggested a probable balancing act between the two systems, leading to the rejection of the Plantade–Halley system in favor of the Mairan system.

But this period presented another historical particularity. Several authors, such as Pierre Brunet (1970) and John Bennett Shank (2008), describe the period that began in 1727 and lasted until the end of the 1730s as a phase of renewed tension between Cartesians and Newtonians. A synthesis of the periodizations of the penetration phase of Newtonianism proposed by different authors appears in a recent book by Crépel and Schmit (2017, pp. 21–40). For Brunet, the Cartesian reaction began at the beginning of the 18th century, with opposition becoming increasingly intense from 1727 onwards, whereas for Shank, the process did not really begin until 1715 with a growing rejection of Descartes by Newtonians, Fontenelle's eulogy of Newton in 1727, contrasting Descartes' "bold flight" to Newton's "timid walk". This was a blow to the self-esteem of the English and exacerbated tensions. Let us judge:

> The two great men, who are in such great opposition, have had great relationships. Both were geniuses of the first order, born to dominate over other minds, & to found Empires. Both excellent Geometers saw the necessity to carry Geometry into Physics. Both of them founded their Physics on a Geometry, which they had almost only from their own lights. But the one, taking a bold flight, wanted to place himself at the source of everything, to make himself master of the first principles by some clear, fundamental ideas, in order to have only to go down to the phenomena of Nature, as to the necessary consequences; the other, more timid, or more modest, began his march by leaning on the phenomena to go up to the unknown principles, resolved to admit them whatever the sequence of consequences might give them. The one starts from what he hears clearly to find the cause of what he sees. The other starts from what he sees to find the cause, either clear or obscure. The obvious principles of the one do not always lead him to the phenomena as they are; the phenomena do not always lead the other to fairly obvious principles. The limits, which in

these two contrary roads could have stopped two men of this kind, are not the limits of their Spirit; but those of the human Spirit (Fontenelle 1727).

We have seen that Halley, notably through his calculations of comet trajectories, was at the forefront of the Newtonian scene, and that Mairan, although admiring Newton and integrating Newtonian elements in his mechanistic constructions, remained attached to the vortices and to the subtle matters invoked by Descartes. He wrote in 1742: "Whatever the destiny of vortices, it is a great and attractive theory which deserves that we make the last efforts to maintain it and to deliver it from the pressing objections with which the partisans of the vacuum have been trying for more than fifty years to overwhelm it" (de Mairan 1742, p. 209). We have underlined the determined and combative character, under the veneer of the honest scholar, of Fontenelle, penetrated by the importance of his role as Secretary of the Académie Royale des Sciences and editor of the *Histoire*, exalting the deeply Cartesian cosmo-atmospheric system of his colleague and friend. The rejection of Plantade's memoir, and through it, of the system promoted by the eminent representative of Newtonian mechanics that was Halley, in a period of exacerbation of the tensions, can thus be conceived as a "political" choice, all the more effective that it was accompanied by the emergence of a system anchored in Cartesian cosmology (Chassefière 2022a). Each country, in the context we have mentioned of an accelerated internationalization of science, needed to take its marks, and the choice of Mairan's system can be seen precisely as an affirmation of the French specificity.

Taking up Shank's periodization, the period 1716–1726, during which Halley's system was never mentioned in the columns of the volumes of the Academy, can be inscribed in the period of development of the anti-Newtonian discourse that Shank started in 1715, the period of "Newtonian war" after 1727 seeing successively the refusal of the publication of Plantade's memoir (1727–1730), this act threatening the French scientific institution and seeming to us to have been motivated only by a "political" imperative of the first order, followed by the finalization and publication of Mairan's treatise on the aurora borealis (1731–1733), which sealed and offered perspectives to the Cartesian doctrine.

2

Joseph-Nicolas Delisle: Grandeur and Vicissitudes of a Newtonian Scientist with Thwarted Ambitions

2.1. Introduction

Most of the observers of aurora borealis of the first half of the 18th century, who provided to Jean-Jacques Dortous de Mairan observations intended to test his aurora system, were in epistolary contact with Joseph-Nicolas Delisle. Delisle, just before his departure for St. Petersburg to take over the new Observatoire Impérial, began to set up a network of astronomers whose observations he tried to coordinate. Delisle's abundant correspondence, a man who established through his network a system of exchange of information and results of observations that he intended in particular for the drafting of a great historical work, *Histoire Céleste*, containing all the observations made by all the astronomers of Europe (Delisle 1738, pp. 3–4), attested to the scope and dynamism of his approach. This was initiated in 1718, on the basis of the archives he had already constituted of the manuscripts, notably, as he wrote to Jean-Paul Bignon (Delisle's letter to Bignon, June 26, 1720, Bibliothèque Numérique – Observatoire de Paris[1]), of the works of Tycho Brahe, Jean Picard and Philippe de La Hire, recently deceased at this point, to whom he succeeded as Chair of Astronomy of the Collège Royal, as well as Jacques d'Allonville de Louville and Sédileau (Bigourdan 1895, p. 39). The originals of Tycho Brahe's observations, brought back from Uraniburg by Jean Picard in 1672 (Picard 1736, p. 67) and preserved in the archives of the Académie des Sciences, were copied by Delisle, before being returned to the King of Denmark in the first half of the 18th century (Bigourdan 1895, p. 24). Delisle also addressed the astronomers of his time. For

---

1 La Bibliothèque Numérique – Observatoire de Paris is available at: https://bibnum.obspm.fr/.

example, in January 1720, he started a correspondence with Christfried Kirch (Jaquel 1976), son of the astronomer Gottfried Kirch, director of the Berlin Observatory and member of the Berlin Academy of Sciences, to whom he also asked for the communication of the volumes of his father's observations, in particular concerning the eclipse observations. Kirch's answer was encouraging, and Delisle wrote to him in the summer of the same year: "I am pleased to find you willing to enter into an epistolary trade with me on astronomical matters […], but", he specified further, "in order to make this trade more useful to both of us, it is necessary to write to each other often, and this until we are perfectly instructed in our ways of observing and our particular designs in the study of astronomy" (ibid., p. 419). In the following letter, Delisle outlined a real shared observation program, to be implemented as soon as his Berlin colleague had instruments as powerful as his own:

> We will then agree on the observations that we will have to make; the first ones will be those that will establish the difference of our meridians and of our height from the Pole, after which we will be able to work more usefully together to establish the theory of all the planets by the most appropriate observations. It is for this purpose that I have gathered all the astronomical observations that I could; as I believe that you have the same purpose in the study that you make of astronomy, I will gladly associate myself with you for this work; I have even spoken about it to Abbé Bignon, president of the Académie des Sciences and who is here the protector of the sciences, of language and literature and of the arts (ibid., p. 420).

Insisting on the observations of Kirch's father, which his son said he had no time to classify and translate into Latin, the French astronomer reassured him:

> We have enough people here who can read the most poorly written German; thus you can send us the very journal of all the observations of your father without taking the trouble to extract from it what is only astronomical nor to translate it into Latin. You will be very pleased to send us at the same time all his other observations which are not in this journal and which you will apparently have collected from printed books (ibid., p. 420).

He repeated with other scholars the same operation of making contact and setting up an ongoing epistolary relationship, notably Johann Philipp von Wurzelbau, then Johann Gabriel Doppelmayr in Nuremberg at the beginning of the 1720s, Edmond Halley and other English astronomers in London in 1723, then, from St. Petersburg, in the 1730s and 1740s: Anders Celsius, Olof Hiorter and Pehr Wilhelm Wargentin in Sweden, Johann Wilhelm Wagner in Berlin, among others. His exchanges with all these correspondents were frequent, and, in addition to the cross-fertilization of observation results, resulted in the support of some for the nomination of others to their national academies, allowing for an internationalization of the constitution of these academies, and a strengthening of the institutional scientific links between European nations. The observations exchanged in the correspondences were essentially astronomical, concerning the eclipses and more generally the events of alignment of the bodies of the solar system: eclipses of the moon and of the sun, eclipses of planets (and fixed stars) by the moon, eclipses of the satellites of Jupiter by the planet, transits of Venus or of Mercury on the sun, so many events whose importance was then considerable, as well for shaping the solar system as to measure the latitudes and longitudes of places for cartographic purposes. In the letters exchanged by Delisle with the major observers of the aurora borealis, such as Kirch, Celsius and Wargentin, scientists to whom he would become very close, we find nothing, or very little, about the aurora borealis. These were observed de facto in the field, or from observatories, by the astronomers of northern Europe, or even of the south even if they were seen more rarely, during their nights of vigil, but the science was completely different from that of the measurement and calculation of the trajectory of stars, which, in a climate of slow penetration of the Newtonian theory of gravitation on the continent, monopolized the attention and energy of astronomers. On the other hand, the meteorological measurements, or the orientation of the magnetic needle, were from time to time the object of dedicated paragraphs in Delisle's letters. But, more than that, it was thermometers that Delisle disseminated within his network of observers, instituting a true European meteorological network, made up of inter-calibrated thermometers enabling direct comparisons of the temperatures measured in various places and under various climates, whose data was used in particular by Father Louis Cotte, one of the most eminent meteorologists of the time in Europe (Fressoz and Locher 2015, p. 65), in his great treatise on meteorology (Cotte 1774). The latter paid a heartfelt tribute to Delisle for his observation portfolios, preserved in the Dépôt de la Marine, but entrusted to him for the occasion, files rich in readings taken in remote countries, and which constituted his main source of data. We will look in more detail at the "Delisle" network and the European academies of sciences involved in these exchanges (Sweden, Germany, Russia) in Chapter 4 devoted to Celsius.

Delisle was the largest contributor to the collection of aurora observations used by Mairan in his treatise of 1754. The only publication made by Delisle of his aurora observations appears in his *Mémoires pour servir à l'histoire et au progrès de l'astronomie, de la géographie et de la physique*, published in 1738 in St. Petersburg (Delisle 1738). This work contains a part entitled *Dessein de cet ouvrage* delivering his great project of a Celestial History containing all the observations made in all countries and at all times, an idea which was the central theme of Delisle's action during his life. Here is how he described his great project:

> Having applied myself by inclination to the study of Astronomy, for more than 25 years, & having collected during all that time all the observations, theories, & other Memoirs which could be useful to the progress of this science, I understood at the end that this science was too extensive for a single Astronomer to hope to be able to treat it in all its parts, in a way which leaves nothing to be desired in the century in which we are, & that thus the project which I had formed for some time, to compose a complete treatise of Astronomy exposed historically, & demonstrated by all the observations made until now, &c; That this project, I say, was above my forces, or that at least it was uncertain if I could complete it alone. I have learned from my own experience how difficult it is to associate with others in such works: thus I believed that not to lose my work & my views for the advancement of Astronomy, I had to publish Memoirs which could serve the same purpose, that is to say to help to compose in the future a complete treatise of Astronomy, when what I give will be joined with what the other Astronomers have already done, & what they will be able to add to it, & to do better in the future (ibid., pp. 3–4).

The pieces published in these *Mémoires* of 1738, namely the reports of the aurora borealis observed in St. Petersburg (1726–1737) and in the region of Archangel by his brother Louis during his cartographic expedition (1727–1730), the astronomical observations and the experiments on light made in his observatory of the Luxembourg Palace before his departure for Russia (in the middle of the 1710s), and the realization of mercury thermometers measuring very low temperatures (1733), needed to be conceived, he said, as the elements of the future complete treatise of astronomy of which he already made his master work. About the two chapters devoted to the aurora borealis, here is what he wrote:

> The northern region, where Russia is mainly located, makes the spectacle of the Aurora Borealis and other similar phenomena much more frequent than in other more southern countries. It is also for this reason that one expects from this country the principal reports of this phenomenon to which one now pays so much attention. This is what makes one believe that one will be very happy to find here gathered all the observations of which the greatest part was made by my brother in the years 1727, 1728 & 1729, in Archangel & in Kola or in the surroundings, & by me in Petersburg, since the year 1734. I will join to it all the other observations of a similar nature, made in Russia, & of which I could have knowledge (ibid., pp. 19–20).

In the chapter devoted to the observations of aurora borealis, Delisle mentions the very great frequency of aurorae, of which all are not listed, because, he said, the observers had other activities. The observations came from Friedrich Christoph Mayer, who observed many aurorae and began to communicate them to the Académie Impériale in 1726. He proposed an aurora system, to which we will return at length in Chapter 3, a system that he perfected until his death in 1729. Georg Wolfgang Krafft took over the aurora observations after Mayer's death, and provided detailed observations of the most remarkable aurorae, for example, the grandiose one of February 15, 1730:

> On February 15, there appeared a boreal light which was one of the greatest that one observed in Petersburg. The whole sky appeared to be on fire. It began at about nine o'clock in the evening, the sky being very serene. At first, it appeared as rods of light that rose from the East, the West, and the Septentrion, up to the Zenith. There were even some which passed beyond the Zenith, & which formed around this vertical point a luminous core, so that all the sky resembled a luminous vault. On the side of the Septentrion rose a luminous arc which passed at a little distance from the Zenith, & whose sides were almost vertical and perpendicular to the horizon. Within this arc, one could see rods of light moving to and fro, although the air was very still, without any noticeable agitation. At the extremities of the rods of light which rose from the East and West, a second arc began to form on the South side, which was soon complete, and whose height was about 7°. The inside of this arc was dark; however, one could clearly see through it the planet Mars, which was rising, and which was near Virgo. This apparent darkness was terminated by a luminous arc which was at first in perfect repose; but towards midnight one saw also rods of light rising from this southern arc, & these rods joined with those which rose from the Septentrional arc, & which passed the Zenith. When the southern part of the whole phenomenon was in its greatest strength, the rods of light which came from the Septentrion seemed to diminish, & the southern part of the phenomenon also seemed to be transformed into several clouds, or heaps of a darker light: but nevertheless this southern part of the phenomenon did not delay in resuming its first form. There also appeared colors in this great Aurora Borealis, & principally in the rods of light which came from the north, which appeared red (ibid., pp. 81–82).

Delisle, for his part, started his observations of the aurora borealis in September 1734. At that time, he started to spend all his nights at the Observatory to observe the variations of the fixed stars and the movement of the planets. He continued these observations in 1735 and 1736. But he judged that his pendulum was not exact enough to provide him the variations in right ascension of the fixed stars and consequently stopped spending his nights there. He explained that he gathered in this memoir the observations of Mayer, Krafft and himself, as well as those of Pierre Le Roy, one of the few Frenchmen of the Imperial Academy of St. Petersburg which at the beginning was mainly composed of German scientists. He was said to have reduced the times to the new style (Gregorian calendar, the Julian calendar being then still used in Russia), as for the observations of his brother Louis

> with the intention of being able to compare them more easily, & to make them serve as supplements to those of which Mr de Mairan gave the catalog which ends with the year 1731 [...] If one joins all these observations made in Russia, with those that Mr Celsius has collected, made in Sweden since the year 1716 until the end of the year 1732, one will have well enough to increase the general size that Mr de Mairan drew up of all those which he was able to collect (ibid., p. 79).

Delisle thus had in mind to complete the catalog of aurorae published by Mairan in 1731 with his theory, and that in complementarity with Celsius, with whom he maintained a regular correspondence.

The isolated character of the *Mémoires* of 1738, of which no sequel was published, and which contains the only part ever published of the great celestial history that Delisle planned to write, namely his introduction, is intriguing. Delisle certainly published, during the first third of his stay in Russia, his observations of the eclipses of Jupiter's satellites in the first six volumes of the *Commentarii Academiae scientiarum imperialis* of the Russian Imperial Academy of Sciences (1726–1733) and in the *Philosophical Transactions* of the Royal Society of London in 1729 and 1736. We also owe him a *Projet de la mesure de la terre en Russie*, published in Amsterdam in 1737, of which an English translation appeared the following year in *Philosophical Transactions*. But this was relatively little for a scholar of his size. Why was the publication of the first volume of the *Mémoires* not followed by any other? Why were his astronomical observations and experiments on light made in Paris 20 years earlier not published as part of the *Mémoires* of the French Académie Royale des Sciences? To which combination of circumstances do we owe the presence of Delisle in Russia in this period, and thus the observations of aurora borealis that he delivered to Mairan for the second edition of his treatise? It is these questions that we will try to answer in this chapter, devoted to the trajectory of an extraordinary astronomer, whose scientific aims were constantly thwarted throughout his life, either because he was confronted with a scientific power that did not share his philosophical options (in this case Newtonianism), or that the political power to which he addressed himself to obtain the necessary means for the realization of his projects could not bend, in its own objectives, to the requirements of rigor and the necessarily long timeframe of the scientific process.

An essential source of information used for this chapter is Delisle's abundant correspondence, for which we use the following sources: Bigourdan (1917) for Leonhard Euler, Benitez (1986) for Count Plélo (Louis-Robert-Hyppolyte de Bréhan), Jacquel (1976) for Christfried Kirch, Olivier Courcelle's online site "Chronologie de la vie de Clairaut" (Clairaut.com 2022) for Lemonnier, the Bibliothèque numérique – Observatoire de Paris for all other correspondents. By default, in the absence of contrary indication, the source used is the latter. In section 2.2, we examine, in order to understand the reasons for Delisle's departure to Russia, what was the personal and professional material situation of Delisle in Paris at the time he decided to leave, and in which context his scientific activity was developing. Section 2.3 is devoted to Delisle's activity in Russia, and the difficulties he encountered there. In the conclusion, we try to restore the overall coherence of Delisle's life path. As we mentioned in the introduction of the book, our aim is to place the observers of the aurora borealis in the scientific, human and institutional context of their time, and we will in the following pages depart from the question of the aurora borealis itself.

2.2. Delisle in the period before his departure for Russia (1710–1725)

2.2.1. *Delisle's beginnings in astronomy and optics: a Newtonian*

Nevskaja writes about Joseph-Nicolas Delisle, to explain his departure to Russia to set up the Imperial Observatory of St. Petersburg: "Moreover, as a follower of Newton, he could not hope to make a career in France where Cartesian doctrine was dominant at the time" (Nevskaja 1973, p. 292), but does not justify this judgment by precise statements of the person concerned. Moreover, nowhere in Delisle's writings and correspondence is there an explicit expression of a conflicting situation that made his life difficult and pushed him into exile. Delisle was an important and recognized scholar of the French astronomical community of the first quarter of the 18th century, as we shall see, whose "dynamic and sometimes aggressive egocentrism" (Jaquel 1976, p. 414) is noted by Jaquel. His correspondence, from the 1720s onwards, reveals a determined man, persevering in his requests for information, easily impatient when he did not receive from his correspondents the answer he expected, leaving no room for affect in his professional relations. He was certainly not a man to complain publicly. However, during the decade 1715–1725, he met with frank opposition from certain Cartesian scholars of the Académie Royale des Sciences, and it is legitimate to wonder if this opposition had an influence on his decision to go into exile. The Tsar of Russia came to France in 1717, with the aim of establishing links with the Académie Royale des Sciences. He proposed to Guillaume Delisle who would receive the title of royal geographer the following year, to come to Russia to work on the establishment of a map of the empire, but the latter declined, and put forward the name of his brother Joseph-Nicolas, also an academician and astronomer, who was very familiar with the measurement of positions by astronomical methods. The official invitation was addressed "by mouth" to Joseph-Nicolas by Johann Daniel Schumacher, the tsar's first librarian, during his trip to Paris in 1721 (Marchand 1929).

Delisle was one of the first academic astronomers, with Jacques d'Allonville de Louville, to take a clear position in favor of Newton's ideas (Schaffer 2014). The chemist and academician Étienne François Geoffroy had been the first major French scientist at the turn of the 18th century to give credence to the Newtonian theory of attraction (Shank 2008, pp. 114–120), although it is unclear whether he was actually a Newtonian (Joly 2012). Geoffroy spent the entire year of 1698 in England, where he was made a fellow of the Royal Society. Knowing English, he was the first to present Newton's *Optics* to the Académie des Sciences in 1706. In May 1715, Louville went to London to observe an eclipse of the sun. That year, he and two other academicians (Claude Joseph Geoffroy, brother of Étienne, and Pierre Remond de Montmort) were the first Frenchmen appointed to the Royal Society since Geoffroy's election (Bonno 1948, p. 124). Delisle was appointed in 1724 on the occasion of his trip to London. Back from the English capital at the end of 1715, Louville diffused in his turn, after Geoffroy, Isaac Newton's optical theory. Delisle read Newton's *Principles* in their second edition of 1713, and worked during the summer of 1716 on his own version of a lunar and solar theory derived from the *Principles*, in addition to his experiments on light, amplifying those done by Newton. Louville published in 1720 a memoir on the celestial mechanics of Newton, whose discoveries he cautiously attributed to Kepler (Schaffer 2014, p. 154). The same year, the French translation of the second edition of Newton's *Optics* appeared, probably under the influence of Delisle (Nevskaja 1973, p. 303). According to John Bennett Shank, the publication of Newton's *Optics*, in 1706, was "perhaps the first catalyst" of the tension that arose at the beginning of the following decade in France between Newtonians and Cartesians (Shank 2008, p. 113). The theory of gravitational attraction was set out very explicitly, in a way that was more physical than mathematical, and therefore more accessible and clear than in the *Principles*. The "pax analytica", resulting from the purely mathematical reading of the *Principles*, which reigned during the 20 years following their publication, thus rapidly crumbled after the publication of *Optics* (Shank 2008, pp. 64–76; Crépel and Schmit 2017, p. 36). Delisle, through his work on the diffraction of light, and his projects for measuring the shape of the Earth, developed at the same time, found himself precisely at the heart of the doctrinal quarrel which was gaining momentum in the 1710s. The analysis of Delisle's case sheds light on this pivotal period in the penetration of Newtonianism on the continent, as we will detail in this section. We base ourselves partly on the very important archive of the Observatoire de Paris concerning Delisle which provides, in particular through the examination of his correspondence, digitized in the *Bibliothèque Numérique – Observatoire de Paris*, significant information on the reception at the Académie Royale des Sciences of Newtonian optics, and more generally of Newtonian attractionism.

Jean-Paul Grandjean de Fouchy, in his eulogy of Delisle (Fouchy 1768), mentions that Delisle began studying mathematics in 1706, and that the total eclipse of the sun that year gave him a taste for astronomy. One of his first subjects was the eccentricity of the orbit of the Earth and the determination of the evolution of the position of the sun in the sky taking into account this eccentricity. He started to work on the tables of the sun and the moon based on the work of Jean-Dominique Cassini, who was blind at the time, and who took on the task of instructing him. This strong link with Cassini would have been partly responsible for the hostility of his son Jacques Cassini to Delisle (Sigrist and Moutchnik 2015, p. 96). It was in 1710, at the age of 22, that Delisle obtained permission to occupy the dome above the main door of the Luxembourg Palace to make his observations. He made, according to Fouchy (1770), a wooden quadrant, which he realized was not sufficiently precise because of the deforming effect of humidity variations. Not being able to carry out observations, for lack of instruments, he established at the request of Jacques Cassini a table allowing him to quickly calculate the eclipses of the planets and fixed stars by the moon. He began to make astronomical observations in Luxembourg only in 1712, and in 1714 he entered the Académie Royale des Sciences as a student of Jacques Philippe Maraldi. In 1715, Delisle observed the occultations of Venus and Jupiter by the moon, and exchanges on this subject formed a correspondence with the abbot Teinturier who made the same observations in Verdun. The observation of a coloration of Venus at the approach of the lunar disk was then debated, Jacques d'Allonville de Louville attributing it to a lunar atmosphere, J. Cassini to a fatigue of the eyes or to an instrumental effect, Delisle leaning towards an effect related to the diffraction of light.

Delisle's correspondence at this time tells us that he lived poorly, and lacked the means to practice astronomy. At the end of September 1715, Delisle had to leave his premises in Luxembourg, where the Duchess of Berry had just moved. He interrupted for more than a year his astronomical observations, in favor of laboratory experiments on the diffraction of light in the vicinity of solid bodies. He resumed his astronomical observations only in December 1716, once installed at the Taranne hotel, in an apartment that Louville had previously occupied for his observations. In the same year, he was appointed Assistant Astronomer at the Academy (Fontenelle 1716a), but this position still did not entitle him to any fixed remuneration. It was through Abbé Bignon that Delisle had asked the regent that this apartment be attributed to him, because, he wrote to Bignon in the fall of 1716, "this help that you would grant me would revive again the ardor that I have for astronomy & I could hope by dint of study & application to make myself worthy of the protection with which you honor me; I count Sir to send you incessantly a small text that I am completing on light" (letter from October 1716). We will return to the text mentioned by Delisle in this letter, which is an account of experiments on the diffraction (or "inflection" in the terminology used by Newton) of light, but this example shows that Delisle used the channel of Bignon, who had direct access to the king, for his requests for financial support. We learn in a letter from Delisle to Bignon from June 11, 1717 that the latter gave him the hope of financing a quadrant for astronomical measurements, following the request that Delisle made to him through Jacques Cassini. We understand that the expected contribution of Teinturier, in exchange for the service that Delisle made him by taking his nephew under his wing to teach him astronomy, was not sufficient, and that the essential need for financing remained to be found to allow him to "follow his genius". The financing was obtained and the instrument realized a few years later, as we learn it from a letter from Delisle to Laurentius Blumentrost dated September 8, 1721. In his letter from May 6, 1718, Delisle asked René Antoine Ferchault de Réaumur who, as we have seen, was then the most influential scientist of the Académie Royale des Sciences, to support the request of Bignon (of whom Réaumur, as we have seen, was particularly close) that he be given the chair of professor left vacant at the Collège Royal following the death of Philippe de La Hire. The obtaining of this post pulled him, he wrote, of "the necessity in which I am reduced, which is if I dare to say it, to calculate for the astrology judicial, & thus to prostitute astronomy to research studies for which I have a sovereign contempt". He added that J. Cassini, who knew of his inclination and his dispositions, could serve as a guarantor if needed, and invited the Academy to recognize "by the good use I will make of the time I will be able to have to myself, what I could have done if I had been in a more comfortable situation". On May 31, 1718, Delisle insisted to Bignon about the chair at the Collège Royal, because he was without resources, with a sick father. To finance the instruments and books he needed, he had only his judicial astrologer's fee of 1,000 francs per year, which was paid with increasing delay. He mentioned, to reinforce his legitimacy, that Cassini, who left then to prolong towards the north the meridian of the Observatory until the sea, entrusted him the task to make at the Observatoire de Paris the observations corresponding to those which he made in his voyage of the meridian heights of the sun.

Delisle was indeed appointed professor at the Collège Royal on August 24, 1718, the same day that his older brother Guillaume was awarded the title of First Geographer of the King. He thus inherited all the works of Philippe de La Hire's library, which completed the works of the King's Library to which Bignon had given him access. The following year, he became Associate Astronomer at the Académie des Sciences, replacing Louville, who was promoted to member, a role that did not bring him any significant income[2]. Nevertheless, in addition to the 36,000 annual francs intended for the salaries of the academician members, a sum of 12,000 francs, also taken from the royal treasury, was used for general expenses, and for the setting up of experiments, also allowing for the payment of aids to the Associates and other non-resident academics (Bertrand 1869, p. 97). Delisle benefited from such aid, and declared that he received overall fees of 3,000 francs per year from the Collège Royal and the French Academy (letter from Delisle to Laurentius Blumentrost dated September 8, 1721). His nomination to the Collège Royal allowed him to receive the equivalent of the salary of an Academy member, a sum which was nevertheless insufficient according to him, considering his modest condition, to be able to devote himself fully to his profession as astronomer. In a letter to Bignon dated June 26, 1720, Delisle asked for the means proportionate to his research to buy books and to obtain instruments which he still lacked. He did not have any working space at the Observatoire de Paris, and asked his interlocutor to intercede with Cassini and his collaborators, who "have all the house to them" so that a room was assigned to him at the Observatory. As always in his requests to Bignon, he put in front of his request a research program that justified it, exposing his great *Histoire Céleste* project that we could qualify as epistemological before the letter:

> Another thing would contribute infinitely to the progress of astronomy, and facilitate its study. It would be to treat this science historically, that is to say, to describe historically the progress and discoveries that have been made, which will more easily and more surely inform those who wish to know this science, and which will put those who feel sufficiently enthusiastic and willing to improve it in a better position. For want of studying in this way, one takes great pains to invent things that one does not know have been said and found before, and often better than one finds them. As I have made a special point of knowing astronomy in this way, the treatises I will put my hand to will be based on this historical knowledge.

Delisle was said to have in mind in the short term a complete theory of the sun treated in this way, and advocated more generally, and in the long term, a coordinated approach, that we would say today of networking of exchanges of observation results and more generally of information between astronomers. It was in this spirit that he collected the observations made by other astronomers, in particular those of Picard, de La Hire, Georg Margraff, some of those by Tycho Brahe, Jean-Matthieu de Chazelles and Louis Feuillée and worked to acquire the volumes of observations and the books produced by the foreign astronomers like Ole

---

[2] Three thousand francs per year for the oldest, and 1,800 and 1,200 francs for his younger colleagues, did not allow for a decent living in Paris (see Hahn 1975, pp. 501–513).

Christensen Rœmer in Copenhagen, Kirch and Wagner in Berlin, Georg Christoph Eimmart and Wurzelbau in Nuremberg, Robert Hooke, William Gascoigne and John Flamsteed in England, "enough to make a complete celestial history". He told Bignon that he planned to write a great treatise of historical astronomy.

Delisle finally obtained the room at the Observatory that he had requested, but he was not allowed to stay there permanently, which obliged him to travel frequently and tiringly between his home and his place of work (letter to Bignon dated February 6, 1721). He attached to his letter a statement of the expenses he had to make to equip the room, and said that he had lost hope of obtaining help from the Regent to acquire new instruments. He wished to be able to occupy the room on a permanent basis. On September 8, 1721, he wrote to Laurentius Blumentrost, doctor of Peter the Great, who became a few years later the first president of the Imperial Academy of Sciences of St. Petersburg, a letter in which he evoked his project of great historical treatise of astronomy and the means at his disposal to carry it out: a library rich in many books, and powerful instruments, of which the great quadrant that he succeeded in having financed, then realized and of which he could dispose as he pleased, according to him one of the best instruments of this type in the world. His financial situation had also improved a lot, since he had been chosen by the States of Brittany to head a team of engineers in charge of establishing an exact map of the region, supported by all the astronomical observations necessary for this purpose. The salary he received for this task was 12,000 francs per year, to which were added his fees of 3,000 francs from the Académie des Sciences and the Collège Royal. On November 1, 1723, Delisle said in a letter to Wurzelbau that he was still waiting to obtain a room worthy of this name that he could occupy permanently at the Observatoire de Paris; an isolated tower was attributed to him the following year. He wrote in the same letter that Abbé Bignon obtained from the Regent "of what to maintain 3 young people who would be destined to the study of Astronomy; and I proposed to answer this establishment and to advance more promptly these young people, and to put them sooner in a state to work usefully for astronomy; I proposed, I say, when my assortment of astron. tables will be completed, to add to it all the explanations necessary to make understand thoroughly the construction and the uses of these tables".

The letters written at this time by Delisle to Cassini were short and dry, and concerned exclusively requests for materials, or books, without any exchange on the results of the observations, as it was the case with his interlocutors of the time, such as Teinturier in Verdun, Kirch in Berlin, Wurzelbau and Doppelmayr in Nuremberg and Halley and Newton in London. The very existence of these letters shows that Delisle was not present at the Observatory on a regular basis; otherwise, his requests to Cassini would have been made orally. In the fall of 1721, Delisle sent Cassini a list of astronomical works of which he had knowledge, some of which were in the King's Library, to which Bignon had given him access, and asked Cassini at the same time if he could have access to his personal library to complete his list (letter from September 6, 1721), which shows that Cassini's library was not open to the astronomers of the Académie des Sciences, at least not to Delisle. At the end of 1723, anticipating a forthcoming installation at the Observatory, Delisle asked Cassini for the authorization to transport some of his instruments from the Luxembourg Palace to the Observatory, and to make his observation of the passage of Mercury on the sun in one of the two towers (letter of October 30, 1723). The following month, one week after the event, he asked Cassini to give him the time of passage of the sun at the wall dial, because he needed the true time, to recalibrate his own observations on this (letter from November 16, 1723). As a testimony of a very probably distended relationship between the two men, Delisle did not obtain an answer, since he repeated his request one month later (letter of December 12, 1723). As we have seen, Delisle must have addressed himself to Bignon, bypassing the board of the Observatory, to obtain the permission to occupy the premises of the Observatory.

As we have seen, Delisle's requests to Abbé Bignon, whether for resources to be obtained from the Regent or for desired appointments in the academic system, were on the whole successful. Delisle was also listened to in his recommendations for recruitment to the Academy, as when he asked Louis Godin in 1722 for a position as assistant astronomer, replacing Claude-Antoine Couplet, who died the same year (letter to Bignon from August 30, 1722). He began by advising his interlocutor that the place should not be attributed immediately, to consider potential candidates, such as Godin and La Lande, a young astronomer of whom Jacques d'Allonville de Louville spoke to. He pointed out that the Academy was not satisfied, when he presented himself, with the recommendations of Cassini and Maraldi, but required from him two works attesting the quality of his work. It was necessary, according to him, to ask the candidates to have kept a journal of observations for some time to be accepted. But, he added, if Bignon intended to fill this place "without requiring anything else than the testimony of an astronomer academician with whom he will have worked and without any other proof of his capacity than some calculations", he could testify to the qualities of Godin, who presented to the Academy the calculation that he made of the passage of Mercury over the sun in November 1723, which had not occurred since 1697. Louis Godin was indeed received as a member of the Academy as Assistant Geometrician in 1725, then as Assistant Astronomer two years later. Delisle was clearly a respected scholar in his community.

Besides the development of Newton's optical theories, and the debates on the shape of the Earth to which Delisle tried to be a close part of, and on which we will return, "one knows less how the celestial mechanics developed in spite of the opposition of the Observatory and Fontenelle" (Schaffer 2014, p. 151). Delisle was in the forefront with his project of tables of the moon and of the sun. In a letter to Teinturier from February 7, 1717, Delisle mentions the numerous experiments he was doing on light. "These experiments", he added, "only serve me as recreation in the more serious study that I make of astronomy. If the system of Mr. Newton could please you, you will perhaps not be upset to learn that I am working to establish it in a way that has not yet been done. We have been content until now to demonstrate the theory, and although the English astronomers assure us that it agrees with the observations better than any other, none of them has shown this agreement: which means that our astronomers have not wanted to take the trouble to prove what they found wrong with it". Delisle thus calculated tables "for the sun and the moon only on the determinations that Mr. Newton drew from his theory of gravity". Newton recognized many more irregularities in the movement of the moon than other astronomers admitted, which had to be attributed to physical causes. Delisle now needed to make the appropriate observations to compare these tables with reality, and these measurements required the means in instruments and observation rooms, which he was on the right track to obtain from the Regent, he explained. On April 27, 1717, Delisle wrote a long letter to Réaumur, which summed up all his efforts to exploit Newton's theory for the purposes of calculating celestial mechanics, and to verify this theory by comparison with observations. Here is a part of it, which summarizes well Delisle's intellectual approach, and the precautions he took so that his project, which he considered to be of interest to Bignon, and that consequently he could use it as an argument to obtain from him the necessary means for its accomplishment, did not fall into the hands of the board of the Observatory, and probably also of Fontenelle, who were hostile to Newtonian attractionism:

> Sir,
>
> I had the honor to see this morning Abbé Bignon on the occasion of a work that I had presented to him last week. He did me the honor of telling me that he would give it to you to return to me. I beg you, to do it without showing it to anyone. These are new astronomical tables of the [sun] & [moon], calculated according to the latest determinations of Mr. Newton in accordance with his theory of gravity. By presenting them to Abbé Bignon, I hoped to have the honor of explaining to him the aim, and to show him my project in full, of which these tables were only the beginning: but not having been able to enjoy Abbé Bignon's company long enough to do so, I found myself obliged, in order to inform him, to leave him a document that I had not written in

order to show him; but only in order not to lose certain views that had come to me, and from which I intended to subtract and add everything that experience would have taught me. Indeed, if you read it, you will find declamations against the ordinary way of treating astronomy, which I will perhaps reduce as I go along […] The main part of the work is still to be done, and this is what I did not have the leisure to make Abbé Bignon understand, not having the honor of speaking to him as easily as you do.

Although Mr. Newton has established on the observ. the numbers on which I have just told you that I had calculated, however, as he has not given an account of the observ. on which these determinations are based, one cannot derive any use from this theory, unless one compares it with the sky. The other astronomers who have adopted Mr. Newton's system have been satisfied with demonstrating the theory, without showing its agreement with the sky […] It is to make this comparison more easily that not finding tables exactly calculated according to the most recent determinations of Mr. Newton, I have calculated those that I have shown to Abbé Bignon. To compare these tables with the sky, observers are needed. I have already collected a good number of them, & I hope to have soon all that has been done so far: but however great the number of these observ. already made, they are not enough for me; for I believe that by following for a few years the course of the moon in my own way, that is, by my own observ., the accuracy of which I will know, and which I will make in the circumstances appropriate to my purpose, I hope by this means to advance much more and much more quickly than by using only the observations of others, whose accuracy it is very difficult to know, and which have not been made in the circumstances in which I might need them.

Delisle, at the end of his letter, asked Réaumur to inform Bignon of his project. This project of "new astronomical tables of all the celestial movements" was central in his activity at that time. He mentions it in his correspondence with several of his interlocutors, such as Teinturier in Verdun, Rast in Königsberg, Wurzelbau and Doppelmayr in Nuremberg. It was part of the increasingly frequent contacts established at that time with English scholars. As we have said, Louville on his return from England, after the observation of the eclipse of the sun, spread in France Newton's optical theory. Shortly afterwards, as we have seen, the French translation of the second edition of Newton's *Optics* appeared, probably under the influence of Delisle. Delisle read Newton's *Principles* in their second edition of 1713, and worked during the summer of 1716, alongside his experiments on light, on his own version of a lunar and solar theory derived from the *Principles*, which he had to assimilate quickly. It was at the beginning of 1717 that he finished his tables, which it seems probable that they incited Bignon to support his nomination to the Chair of Astronomy at the Collège Royal. In a letter to Georg Heinrich Rast from July 1718, having recently met Halley in London, he thanked him for the detailed information that he communicated to him on the activities of Halley and the other English astronomers (letter from July 16, 1718). He asked Rast for Halley's perpetual ephemerides of the movement of the moon. He indeed made print tables of the moon, the planets and comets according to Newton's theory. These tables were accurate to within 4' of arc according to Halley. He wanted to be able to compare them with his own tables of the moon, calculated similarly according to Newton's theory, but which he had not yet had time to compare with observations.

Delisle himself went to London in 1724 to meet Newton and Halley, as well as other members of the Royal Society, like James Pound and his nephew James Bradley, who discovered the aberration of stars (Schaffer 2014). On this occasion, he made known to Halley the observations of the members of his network of observers in France, Germany and Italy. Upon his return to Paris, he actively disseminated Halley's work. He showed himself in a letter written to Grammaticus (letter from October 1724) to be extremely critical against Cassini's tables of which only "those of the sun & satellites of Jupiter & of Saturn" were printed, Cassini and his close circle making the other tables "a great mystery". He reproached La Hire's tables for being constructed "without hypotheses & by the only observations". "As it happens quite often that these tables deviate from the sky", wrote Delisle, "one does not know where to reject the error, because one does not know the foundations, & that they are not built on a regular theory [...] & it is what disgusted me to use them". In the same letter, Delisle said that he was building better tables, based on "a regular & uniform theory, both geometric and physical, which [is] that of the English". Halley had given him a copy of his tables while in London, the basis of which he had explained to him, and this prompted him not to publish his own tables constructed in 1716, since Halley's were exactly as he had intended (Schaffer 2014). The bookseller of the Société Royale sent him a definitive copy of Halley's tables in 1724. Delisle, on his side, dissuaded the Parisian booksellers to republish La Hire's tables, and Claude Jombert, one of the principal scientific booksellers of Paris, ordered several copies of Halley's tables. Thus, Delisle's tables project lost their relevance with the publication of Halley's tables, which played exactly the role that Delisle wanted to assign to his own tables.

Delisle played an important role in the propagation of Newtonian mechanics, and its validation by astronomical observation. His departure for St. Petersburg delayed somewhat the development of celestial mechanics and cometography in Paris (Schaffer 2014). Until the 1740s, Halley's methods, which Delisle worked to make known in his Parisian period, remained little known in France. The main debates in the 1730s revolved around geodesy and the shape of the Earth, not cometography.

During his stay in Russia, Delisle developed his program in celestial mechanics and cometography. He coordinated there the observations of eclipses and transits in Germany and Russia, and organized a series of regular lectures, some of them public, on celestial mechanics at the Imperial Academy of St. Petersburg (Schaffer 2014, p. 174). His correspondence from the years 1741–1746 with Mairan in France, Euler and Wagner in Germany, and Celsius, Hiorter and Wargentin in Sweden gave a large place to comets, several of which were observed during this period. Thus, in his letter to Euler of June 23, 1742 (Bigourdan 1917, pp. 265–270), he asked Euler to send him all the Berlin observations of the comet that appeared in March of the same year. He communicated to him in return his own observations, in which he positioned the trajectory of the comet compared to the stars of Flamsteed's catalog. He said that he tried to determine the trajectory of the comet by applying Newton's laws, and asked Euler to improve his calculations. Euler answered him the following month (ibid., pp. 270–275) that his observations were the first to enable a precise determination of the comet's orbit. In August 1744, Delisle congratulated Hiorter for his representation to one minute of the trajectory of the comet of the beginning of the same year by a parabola, and asked him to communicate to him all the elements relating to this comet, in order to increase Halley's table of comets (letter of August 21, 1744). The same Hiorter, one year later, indicated to Delisle that he was going to establish the true orbits of seven comets: 1698, 1702, 1706, 1707, 1718, 1729 and 1739, and to determine by calculation the periods of these comets (letter of June 16, 1745). At the beginning of 1746, Wargentin wrote to Delisle that Hiorter continued his research on the last comet, whose parabola he converted into an ellipse and estimated its period at 345 years (letter of January 19, 1746). Thus, Delisle remained all his life very involved in the observation of comets and the validation of Newtonian mechanics.

In spite of his eminent position in the French community, Delisle suffered two significant setbacks in the years 1715–1725 from the Cartesians of the Academy that we will now detail (see also Chassefière 2022b).

2.2.2. *Delisle's setbacks at the Académie Royale des Sciences*

2.2.2.1. *Light diffraction experiments*

The experiments that Delisle carried out in 1716 on the diffraction of light were rooted in the observations of several eclipses made the year before, in which, as an astronomer, he was closely interested. On May 3, 1715, a total eclipse of the sun took place, which Louville observed in London, a place where it was the most marked, followed by eclipses of Venus (June 28) and of Jupiter and its satellites (July 28) by the moon. In the debate that accompanied these observations, several hypotheses clashed. For Louville, the luminous ring visible around the moon during the eclipse was caused by an atmosphere surrounding the moon (Nevskaja 1973). Moreover, Delisle wrote in a letter to Teinturier, when there was, according to Louville, a small crescent of the sun emerging from the moon, this one was red, "as it should happen by the refraction through an atmosphere" (letter of July 12, 1715). But in Paris, Maraldi was not in favor of this idea of a lunar atmosphere and attributed the red color to a fatigue of the eyes. Delisle said that, during the occultation of Venus by the moon followed from his Observatoire du Luxembourg by himself, Louville, and an English scientist of the Royal Society, the planet also took a red color on the side of the moon just before occultation, and that they saw it enter on the apparent disc of the moon of almost all its diameter. J. Cassini, Maraldi and Nicolas de Malézieu observing the same occultation at the Observatoire de Paris saw nothing of all that, as well as François de Plantade in Montpellier. Cassini did not deny, although he said not to have observed it, the red coloration, but attributed it to a chromatic effect of the telescope at the edge of the field. Of the report of the eclipse read by Cassini to the Academy, Delisle affirmed in a letter to Teinturier of August 27, 1715 that it was incomplete, and asked his interlocutor in Verdun for a copy of his own observations:

> I do not believe, however, that all those who extract observations sent to the Academy by foreigners, cut out what does not agree with their opinions, but whether by malice or not, it is very certain that in these kinds of extracts the authors are often made to speak quite differently than they did. Sometimes their discoveries are attributed to themselves; and it is very rare that one does not cut out things that one believes to be useless, which would not fail to be useful to others, when it would only be to give rise to thoughts.

> I told you in my last letter that Mr de Plantade had told Mr Cassini that in the last eclipse of Venus by the moon, he had not noticed anything that could serve to establish or destroy the moon's atmosphere: however, since then, he has told Mr le Chevalier de Louville that he had seen the same colors as us, arranged in the same way with respect to the moon; and moreover, that he had been assured that these colors were not caused by the telescope.

The allusion to Plantade's double discourse gives Jacques Cassini the image of an intimidating character, in front of whom nothing could be said. Delisle mentions in the letter that the occultation of Jupiter by the moon, observed as well by Cassini and his colleagues as by himself, did not give rise to the same effects as that of Venus in terms of coloration or apparent entrance of the planet on the disk of the moon. He mentions that Father Feuillée in Marseille noticed that at emersion, Jupiter's disc lengthened parallel to the edge of the moon, and the satellites appeared larger, and pointed out in passing that Cassini gave only little information on Feuillée's observation, "perhaps because this Father is in the opinion that the moon has an atmosphere", suggesting the scientific partiality of the director of the Observatory. Delisle seemed for his part open to the possibility of a lunar atmosphere, even if his interpretation was quite different, appealing to the diffraction of light by the edge of the lunar disk. He mentioned the experiment he made at the Academy, consisting of letting the sunlight enter through a hole in a dark room and receiving its light on a cardboard circle. By placing white paper behind this circle, we could observe on the paper the projected shadow of the opaque circle, bordered by a luminous ring similar to that seen around the moon during a total eclipse. We could also hang a ball in front of a window and place it so that it exactly occulted the sun, and then a white ring around the ball could be seen.

The experiment of which Delisle spoke was discussed, with other elements, by Fontenelle in the *Histoire* of 1715 concerning the eclipse of May 3 of the same year (Fontenelle 1715). The author noted that, as in the 1706 eclipse, a luminous circle of silver color appeared around the moon during total darkness, having in width 1/12 of the moon's diameter. Jean-Dominique Cassini had attributed it to the sun's atmosphere, while Louville thought that it was due to the moon's atmosphere, because if it was the solar atmosphere, the ring should not appear concentric to the moon at the beginning of the eclipse, the apparent diameter of the moon clearly exceeding that of the sun during the eclipse of 1715, which was, however, the case. Louville added other arguments, such as the red coloration of the edge of the moon which had not yet left the sun at the end of the eclipse, the progressive dimming of the sunlight in front of the edge of the moon which entered the sun at the beginning of the eclipse (which he attributed to a lunar atmosphere charged with vapors and exhalations), or of the lines of light coming out of the moon's disk at the beginning of total darkness, which could be lunar atmospheric flashes. From the width of the luminous circle, Louville deduced a thickness of the lunar atmosphere of 300 km, thus three times higher than that of the terrestrial atmosphere. This was not surprising since gravity is three times less on the moon. But Fontenelle judged the hypothesis of the atmosphere doubtful, and mentioned the phenomenon of the diffraction of light observed by Francesco Maria Grimaldi in the previous century, different from reflection as well as from refraction. This phenomenon means that if a light source is placed behind an opaque body, so that the light is normally hidden from the observer, the latter nevertheless sees a luminous crown around the body, as if the rays were deflected in the vicinity of the body. Fontenelle related the experiment made by Delisle. He also quoted La Hire who did the same type of experiment using a stone ball, but who invoked an effect of the multiple reflection of the light rays by the asperities present on the surface of the interposed body.

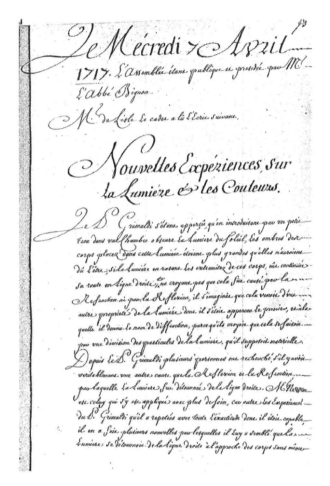

Figure 2.1. *Extract from the minutes of the Académie Royale des Sciences for the year 1717 (Bibliothèque numérique en ligne Gallica)*

It was in 1712 that Delisle discovered, or rather rediscovered after Grimaldi and Newton, the diffraction of light (Delisle 1738, pp. 205–266). He indeed realized in Luxembourg a gnomon allowing him to more easily adjust the true time of his pendulum. For that purpose, he stretched in the air a meridian made of a single thread, consisting of long hair tied together. The image of the sun was received in a dark room on a board coated with white. He was struck by the particularly sharp shadow of the thread, these two bands being themselves bordered by two narrower bands, from which they were separated by a black line. A third, weaker band was visible. At that time, he was not yet aware of Newton's experiments in the third part of his *Optics*, nor of Grimaldi's. Following this first, fortuitous observation, he showed that the shadow of all kinds of bodies was similarly bordered by luminous bands, and attributed to this phenomenon the luminous ring seen around the moon during the total eclipses of sun. He quoted Vittorio Francesco Stancari who, in the Comments of the Academy of Sciences of the Institute of Bologna, proposed the same explanation. Delisle said that he then carried out numerous experiments during the year 1716 (during which he had no place to observe the sky), before moving to the Hotel de Taranne at the end of the same year. Nevskaja (1973) mentioned the publication of the second edition of Newton's *Optics* on June 16, 1717, dealing with all the questions of diffraction. She notes that in 1720 a French translation of this second edition of *Optics* was published, and that this publication was ordered by the Princess of Gallie, who regularly attended Delisle's experiments. This suggests that Delisle was indirectly responsible for the French translation. Delisle read the results of his observations in the Public Assembly of the Académie Royale des Sciences of April 7, 1717, presided over by Bignon and open to the public outside the Academy, as indeed is attested by the handwritten minutes of the Assembly of that day (Bibliothèque numérique en ligne Gallica), the printed text of which, along with a detailed table of the results of his numerous experiments, can be found in the *Mémoires* of 1738 published in St. Petersburg (Delisle 1738). He said that he had postponed the publication of these data, but that he did not intend to carry out any more experiments, so he decided to publish them. He explained that his speech to the Academy was not mentioned in the *Histoire* of the Academy of that year, and that it was not published in the *Mémoires* of the same year.

Delisle then summarized the speech he read on April 7, 1717. He began by describing Grimaldi's discovery: by exposing objects placed in a dark room to the sunlight that he introduced through a small hole, he found the shadows to be larger than what they should be geometrically. This could not be an effect of reflection or refraction, and he imagined a third property of light, which he named diffraction. Newton repeated Grimaldi's experiments, and carried out others, showing, writes Delisle, that "light turns away from the straight line when approaching bodies without even touching them", as he detailed in his third book of *Optics*, which he judged to be "imperfect" (Newton 1722, First Author's Warning). Newton did not find the time to carry out more experiments than he did, said Delisle, "& is still trying today to excite someone to pursue this subject, convinced of its usefulness for Physics". Thus, Delisle felt himself to be in some way invested with the role of Newton's successor in the exploration of the properties of diffraction, or "inflection" as Newton called it, of light. Grimaldi having made only general observations, and Newton having limited himself to the study of the shadow of a hair, Delisle explained that he set out to precisely measure the shadows of all kinds of bodies, realizing in the course of his experiments "that the shadow of bodies took on different figures and different colors, so that in certain circumstances it was almost unrecognizable from what it ordinarily is". He noted that the variation of the size and the figure of the shadow of the bodies was intimately related to "the play of the light which is made inside the shadow itself" and interrupted his experiments on the size of the shadow to "pursue these appearances which are made in the shadow; because before proving the inflection by the size of the shadows, it is necessary to agree on what one must take for the shadow". He examined the shadow of several kinds of bodies: hair, needles, brass or silver wire, small copper or lead blades, and noted that this shadow, very black and clear when the plane of projection was close to the body, was less so when one moved away from this plane, the edges of the shadows becoming mixed with a little white. He observed that the shadow was divided into dark bands, whose number decreased when the projection plane was moved away, so that it became possible to count them. He noticed that the dark bands were even in number, unequally distant from each other, the intervals between bands being all the greater as one was close to the center of the shadow. Moreover, these intervals were all the wider that the body was small. He described the chromatic character of the diffraction pattern:

> All these appearances are made on the shadow, as I have said, and are much more marked than the shadow itself, which appears only as a tint of India ink spread equally over the whole, which being very black close to the body, prevents one from distinguishing anything but it to the simple sight, instead of which further away from the body, this shadow becoming lighter, does not take long to fade away. And it is then that one begins to see in each of these intervals colors similar to those of the rainbow, arranged in the same way, & which are all the more sensitive as the shadow is brighter. These colors are thus arranged from the middle of the shadow to the edges; violet, then blue, then green, yellow, and finally red: after which these colors begin again in the same order, as many times as there are intervals in the shadow; so that one sees that the dark lines which separate each of these intervals, are formed of the extreme colors of two of these contiguous suites; namely by the violet on the outside leaning against the red on the inside; instead of the less dark intervals that the dark lines leave between them, are composed of the lighter colors, namely blue on the inside, yellow on the outside, & green in the middle (Delisle 1738, pp. 214–215).

Of the 257 observations described in the table published in 1738, it is not known which were made in 1716–1717, and which possibly later, including in Russia, but the precision with which Delisle described the geometrical and chromatic effects of diffraction at the Public Assembly of 1717 already shows a detailed experimental exploration of the phenomenon. Delisle's speech, however, was completely ignored by Fontenelle in the *Histoire* of 1717, which cannot be considered as insignificant, considering its quality and the importance of the subject, underlined by Newton's interest in it. The non-contact deflection of light rays by a solid body was inexplicable in the eyes of Cartesians. They therefore only accepted such results with extreme circumspection. Let us specify in passing that the experiments carried out on light by Mairan in 1716–1717 (Guerlac 1981, p. 65) were not about diffraction, but about the validation of the *Fourth Proposition of the First Book of Optics* dealing with mixtures of homogeneous and heterogeneous colors, according to whether, looked at through a prism, they remained identical or were resolved into their elementary colors (Newton 1722, pp. 149–151).

Six years later, in 1723, Maraldi published a treatise in the *Mémoires de l'Académie Royale des Sciences* in which he presented the results of his own experiments (Maraldi 1723), similar to those carried out by Delisle in the previous decade. Fontenelle, who reported on these results in the *Histoire* of 1723, only briefly alluded to Delisle's original work, which he had entirely ignored six years earlier. "Mr Maraldi", he wrote,

> has carried out a great number of experiments to clarify this matter, not only with respect to astronomy, but even more in relation to optics, which, as it will be studied more and more, will always give more curious phenomena, & which one would have expected less. M. Delisle the younger had already given in 1717 his experiments and reflections on this same part of optics (Fontenelle 1723, p. 91).

As for Maraldi, he did not quote Delisle, whose experiments he seemed to have taken up. Delisle was sickened. On August 5, 1723, he wrote to Réaumur, a very influential personality, who by the way had no alliance with the Cassini camp (Schaffer 2014, p. 158), the following letter:

> Sir,
>
> I was flattering myself to have the honor of seeing you this morning, when my brother told me yesterday that you had to go to Meulan. I could not find a better opportunity, Sir, for the matter at hand, since you can discuss it with Abbé Bignon, & that it is a matter of academic discipline, & of the justice I ask for against one of my colleagues, which I believe you cannot refuse me. About a month ago, Mr. Maraldi read at the Academy a large memoir on optical experiments, from which he drew all the material from a small memoir that I read at a public meeting six years ago, and which is inserted in the registers of the Academy, but without having been printed, nor even without having been mentioned in the *Histoire* of the Academy. For 2 or 3 years before the reading of this memoir, I had almost exclusively applied myself to making the experiments of which I have only given the details. The Memoir of Mr. Maraldi contains only the same facts of experiments, only more extended in words, instead of that in the

small number of pages in which I had contained all the history of the phenomena that I had observed, I had announced more than Mr. Maraldi has reported in all his great Memoir. It is true that he added some good or bad explanations: I say good or bad, because you know better than I do, Sir, that they do not conclude anything; there is even manifest falsity. But whatever these explanations are, I ask if it is fair to let Mr. Maraldi's experiments be printed without also publishing mine, which are several years older. Have these experiments become better since they passed into the hands of Mr. Maraldi? When I gave the Academy the details of my experiments, I reserved for myself all the details which are quite different from what Mr. Maraldi seems to have done; that is to say, in addition to the general experiments that he reported, which are similar to mine, he does not seem to have entered into such a particular examination as I have done. He reports in his Memoir some circumstances which make me judge that he did not take the care to measure as I did the size and proportions of all the appearances which are seen in the shadows. I have a portfolio filled with all these measurements and details: but I do not know when I will be able to communicate the whole of it to the Academy. In the meantime, I hope that you will not refuse me, Sir, to indicate to me an expedient so as not to lose in the eyes of the public the part that I have in this work, 7 or 8 years before Mr. Maraldi (letter from August 5, 1723).

The explanation given by Maraldi in his memoir, and taken up by Fontenelle in the *Histoire*, was indeed hardly convincing, even obviously false, in contradiction with the results of the experiments. It was clearly only intended to provide an explanation that did not violate the beliefs of the mechanists of the Academy. As Fontenelle summarized it, "Mr Maraldi conceives that an infinite number of rays must enter through the hole, reflected by the particles of the outside air, and that the shadows they cause, which would not have been noticeable in a place that is usually lit, will be noticeable in the darkened room" (Fontenelle 1723, p. 97). Thus, to Newton's idea of an inflection of light rays in the vicinity of solid bodies, judged by Maraldi and Fontenelle to be not very "simple" and not very "physical", the two scholars substituted an explanation attributing the enlargement of the shadow to the juxtaposition of numerous secondary shadows caused by the solar rays reflected upstream from the hole where the sunlight enters, superimposed on the main shadow. As Delisle details in the *Mémoire* of 1738, this explanation was in direct contradiction with the results of his experiments, and indeed with the results of those of Maraldi himself.

2.2.2.2. *The project to measure the shape of the Earth*

The measurements made in 1718 for the extension of the meridian confirmed, according to Jacques Cassini, that the Earth was elongated along the axis of its poles

(Cassini 1718, p. 243), and not flattened on its equator as Newton predicted it, and justified it from the measurements of the length of the pendulum beating the second under different latitudes (Newton 1759, pp. 57–59). Delisle, in a memoir entitled *Nouvelles réflexions sur la figure de la Terre*, communicated in April 1720 to Jean-Paul Bignon (Delisle 1720), asserting that "nothing seems more appropriate to decide the question of the figure of the Earth & to know surely the extent of France than to measure the parallel of Paris at least in all the extent that it occupies in France". He recognized in his memoir the seriousness and the precision of Cassini's observations, and took note of the inequality, which "is reduced to 23 [seconds of arc] of which the difference of the parallels of Paris and Dunkerque was found greater by observation than it was supposed to be in the spherical hypothesis", and which he estimated probably exact to a few seconds of arc, which led Cassini to conclude to the elongation on the poles. But this, to extend to the entire Earth the results of the measurements made on France, appealed to hypotheses, of which we understand that Delisle found them debatable but did not wish to discuss them in his memoir, supposing "to be well established only the part of this meridian which was observed because it is known independently of any hypothesis". Delisle calculated the deviation that Cassini's hypothesis on the shape of the Earth would imply, compared to the spherical hypothesis, for the length of the parallel of Paris in its French part, and found it of nearly one minute of arc, easily measurable, "the eclipses of the satellites of Jupiter & of the fixed stars & of the planets by the moon providing us [with] a way to know the difference of the meridians much more exactly than one could do until now before the use that the academy drew from these kinds of eclipses". Delisle concluded his memoir as follows:

> If we had fairly detailed maps of the different provinces of France located on the same parallel & that we could link them together in such a way that by their means we could know approximately the distance of two fairly distant places of which we would know exactly the difference of the meridians, as it would not be difficult that it could be found by the exact observations of the differences of meridians which were made by order of king in all the borders of France, one could perhaps by this means in draw some sort of proof from the figure of France while waiting for it to be done more exactly by the same current measurement of the parallel of Paris made only in this view and executed with the same accuracy as the meridian was measured (Delisle 1720).

Aware of the important means required for a complete triangulation of France along the parallel of Paris, he proposed, in a second note, and also communicated to Bignon, a less expensive way to test Cassini's hypothesis on the shape of the Earth, namely to use the fires method, that Jean Picard had already used in the previous century. It was a question of replacing the common signal constituted by the occultation of a satellite of Jupiter, or of a fixed star by the moon, for two distant observers wishing to measure the difference of the true hours between two meridians (giving the difference of the longitudes), by a visual signal temporally more precise, in this case a fire in a place sufficiently elevated to be seen simultaneously by the two observers, which was masked or uncovered by means of a screen raised or lowered in front of it. Charles Marie de La Condamine, in his memoir of 1735, detailed this method, which he planned to use for the measurement of the parallel of the equator, and showed that we could thus reach a precision of approximately 1 second on the instant of the signal seen simultaneously by the two observers, 10 times better than that obtained by the observation of the occultation of Jupiter's satellites (de La Condamine 1735). Taking into account the precision of the same order that it was possible to reach on the true time in a place, La Condamine estimated at 1 second of time the order of magnitude of the precision on the difference of the longitudes between two places that we could obtain using the fires method (i.e. 360 degrees divided by 86,400 seconds of time, i.e. 15 seconds of arc). Delisle proposed a measurement of the difference in longitude between a tower, that of La Ferté Alais and the cathedral of Chartres, separated by 13 leagues (62 km) and whose line which connects them is directed almost east–west, by making a light on this tower, or by using a third place (Montmartre, in visibility of Chartres as well as of La Ferté Alais) for the signal of the light. He provided in particular the sizing of the fire required for it to be seen at such a distance. We can deduce from the calculations made by Delisle in his memoir that a precision better than 3 seconds of time on the difference of longitude between la Ferté Alais and Chartres would enable verifying, or invalidating, Cassini's hypothesis on the shape of the Earth. This time was thus compatible with the uncertainty of 1 second provided by the method, and we cannot doubt that it was in perfect knowledge of the facts that Delisle proposed such a device, which made it possible a priori to decide in favor of, or against, Cassini's hypothesis.

In his memoir of 1737 entitled *Projet de Mesure de la Terre en Russie* (Delisle 1737), Delisle returned to this idea expressed in the unpublished memoir of 1720, imagined as early as 1718, to decide the question of the shape of the Earth by the observation of the degrees of the parallel, compared with those of the meridian. He specified that the observations he proposed to make consisted of forming triangles along the parallel of Paris, and to observe at both ends the difference of the meridians in the most precise way possible. The difference was even, he said, so considerable, that he thought of using the fires method, as just described. And here is what he said retrospectively: "I went for this purpose in April 1720 at some distance from Paris towards the South, in the places which seemed to me the most suitable for this purpose: but it was not carried out, for lack of help, & by some other reasons which I will not mention" (Delisle 1737, p. 183). Thus, Delisle's choice to bypass the Observatory and Fontenelle by addressing himself directly to Bignon, no doubt justified by the necessary financial support of the royal power, ended in failure, the approach having turned out to be without effect. Maupertuis did not do so otherwise 15 years later by bypassing also the scientific institution, with success in his case, to obtain the support of the State for the Lapland expedition, which we know decided definitively the question of the shape of the Earth. Concerning the measurement of the parallel of Paris, it is from a political decision to establish a complete map of France taken in 1733 that the means put at the disposal of Jacques Cassini for this objective were achieved. And it was nothing less than the project initially sketched by Delisle which, according to him, would be implemented 15 years later, Cassini taking credit for "the method of determining the figure of the earth by the comparison of the degrees measured from the parallel of Paris different from what they should be in the spherical hypothesis", as mentioned in a non-autographer's bibliographic note, perhaps by Charles Messier, found in the archives of Joseph-Nicolas Delisle at the Observatoire de Paris (Delisle 1720).

Figure 2.2. *Sketch representing the tower of La Ferté Alais mentioned by Delisle in his essay (Delisle 1720)*

However, things were not so simple, because the method used by Cassini, described in the *Histoire* of 1733, did not use a measurement of the difference in longitudes of two distant points on the parallel. The geometrical method used by Cassini relied only on latitude measurements, more accurate according to Cassini than longitude measurements. The method consisted of measuring the difference in latitude between a point of the parallel far from Paris and its projection on the circle tangent to the level of Paris at the parallel, this circle passing through the center of the Earth and thus cutting the Earth's surface along a straight line, precisely what is provided by the triangles method. The deviation, which was zero at Paris (point of tangency), increased when we moved away from Paris towards the east or west, the parallel not being, in projection on the surface of the Earth, a straight line, and the law of variation of this deviation giving the shape of the Earth by comparison with the expected deviation, calculable, in the case of a spherical Earth. The method used by Cassini was not strictly speaking plagiarism of the one proposed by Delisle in 1720, even if the idea of measuring a parallel could probably be attributed to Delisle. We read, still in the same archives of Delisle, that he protested, by a letter dated July 31, 1734, to Fontenelle, and that he also wrote the same year to the Count of Maurepas, thus bringing the matter to the political level, a letter to which Maurepas replied the following year, giving him his opinion and that of Cassini on the question. In June 1735, Delisle had Father Souciet send a "letter for the author of the Trévoux journal with the original of his memoir presented to Abbé Bignon in 1720". At the same time, Delisle's sister (the latter being then in Russia), probably under the influence of the boards of the Observatory or of the Academy, in order to avoid a scandal which could have been prejudicial to the institution, urged the editor of the Trévoux journal not to publish the dissertation of his brother, "to keep it secret and not to say a word to anyone". This one, indeed, never appeared.

Following this failure, Delisle turned to the Russians. In a very long letter to Blumentrost dated September 8, 1721, he presented his project of tracing a meridian comparable to the one realized in France, but running further north because of the high latitude of St. Petersburg, and evoked the question that interested him most, that of the shape of the Earth. He took up the idea which was dear to him of the crossed use of the measurements of the degrees of the meridian and the parallel to determine the shape of the Earth, which he transposed from Paris to St. Petersburg:

> In the assumption that the earth is spherical, the degrees of the 60th parallel which is about that of Petersburg are precisely half of the degrees of the meridian. In the other hypotheses of the earth elongated or flattened by the poles, the degrees of the 60th parallel are more or less than half the degrees of the meridian; thus the measurement of the degrees of the parallel of Petersburg compared with the degrees of the meridian, would give new knowledge about the figure of the Earth,

which is absolutely necessary to know for the exact use of nautical charts, especially in these northern countries.

In the continuation of the letter, he details his knowledge in astronomy and, on the basis of his competence in the matter, offered his services to help set up an observatory in St. Petersburg; we will return to it in section 2.3 dedicated to the Russian period of Delisle's life.

He ended the letter by giving the broad outline of his *Histoire Céleste* project. We must, he said, begin by historically describing the progress of the discoveries, which is the most natural way to lead to the ideas that have gradually emerged. Then, we must compare the methods and discoveries of the different astronomers in order to identify their advantages and disadvantages. And, from this, the author should enumerate all the possible methods to allow astronomy to progress; and then enumerate all the possible observations, whose history would be described and the circumstances and verifications they carry with them would be discussed. It would then be necessary to describe the instruments and the way to use them, and then to show the use that we make of the observations to establish the rules of the celestial movements with the help of the tools of geometry. This theoretical part could also be treated historically. He criticized the existing works for describing only particular methods, results and opinions. We find again in his work the objective of producing a complete and universal work that would serve as a basis for further progress in astronomy.

2.2.3. *Delisle's great project:* Histoire Céleste

In a letter written two years later to Wurzelbau, Delisle describes the pedagogical use he wanted to make of the set of astronomical tables he was preparing, to which he added "all the necessary explanations to make the construction and the uses of these tables understood in depth" (letter from November 1, 1723). He conceived them as a tool, in the same way as the instruments of astronomy, of which he planned in his great *Histoire Céleste* to describe "the construction and the uses", as well as "their verifications", to make the public aware also of the observations. He added: "I will be able to join to these two parts [on the tables and the instruments], geometrical elements of the sphere, which I already composed and dictated to the Collège Royal; what will make more complete this astronomical institution that I will publish as soon as it is made, in order to facilitate the study of astronomy". And it is on this basis that the taste for astronomy could be born in some people. And, added Delisle, for those "who will want to devote themselves to this study for its own sake, I will gladly communicate to them all the observatories they will need; if they want to work on the particular theories, they will find at my place all the observatories collected & arranged, which will advance them a lot". He explained that he would put at the disposal of whoever wished it all the observations that he collected, "persuaded that I am", he said, "that it will be able to find many people who will be able to make a better use of it than me". Delisle encountered obstacles in the realization of his project, the provision of the community of the observations carried out by each one not being to the taste of all astronomers:

> However, I will try to do my best. I believe that it is necessary to do so for the advancement of science. Astronomy especially needs a lot of help and advice; and the work that needs to be done is so immense that one cannot call upon too many workers: but it would be necessary to do it well if all those who meddle in it were as disinterested in their particular reputation as I am in mine, & that they all satisfy the advancement of science, instead of finding too many who do the opposite, & who out of jealousy of the progress that others could make, refuse them the help of observ. & advice that they could give them. These kinds of people do this most often only because of a well-founded apprehension that they have of the progress of their inferiors. As they are not hard-working enough to prevail over them on equal terms, they try to weaken them by hiding this help from them: but they are mistaken, for one sees most often that by assiduity in work combined with a good mind, one overcomes many difficulties, and that one goes further than with more help and less application. This is said in passing.

We know that Delisle, not more than Pierre-Charles Lemonnier, who would be chosen 15 years later, at the age of only 23, by the Académie Royale des Sciences to write "the celestial history of all the astronomical observations made in France since the establishment of the Academy", as he said in a letter to Delisle from May 12, 1738 (CVC), was not able to obtain communication of all of Cassini's observations, and there is no doubt that the preceding criticism was directed against the Cassinis. In a letter to Cassini de Thury, son of Jacques Cassini, dated February 5, 1746, Delisle tells him that he was a pupil of his father and his grandfather. He informed him that his father did not answer his requests for results of observations, pretexting a lack of time to deal with it. He formulated in his letter the following request, which implicitly expressed his constant failure in the past to obtain from Jacques Cassini and Jacques Philippe Maraldi their observations:

> Can we not also hope that you would be willing to put in order to procure the printing of the beautiful part of the celestial history that the journals and observations of your late grandfather, your father and the late Mr. Maraldi can provide. It is a work that you should undertake following the example of the one that Mr. Lemonnier

undertook on the observations of Mr. Picard, La Hire & others. You would have for that, above Mr. Lemonnier, the advantage of benefiting from all the knowledge and lights that your father could give you, who has them from his own, which to the great advantage of astronomy makes us go back to the time when this science took a new form for the Cassinis. Without this living help, we often remain with uncertainties very harmful to the progress of science.

Delisle had a high opinion of his role as a collector of observations, which he carried out through his sustained action of network building, which did not weaken throughout his life as an astronomer. He describes thus to Wurzelbau, in the already mentioned letter of November 1, 1723, his motivations to collect observations "of all times & places":

> I gather all that I can of Astron. observ. of all times & places, to be in a better position to establish solidly all the foundations of astronomy I propose to discuss & fix each one separately & independently of the others, & that by the most appropriate observables, & by all those which can be used for it. For I imagine that the theory of each planet being composed of several elements, some theories can pass for good, although the different particular elements of which they will be composed are not well established, and this because the defects will be compensated. This is why one can only be sure of having reached the truth, after having established each particular element independently of all other knowledge, that is to say without assuming anything, as far as it is possible. The theories whose elements will thus be fixed each in particular by the most proper observ., seem to me to be as perfect as it is possible to have them. But the difficulty is to establish thus each element of astronomy. There is only the quantity of observations by which one can hope to come to an end, because by comparing all the different observ. made at the same time by different persons, it is a means to recognize the exactitude & the defects of some & others; & secondly, as in each astron. phenomenon some apply themselves to observe one thing while others observe another, one has by the assembly of all the observ. on the same subject all the circumstances that it is possible to have by observ., and this quantity of different circumstances is the only means by which one can be sure of each of the particular elements that enter into the composition of the observed phenomenon. These are my views in general in the collection that I make of astron. observations of all times & places.

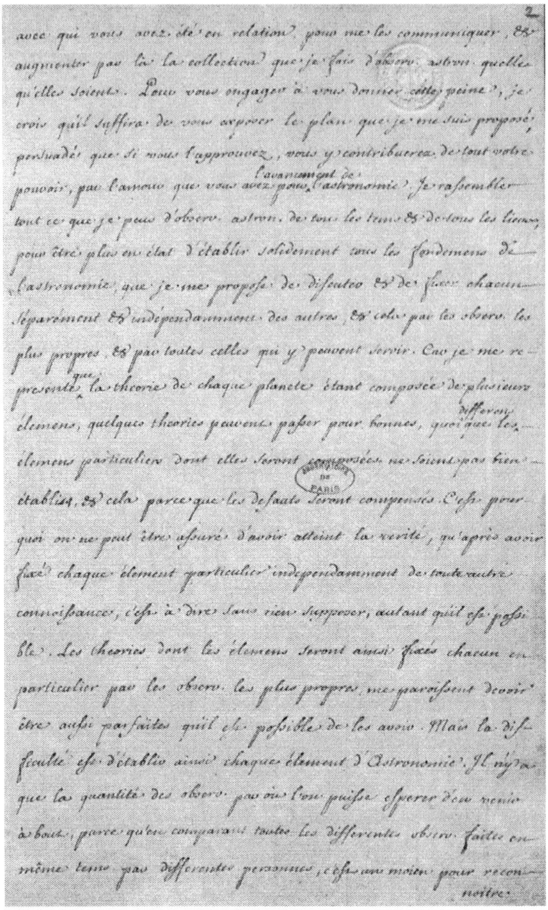

2
avec qui vous avez été en relation, pour me les communiquer, &
augmenter par là la collection que je fais d'observ. astron. quelles
qu'elles soient. Pour vous engager à vous donner cette peine, je
crois qu'il suffira de vous exposer le plan que je me suis proposé;
persuadé que si vous l'approuvez, vous y contribuerez de tout votre
pouvoir, par l'amour que vous avez pour l'avancement de l'astronomie. Je rassembler
tout ce que je peux d'observ. astron. de tous les tems & de tous les lieux,
pour être plus en état d'établir solidement tous les fondemens de
l'astronomie, que je me propose de discuter & de fixer chacun
séparément & indépendamment des autres, & cela par les observ. les
plus propres, & par toutes celles qui y peuvent servir. Car je me re-
presente que la theorie de chaque planete étant composée de plusieurs
élemens, quelques theories peuvent passer pour bonnes, quoique les differens
élemens particulier dont elles seront composées ne soient pas bien
établis, & cela parce que les defauts seront compensés. C'est pour
quoi on ne peut être asuré d'avoir atteint la verité, qu'après avoir
fixé chaque élement particulier independamment de toute autre
connoissance, c'est à dire sans rien supposer, autant qu'il est possible. Les theories dont les élemens seront ainsi fixés chacun en
particulier par les observ. les plus propres, me paroissent devoir
être aussi parfaites qu'il est possible de les avoir. Mais la dif-
ficulté est d'établir ainsi chaque élement d'Astronomie. Il n'y a
que la quantité des observ. par où l'on puisse esperer d'en venir
à bout, parce qu'en comparant toutes les differentes observ. faites en
même tems par differentes personnes, c'est un moien pour recon-
noître

Figure 2.3. *Excerpt from Delisle's letter of November 1, 1723 to Wurzelbau*

In the letter by Lemonnier to Delisle previously quoted, this one specifies that "the project when it was read at the Academy was approved by everyone, except Mr. Cassini who did not say a single word, and who did not seem happy with such a work". In his answer to Lemonnier of June 28, 1738 (CVC), Delisle expresses his surprise that in Lemonnier's *Histoire Céleste*, it does not make "any mention of Mr. Cassini, although it does not seem possible to treat the celestial history of the astronomy in France, without the main part coming from Mr. Cassini; and […] without that being done with his consent, and with his help". He reminded Lemonnier that the observations of Jean-Dominique Cassini, continued by his son and by Maraldi, were "the foundation of all that has been done in astronomy in the Academy", and wrote that it did not seem possible to him to give a celestial history of the observations of France without including those of Cassini and Maraldi. In his answer of September 11, 1738 (CVC), Lemonnier reaffirmed the agreement of the Academy with his *Histoire Céleste* project, which according to him was worth by way of consequence Cassini's agreement, but conceded that he could not obtain Cassini's and Maraldi's observations, of which Cassini said that he would take care to print them "one day to come". He added in post-scriptum: "If you write again to Mr de Maurepas, I beg you to speak to him about this celestial history with some eagerness for me and not for Mr Cassini". Lemonnier, supporter of Newton and participant in the Lapland expedition whose results had just been communicated and confirmed that the Earth is flattened on its equator, as Newton predicted, was caught in the climate of extreme tension that reigned at the French Academy. He confided to Celsius, also a member of the recent expedition, that he was "very uncomfortable with the faction of the Academy that holds for Cassini. This faction or cabal prevails for the present, so that Mr. Cassini is more imperious than ever: and one had even made some threats to me, something very unpleasant to hear although I fear no considerable effects". Delisle wrote to Cassini on September 24, 1739 (CVC), informing him of the letter he received from Lemonnier shortly after his return from Sweden in which he announced his project "to publish a celestial history of all the observ. made by the astronomers of the Academy, since the beginning of its establishment until now". He said to Cassini to have testified his surprise in his response to Lemonnier, as well as to the Count of Maurepas, that it was not mentioned in Lemonnier's project of his observations and those of Maraldi. We learn in a letter written by Delisle to Celsius on July 17, 1739, in which he gave him the chronology of his exchanges with Lemonnier on *Histoire Céleste*, that he conditioned the supply of his observations, that Maurepas required of him for the *Histoire Céleste*, by that of Cassini's observations, which in practice amounted to saying that he would not supply them, since Cassini kept his. He added:

> I then told Mr. Le Monnier how I thought this collection should be made for the best. I would have liked him to have read my letter to the Academy, but he did not think it appropriate to do so: so I do not know what the Academy's feelings are towards me in this matter. However I received shortly afterwards letters from the Count of Maurepas and from Mr. du Fay who was last year director of the Academy, who invited me on behalf of the Academy to send him the observ.. But I took another course of action which seemed to me the most appropriate, which is to have them printed here under my eyes, in a separate collection, under the title of *Mémoires* to serve the history and progress of Astronomy, Geography and Physics [the memoir of 1738 printed in St. Petersburg]. There is already a printed volume, of which I will soon send you a copy, and on reading it you will easily judge the reasons which made me take this decision. I will

also send it to France to those to whom it is appropriate, by which I will perhaps learn the judgment that the Academy will make of it.

Delisle clearly refused to participate in Lemonnier's work, using as a pretext the fact that Cassini refused to participate by transmitting his observations, publishing on his side in St. Petersburg the first volume of his own *Histoire Céleste*. This refusal must be seen in the context of Delisle's frustrated ambitions to gather all observations, including those of Cassini and Maraldi, for his project, more ambitious than the one that was realized under Lemonnier's leadership, since it included all observations, and not only those of French astronomers. Delisle's geographic distance at the time when Lemonnier was entrusted with this project, of which he was the initial craftsman, could not fail to accentuate his bitterness to thus feel dispossessed. Delisle's refusal revealed above all the value he attached to this project, and his disappointment at not being in a position to play the role he felt he should have. And he was all the less rewarded for it, that the quarrel which resulted from it with Lemonnier meant that the latter was not inclined to recognize him the role which was his in the establishment of the spirit which presided over the *Histoire Céleste* project.

The second volume of *Mémoires pour servir à l'histoire & au progrès de l'astronomie, de la géographie & de la physique* mentioned by Delisle never appeared. A first volume of Lemonnier's *Histoire Céleste* was published in 1741, limited to 20 years of observations, namely those of Picard from 1666 to 1682, and those of Philippe de La Hire from 1678 to 1686, and thus not containing the observations of Cassini and Maraldi (including Jean-Dominique Maraldi, nephew of Jacques Philippe). About them, here is what Lemonnier wrote in the introduction of the *Histoire Céleste*.

> The late Mr Cassini also made a very large number of Astronomical Observations at the *Observatoire Royal*, a large number of which has already been published in the *Mémoires* of the Academy but there is undoubtedly still a great number of them, in his Registers, which were not communicated to me; this is why as I believe them to be in very good order, I do not doubt that the Academy will include them in the *Histoire Céleste*, of which I have formed the Project. (Lemonnier 1741, project read at the Academy on May 10, 1738)

In this introduction, Lemonnier quotes Louville's observations, but Delisle's name is not mentioned anywhere. It was not until 1801 that Jérôme Lalande published the rest of *Histoire Céleste*, but he indicated in his foreword, which mentions the "numerous observations of Cassini, Pierre Lemonnier, Joseph Delisle, Charles Messier &c" (Lalande 1801, p. i) as being included in a new *Histoire*

Céleste, preferring to start with the "most recent observations, and especially with the observations of the stars, which are the first foundations of astronomy".

Cassini's hostility to the *Histoire Céleste* project, carried in the years 1710 and 1720 by Delisle, constituted an important aspect of the conflict which opposes him to his father's pupil, and more generally to a young generation of astronomers more concerned with preserving and gathering observations in permanent archives. During the years preceding Delisle's departure for Russia, it constituted an additional bulwark against his ambitions, which we have seen were already severely affected by the treatment that his research was subjected to by stifling it, and/or by plagiarizing in a more or less obvious way some of his discoveries that were passed over in silence by the scientific authorities of the time.

2.2.4. *Epilogue concerning the Parisian period*

The censorship of which Delisle was a victim by the Académie des Sciences concerning his work on the diffraction of light, which he considered as complementing those carried out by Newton, went well in the direction of the periodization proposed by Shank, namely of a significant role of the publication of *Optics* in the beginning of a phase of increasing tension between Newtonians and Cartesians in the Académie des Sciences in the 1710s, reaching its climax around 1730 (Shank 2008, pp. 105–231). The edition of the French translation of *Optics* in 1720 was ordered by the Princess of Gallie, who regularly attended Delisle's experiments, and Delisle thus played a very significant role in the diffusion of Newton's ideas on light in France (Nevskaja 1973, p. 303). The experiments carried out a few years later by Jacques Philippe Maraldi, proposing an interpretation which did not call upon the action at a distance of matter on light, constituted in their experimental aspect a form of plagiarism, and, for the theoretical aspect, an attempt to defuse Newton's corpuscular and attractionist theory of radiation. Delisle's adherence to Newton's *Optics* was coupled with his involvement in the construction of astronomical tables based on Newton's laws of attraction as set out in *Principles*, and in the proposal of a project to measure the shape of the Earth intended to settle the debate between the supporters of Cartesian vortices and the supporters of Newtonian mechanics. We must emphasize that Delisle's Newtonian convictions, clearly claimed with respect to Réaumur and Bignon, did not prevent him from being appointed by the king as professor at the Collège Royal, under the recommendation most certainly of Bignon. Thus, Delisle did not have a recognition problem within the French scientific community. Concerning his proposal to measure the shape of the Earth addressed to Bignon, and through him to the government, thus bypassing the Academy, we cannot doubt that the opposition came from Jacques Cassini, director of the Observatoire de Paris and member of the Académie Royale des Sciences, which possessed at the time the monopoly of the geodesic measurements. The relations between the two men, as we have seen, were cold, either out of jealousy that Delisle had assiduously assisted Jean-Dominique Cassini in his work at the end of his days (Sigrist and Moutchnik 2015, p. 96), or out of dogmatic opposition, Cassini being a fervent follower of Cartesian vortices. Moreover, Cassini, whose investment in this question was considerable, as attested by the size of the 1718 treatise (Cassini 1718), obviously did not wish to lose the monopoly of geodetic measurement campaigns, which generated means for the Observatoire de Paris. The silence of the Académie Royale des Sciences on Delisle's experiments on light diffraction characterizes a clear distancing of the institution from Newtonian attractionism, directly affecting the researcher in the exercise of his profession, and we cannot doubt that this setback, amplified by the other disappointments suffered, was an element entering in Delisle's decision to leave France (Chassefière 2022b). These difficulties had a strong impact on his rate of publication in the *Mémoires de l'Académie Royale des Sciences*, which went from three to four articles per year between 1715 and 1720, to less than one per year between 1720 and 1724. Oppositions within the Academy seem to have had the better of his ambitions, which he tried to realize elsewhere.

2.3. The invitation to St. Petersburg and Delisle's Russian period (1726–1747)

2.3.1. *The cartographic objective of Delisle's mission*

In a letter to Bignon dated May 31, 1718, Delisle informed him of a new method he had developed to measure the difference of meridians, a central question for the determination of longitudes applied to cartography. The measurements of longitude differences consisted, knowing the true time on two meridians, of measuring the difference of the true times by the observation of a common signal, which at short distance could be a fire, as we have seen, but at longer distance could only be an astronomical event visible simultaneously by the two observers. He explained to Bignon that the eclipses of the satellites of Jupiter by the planet, that we must prefer to the eclipses of the moon, were still uncertain to several seconds, which did not happen in the eclipses of the fixed stars by the moon, which were precise to the second. This method was the one recommended by Jacques Cassini. But he, Delisle, had "certain evidence of an inflection at the edges of the moon, from whatever cause it comes", which meant that the measurements of the differences of meridians using Cassini's methods were "all defective, & even in certain cases more defective than those which one drew from the eclipses of the moon". He explained that it would take him "several whole months of calculations to ensure the rules of this inflection, & to rectify all the differences of the meridians that one concluded from it without making this attention". He wrote that when he would have finished this work, "& that the academy will be assured by the demonstrations & the examples that I will give, of the precision that one must wait for, for the determination of the longitudes, one will be able to make use of it in the project that you formed to make trace a parallel from Brest to Strasbourg". This message must of course be placed in the context of the debate on the shape of the Earth, the month of May 1718 being precisely the one when Cassini left to extend towards the north the meridian of the Observatoire de Paris, observations which confirmed him in the opinion that the Earth was lengthened on the axis of its poles.

At the beginning of the 18th century, in a period of rapid development of geometrical methods of mapping by triangulation, completed by astronomical observations of latitude and longitude, the states committed large resources to the mapping of their territories. This was the case in France, as well as in other countries, such as the immense Russia. In 1698, the tsar Peter the Great, who visited the Royal Society and the Greenwich Observatory, hired the Scotsman Henry Farquharson to direct the Moscow School of Mathematics and Navigation (Appleby 2001). This School was officially created in 1701, and Peter the Great endowed Farquharson with books and scientific instruments. The English scholars were very interested in Russia, as much on the questions of cartography itself as on many other subjects of interest such as thermometry in cold countries, the variations of the magnetic needle, the aurora borealis, etc. In 1713, a list of 53 scientific questions was officially addressed by a Royal Society committee of 16 members including Isaac Newton and Edmond Halley, specially created for the occasion, to Farquharson and to the chief physician of the tsar, Robert Erskine, predecessor in this position of Laurentius Blumentrost, who came to be one of Delisle's main contacts a few years later. In 1715, the School was renamed the Naval Academy and was transferred to St. Petersburg, the new capital of the empire. It was built in the winter palace, and during the first third of the 18th century, it played the role of the first Russian scientific center, training generations of explorers, cartographers, mathematicians, surveyors, astronomers and engineers in all fields of interest for the modernization of Russia (Appleby 2001). Between 1717 and 1719, all the collections of academic works of the summer palace of Peter the Great were transferred to the Naval Academy, before joining the cabinet of curiosities of the Imperial Academy of Sciences, which was founded in 1725. Many geodesists were trained at the Naval Academy in the years 1710 to 1720, and sent to carry out field surveys. A map of Yakutia and Kamchatka was made in 1717, and a map of Siberia from astronomical observations in 1722. Two maps of the Caspian Sea were sent in 1720 and 1721 to the Académie Royale des Sciences and won the Guillaume Delisle prize, named after Joseph-Nicolas' older brother, who was a member of the Academy and First Geographer to the King. Delisle referred to these maps in a letter to Wurzelbau of October 3, 1722, in which he pointed out to his interlocutor that his brother had just had the map engraved in true size. This work took time, because it was necessary to translate all the names, and also to analyze what the map brought to the available information on the Caspian Sea (Ptolemy, Albuseda, Struys, etc.).

His brother's dissertation on the question appeared in the *Mémoires de l'Académie Royale des Sciences*, he told him. He sent him as a present this map, with another of the meridian of Paris, also drawn up by his brother Guillaume on the basis of astronomical observations and geometrical operations of astronomers.

Wishing to establish scientific relations between Russia and France, as he did with England, Peter the Great came to France in 1717, with the specific purpose of establishing links with the Académie des Sciences. He arrived in Paris on May 7, and stayed there for more than a month. He visited the Gobelins factory, the Hôtel des Monnaies, the Imprimerie Royale, the Sorbonne, and went to the Académie Royale des Sciences on June 19 (Chabin 1996). He attended a session of the Academy, and talked with several scholars. He met Abbé Bignon, President of the Academy and future Librarian of the King, and was proclaimed a few weeks later "academician out of all rank". A major fact of his visit was a map of Muscovy that he noticed in the king's home, a map published in 1706 by Guillaume Delisle. Struck by the quality of this map, on which he nevertheless corrected some errors as to the position of St. Petersburg and the Caspian Sea, he wanted to meet the scholar. The exchange was fruitful. The geographer exposed to the monarch the collection and the methodical criticism of the materials, previous maps and accounts of voyages, while waiting for the fruit of the scientific observations of the new travelers thanks to the compass and triangulation (Chabin 1996). Peter the Great first proposed to Guillaume Delisle to come and work in Russia, but he declined, and put forward the name of his brother Joseph-Nicolas, also an academician and astronomer, who was very familiar with the measurement of positions by astronomical methods. Thus, at first, it was the astronomer-geographer Joseph-Nicolas Delisle who was invited to Russia as a geographer rather than as an astronomer, thanks to the expertise and advice of his brother Guillaume, a professional geographer. The invitation was addressed to him "by mouth" by Johann Daniel Schumacher, the tsar's first librarian, during his trip to Paris in 1721 (Marchand 1929). At the same time, Schumacher gave Delisle a letter of recommendation and confirmation from Laurentius Blumentrost, the tsar's first physician, who was to become the first president of the Imperial Russian Academy of Sciences when it was founded in 1725.

In the already quoted letter to Blumentrost of September 8, 1721, in which he presents his project of drawing a meridian comparable to the one realized in France, Delisle declared to have learned from Schumacher that Peter the Great wanted to find an astronomical observatory in St. Petersburg on the model of the one in Paris. "On that," he wrote to Blumentrost, "I can instruct you in many things, and offer to His Majesty the Tsar much more capacity and experience than he may imagine to find in me." Thus, it was Delisle who offered himself his services to help set up the Observatory of St. Petersburg, while he was invited there as a priority to carry out a work of cartography of Russia. In the rest of the letter, he put forward his knowledge in astronomy as well as in geometry and algebra. He knew precisely, he said, what was missing in the existing observations, of which he collected a great part, to advance in astronomical knowledge. Here are the words in which he expressed his passion for his astronomer profession and proposed to put it at the service of the tsar:

> Among a dozen astronomers who are born in a century, and who spend the whole of their lives cultivating this science, there is only one Kepler. Animated by the example of such a great man, & feeling an ardor beyond all expression to discover the true rules of celestial movements, I began by collecting as much as I could Astron. observ. in which I was much happier than Kepler who only had those of Tycho's; because besides all the observ. of this great astronomer of whom I have made a copy on the true originals, I have analyzed all the great developments that have been made in the most renowned observatories of Europe, which have the advantage of being more exact than those of Tycho's, being made since Practical astronomy has been brought to such a high point of perfection, by the discovery of telescopes & the application that has been made of them to astronomical instruments, as also by the perfection of pendulum clocks & it is a treasure for to found a new observatory like that of Petersburg to bring there all the heap that I have made of Astron. observ. in nearly 15 big vol. in folio, and which I will not delay increasing again considerably, having already taken all the measures for this with the foreign astronomers with whom I maintain correspondence.

After describing the main themes of his *Histoire Céleste*, he mentioned the responsibility entrusted to him by the States of Brittany for cartography of this region, which granted him an annual income of 15,000 francs (of which only 20% came from his "academic" activities), of which he asked for the equivalent for the job he would have in Russia. He also asked for an advance of 10,000 francs before his trip to St. Petersburg, to arrange some business, to buy new books and to organize the transport of the instruments that would accompany him from Paris. Delisle said he would reserve his answer to the States of Brittany while waiting for a message from Russia. At the end of 1722, the States of Brittany urged him to decide. On October 8, 1722, he asked Blumentrost for a firm answer on the party to take, a letter to which Blumentrost answered only in March of the following year. The Russian–Persian war of 1722–1723, known as the Astrakhan, against the Ottoman Empire, of which Peter the Great feared that it extended its territory in the areas of the Caspian Sea and Transcaucasia, froze indeed the progress of the Observatory project for two years. Besides, Delisle wanted to keep his pensions and prerogatives at the Académie Royale des Sciences and at the Collège Royal during his absence, these pensions having to be paid to his mother and his sister, whom he did not want to leave in need. On his side, the French political power was divided on the fact to let him, or not, leave for Russia. There followed a long game of negotiations between the Russian State, the French State and Delisle, which ended in the signature of a contract on July 8, 1725, a few months after the death of Peter the Great, who was succeeded by his widow Catherine I, stipulating the sending of Delisle to Russia for four years, accompanied by his brother Louis Delisle de La Croyère and the engineer Vignon in charge of the realization of astronomical instruments. His annual salary was fixed at 1,800 rubles per year, that is, the equivalent of 9,000 francs, and housing, as well as heating, were provided to him. The "works and books which will be made for the Academy and which he will order" were entirely with the load of the Court. The contract stipulated that "Delisle being in Russia, he will be free to work on the astronomical observations in the times and places of Russia that he will like, chosen with the approval of the Court and will be able to send his astronomical observations to the Académie de Paris without that one can prevent them nor that one can retain them or delay". Concerning the "works and services that the aforementioned Delisle proposed to do and render in all that concerns astronomy to Her Majesty the Empress, she will do him the grace to give him authentic attestations of the satisfaction that she will have received" (Marchand 1929, pp. 388–389).

Delisle, when he left for Russia, accompanied by his young brother and the craftsman Vignon, intended to rely on his geographer brother for the proper geographical subjects that he would have to deal with in Russia. Unfortunately, Guillaume died in Paris during the trip to St. Petersburg. At the end of the warning placed at the head of his treatise on the *Nouvelles cartes des découvertes de l'Amiral de Fonte*, published in 1753, Delisle acknowledged his lack of knowledge in geography, having not yet had time to read since his return to France in 1747 the principal authors, which his brother knew, "so to speak, by heart, & on which he had left in writing only extracts for his use" (Delisle 1753, p. 10). He said his difficulty in reconstructing, from the maps of Russia and Northern Asia that he found in his brother's cartographic fund, damaged and divided after his death, the geographical vision that his brother had of these regions. "These are all reasons," he said, "which have made me defer until now to publish what I have collected on Geography, both of Russia & Northern Asia, and of America, & which must make me excusable, if I am not as well instructed on these parts as clever people, who have made all their life no other study than that of Geography, with more talent than I have" (ibid., p. 10). This deficit of knowledge in geography, which Delisle could not compensate by frequent exchanges with his prematurely deceased geographer brother, was at the origin of serious errors in the instructions given by Delisle to Vitus Bering May 4, 1741 for the research of the American continent; the excessively long distance traveled by the two ships from Kamchatka was probably responsible for the death of most of their crews, including Bering himself and Louis Delisle de La Croyère at the end of 1741. The inaccuracy of the maps established at the beginning of the 18th century by Guillaume Delisle and the blind trust that Joseph-Nicolas placed in the known apocryphal report (Laboulais 2006, p. 99) of the voyage of a certain Admiral de Fonte, a Spanish navigator who went to the north-west of the American continent in 1640, which he said he had received from London in 1739, led him to postulate the existence of imaginary lands, such as Gama Land, towards which he directed the navigators, leading them far away from the north-eastern direction they should have followed (Lauridsen 2014, p. 66), and which they would have followed naturally if the authority of the academic body had not been so inescapable for the navigators, who feared reprisals from the political power against them in case of disobedience. On his return to Paris, Delisle communicated to Philippe Buache the outcome of Admiral de Fonte's voyage, what constituted the departure of a polemic between Buache and Robert de Vaugondy. The two geographers disagreed completely on the veracity of Admiral de Fonte, which de Vaugondy did not believe (Laboulais 2006, p. 105). At the same time, two unsigned articles very critical of Delisle appeared in 1753, one in the *Nouvelle Bibliothèque Germanique* (Anonymous 1753a), relatively substantive and objective, which can be attributed without any doubt to the historian Gerhard Friedrich Müller, who participated in the great northern expedition, the other, much more polemical and insulting, in *L'Épilogueur*, a magazine published in Amsterdam (Anonymous 1753b). In these two articles, Delisle was accused of incompetence in geographical matters, and also of having taken the credit for having convinced, by the only presentation of his map in 1731, the Empress Anna Ivanovna to start the second great expedition of the north, a colossal operation at more than one million five hundred thousand rubles, employing 3,000 direct participants, which extended over more than 10 years, and finally resulted in rather meager cartographic as well as scientific results. Although tinged with patriotism, especially that of *L'Epilogueur*, these texts underline, probably rightly, the excessive benefit that Delisle tried to draw from this expedition for his personal prestige, with the tragic consequences that we know. It took 20 years after Buache's death for Vaugondy to recognize not being able to decide on the veracity of the account of Admiral de Fonte, while insisting on the impossibility for a professional geographer to be satisfied with the approximate data gathered by de Fonte.

The objectively rather cartographic character of Delisle's main mission in Russia was the argument that convinced the French authorities to let him go to Russia. The Count of Maurepas, Secretary of State for the Navy of Louis XV, exerted a constant pressure on Delisle to communicate the maps on which he worked in Russia. A letter sent by Delisle to Maurepas in May 1729, when his four-year contract was about to expire, was explicit on this issue:

> Whereupon I beg Y.G. [Your Grace] to consider that, as for the three years that I have been here, I have only been able to collect memoirs

for geography, without having put anything in order, for the usefulness of the country or of the Academy here, there is every reason to fear that, if I were to leave now, the Academy would not want to withhold my memoirs so as not to lose all the fruit of my work, and that thus I would only be able to take to France, of all that I already have, only the part that I have done in secret, instead of that by a longer stay in this country, having the leisure to make the two general maps, of which I spoke before, I would have perhaps the liberty to take with me all the memoirs that one knows here that I have copied, which are ten times more detailed than I can execute in the general maps that I have projected: Or at least, during an extension of leave for two years, I might find leisure to take a second copy of these memoirs, which I suspect they would not want me to carry away (Omont 1917, pp. 147–148).

Thus, Delisle informed the French government about the geography of Russia, which was discovered and constituted one of the major causes of the difficulties which he encountered in Russia. Abbé Bignon was eager to provide the King's Library with works from Russia, which he regularly reminded him of in his correspondence. At the beginning of 1730, Delisle learned about the death of Jacques Philippe Maraldi and changed his mind. He wrote to Bignon to tell him that he wished to return to France, doing according to him in Russia only geography, whereas he came to carry out astronomy there and asked for the role of member left vacant (his place was then that of an associate astronomer) to be able to devote himself to his passion (letter from January 3, 1730). Bignon answered to Delisle on February 5 of the same year that he himself had recently expressed the wish to extend his stay in Russia, having told Maurepas so the year before, and that it was too late not to give the place promised to Jacques Lieutaud, of whom he told him that he was ill and should not live more than three years, thus explicitly giving him the hope to reclaim this place in a few years. And Bignon added:

If you would be so kind as to allow me to tell you today what I advise you to do, I will confess that if I were you, I would not be disturbed by this event, that I would remain until the end of the seven years in which you are, in order to gather together so many precious things that we must expect from you: I would even redouble my efforts to show that nothing can slow down the zeal that has made you undertake such a long journey, and persevere in such peaceful occupations, in a country where I imagine that you will not have much pleasure. Please give some thought to this advice which I give you and which I would not give you, if I were less your friend.

In a letter to Maurepas in June 1730, Delisle seemed to have accepted, willy-nilly, the idea of staying for three more years, and asked that the salary of his chair at the Collège Royal continue to be paid to his family. He reiterated his offer of service to the Secretary of State in these terms: "I have not ceased to make all the observations that have been made here and to continue at the same time to collect all the memoirs for the geography of the country, and I hope, by the extension of my leave that H. M. [His Majesty] is granting me, to leave none of them of which I am not taking a copy to France" (Omont 1917, p. 160).

With these exchanges, the motivations of the French political authorities are clear, who agreed, after reflection and in spite of oppositions which appeared initially, to send Delisle to Russia only on the condition that he delivered them the strategic information that constituted the maps on which he would have to work. Delisle's unavailability to practice astronomy in Russia did not enter in account, successfully showing that Delisle was used by the French authorities for a completely different purpose than the one for which the astronomer was intended, namely to direct an observatory and to gather all the observational data of its network in a vast and universal work. As for Bignon, who supported the young and brilliant astronomer in his requests for means to devote himself to his passion, he found his account in the prolongation of Delisle's stay, not as president of the Académie Royale des Sciences, but as librarian of the king eager to increase his content. And, of course, in spite of the growing vicissitudes of the defense of Cartesianism at the Observatory and at the Academy against the Newtonianism that little by little settled down, before triumphing some years later with the resolution of the question of the shape of the Earth, the absence of Delisle could only satisfy the director of the Observatoire de Paris and the mechanists of the Academy, who had stifled the projects he had tried to set up in the decade before his departure for Russia.

2.3.2. *Delisle's means at the St. Petersburg Observatory*

The means attributed to Delisle for his daily life in St. Petersburg, both in salary and in kind, were very comfortable. His salary of 1,800 rubles per year (Marchand 1929, p. 388), the equivalent of 9,000 francs, which rose at the end of his stay to 2,400 rubles (letter to Mikhaïl Illarionovitch Vorontsov from December 29, 1755), was much higher than the average salary of academicians, 300 rubles for assistants, 600 rubles for full professors (Bradley 2007, p. 62). It was almost a hundred times that of an engraver, who earned between 24 and 36 rubles (Shafranovskij 1967, p. 606). This high salary constitutes one of the sources of discord with Johann Daniel Schumacher, the secretary of the Academy at the head of the Chancellery, who earned 800 rubles and then 1,200 rubles (Werrett 2010, p. 115). He tried to drastically limit the salaries of academicians. The annual global endowment of the Imperial Academy at the beginning was only 25,000 rubles (Léger 1919, p. 204), and the sum available for the library and the workshops, for which Schumacher was responsible, was all the more important as the payroll was low. Thus, in 1731, Leonhard Euler, who was then 24 years old, was competing with Johann Georg Gmelin, Josias Weitbrecht, Georg Wolfgang Krafft and Gerhard Friedrich Müller for promotion (Calinger 1996, p. 128). He then wrote to the president of the Academy that there were few scholars in Europe who had brought mathematics to the same level as he had, and that none would come for less than 1,000 rubles. Schumacher asked the president not to give in, otherwise, he said, Euler would become "impudent". The president finally granted him only 400 rubles (ibid.). Delisle had several servants, four as early as 1726 and up to seven in 1737 (Chabin 1983, p. 79), to take care of his house and serve his family (his brother, as well as his wife and a family friend, who joined him a few months after his move in spring 1726). Delisle, first professor in astronomy of the Imperial Academy, was after the ambassador and the consul, the most important Frenchman in St. Petersburg (ibid., p. 80).

These luxurious living conditions contrast with the means at his disposal for the exercise of his profession of astronomer, of which Delisle complained regularly in his correspondence. Upon his arrival in Russia, he had at his disposal the instruments he brought from Paris, and could soon after use instruments bought by Peter the Great in London, which were transferred for his use to the Observatory. The instrumental stock at Delisle's disposition at the beginning of his stay was limited:

> When he arrived in St.-Petersburg, De l'Isle had at his disposal only a small Chapotot quadrant of 18 inches radius, which he had brought from Paris. However, he immediately began his observations with this instrument; but not being able to meet the requirements of science, he proposed to the Academy to buy a large quadrant that he had had built in Paris under his own inspection. This proposal was not accepted. However, it had the desired result, in that larger instruments, already bought by Peter the Great during his stay in England, and which were in the Deposit of the Imperial Navy, were assigned to the use of the Academy's Observatory. These instruments were:
>
> 1) a large quadrant made by Rownley, of three feet etc. radius;
>
> 2) a sextant of 4.5 feet radius, also built by Rownley, but called a wall sextant;
>
> 3) some ordinary glasses of different lengths.
>
> [...]
>
> To complete the collection of instruments of the Observatory, De l'Isle repeated in 1735 the proposal, to bring from Paris the

quadrant of his own construction, and this time the Academy consented to these desires (Struve 1847, p. 83).

In a letter to Maurepas in May 1729, Delisle explained that the building of the Observatory was not finished, and would only be the following year, thus four years after his arrival in St. Petersburg (a delay which corresponded to the duration of the mandate written in his contract):

> This workman [Vignon] could not execute here the instruments of astronomy, of which I gave the plan for two years and of which he made the models, and that because the Academy has not been able until now to make the expenditure of it, because of the considerable expenses of its printing office; in addition to that the observatory, for which these instruments are intended, is not yet completed, but one started again for one month to work on the observatory and one assures me that it will be completed this year. There is talk of undertaking the large instruments soon, but they cannot be done before a year or two from now (Omont 1917, p. 147).

This delay in the setting up of the means also concerned the human potential, and not only for the astronomical objective. In a letter addressed to Count Plélo, the French ambassador in Denmark, in June 1732, Delisle complained about the lack of draftsmen to help him in the establishment of the general map of Russia, from the memoirs which arrived to him from all the regions of Russia, which moreover, he wrote, are not always of good quality (Benitez 1986, pp. 192–197). He said that he lacked the necessary time for cartography, taking into account this lack of means, and also the fact that he had to carry out astronomical observations. As early as 1726, Delisle determined the longitude of the Imperial Observatory of St. Petersburg, then he undertook the first steps to establish a general map of Russia. At his request, 33 maps from the Russian Senate were given to him at the Academy in December 1726. On February 16, 1727, he read a memorandum on the criteria that would govern the establishment of this map. He quickly proposed an expedition to take advantage of the eclipses of the satellites of Jupiter. His objective was to map the positions of the main Russian cities such as Moscow, Kazan and Astrakhan, as well as the cities located along the Volga and Oka rivers up to St. Petersburg, "by means of the geometrical methods used by Mr. Farquharson between Moscow and St. Petersburg, which he communicated to me" (Appleby 2001, p. 195). In 1731, following the three-year expedition (1727–1730) of his young brother towards Arkhangelsk and the Kola peninsula on the White and Barents Seas (Klein 2001), during which he measured the longitudes and latitudes of about 15 places, he asked the Senate for support and obtained the attribution to his project of two geodesists, Andrey Krasilnikov and Nikita Popov (ibid., p. 195).

Delisle's appreciation of his material working conditions was clearly more positive in 1735, perhaps because he obtained the purchase by Russia of his large quadrant made in Paris, when he wrote to Kirch in Berlin to praise the merits of his St. Petersburg Observatory and to convince him to accept to replace him at the board, which would allow him to return to France:

> It will be enough for me to tell you here that this observatory is in such a state as well by the construction of the building as by the instruments which are there that one can make there all the observations which one can make only in few observatories of Europe i.e. that one can make there all the observations which will be to be established again or to be checked and that one can also work there by new obs. to the theory of all the planets as advantageously as the greatest astronomers of Europe can do with the help they have in France, in England or in Italy. The help that one can have in this country as for the sciences surpasses even those that one can have elsewhere (letter from June 16, 1735).

At that time, the instrumental stock was much more extensive than it was in the first years. In 1738, it was necessary to add to the instruments quoted above a second quadrant (probably that bought from the French), a second sextant, three Newtonian telescopes with reflection, of eight, six and three feet of focus, several glasses, a parallactic machine whose telescope was provided with a wire micrometer, several pendulums of French and English manufacture, finally meteorological instruments established on a gallery in the part exposed to north (Struve 1845, p. 7). Nevertheless, in a letter written to Celsius on December 27, 1743, we find a negative judgment on the means placed at his disposal since his arrival in St. Petersburg, certainly exacerbated by the multiple difficulties encountered at the Imperial Academy in the recent years:

> For me, in spite of the effort I have made since I have been in Russia, I have not been able to contribute by similar means to the advancement of astronomy as much as I would have wished. The little help I have had from those who have had until now the board of the academy have caused me to disrupt my plans; and made me take the resolution to return to France where I hope to better use my time than here.

In the same letter, Delisle said that he took the firm decision to return to France four or five months previously, that is, in the summer of 1743. This decision was obviously motivated by the difficulties that he encountered with the Imperial Academy over several years, in addition to the budget cuts practiced by Schumacher of which the Observatory was regularly the object. In particular, he was relieved of his roles of the Director of the Department of Geography at the Academy, and thus responsible for the establishment of the map of Russia on his return from his 10-month trip in Siberia, at the end of 1740, following the progress judged too slow of this map (Chabin 1983, p. 90), and also to the suspicions of espionage which weighed on him. His project for measuring the shape of the Earth in Russia (Delisle 1737), which he submitted to the Russian Academy of Sciences in January 1737, and which was extremely ambitious, proposing a triangulation of the whole of Russia from a large base that he measured completely in the same year, hoping to obtain by an operation that he judged to be of great prestige for Russia, means that he could not otherwise mobilize, was buried in April 1739 (Appleby 2001, p. 198), either for lack of sufficient political support considering the distant term of the project's completion (several decades), or because the question of the Earth's shape was completely elucidated in 1738 following the Lapland expedition, the first reason seeming the most realistic. Let us also note the unfavorable reaction of the French team of Maupertuis then in operation in Lapland, which saw in Delisle's project a competing action likely to deprive it of the first place of its results, as expressed by Celsius, member of this team, in a letter to Delisle in March 25, 1738:

> I first showed your project to my companions, but they did not seem very happy about it. Perhaps because Russia wanted to share the glory of deciding this famous question, which was thought to be due solely to France. As for me, I strongly approve of your undertaking, and I hope that your work will soon be finished; since I have no doubt that it will confirm our measure of one degree towards Tornea. But I do not see the need to measure all of Russia […] The details of our observations at Tornea, you [will] soon see printed in Paris; we are working on it at present (CVC).

It was perhaps Delisle's immense disappointment, following the definitive failure of his project of measuring the shape of the Earth in Russia in 1739, which followed a first failure on the same question suffered 20 years earlier in France, as we saw, which pushed him to undertake the following year a voyage of Siberia, probably the only one that he made during the 22 years that he spent in Russia, whose goal was the observation of the transit of Mercury in Berezov. Tobias Kœnigsfeld, who participated in the expedition, financed by the imperial court, provided a detailed report (Kœnigsfeld 1768). Delisle and his three companions left St. Petersburg on February 28, 1740, regularly escorted by soldiers provided by the authorities of the provinces they crossed, in an unsafe context. Along the way, they regularly installed their instruments to measure the height of the sun or to carry out observations of eclipses to precisely determine the positions of the places they crossed, and the resulting differences in meridians. On April 22, date of the expected transit of Mercury, the weather was overcast and they were unable to observe. They also proceeded with temperature measurements with the help of thermometers that Delisle took with him. Kœnigsfeld reported Delisle's training activity during a large part of the voyage: "Mr de Lisle gave almost daily lessons of practical Astronomy to the Geodesists that the Admiralty sent to him, & I served as his Interpreter". They also took advantage of the opportunity to survey the geographical riches of the regions they visited: "On the 29th we visited the Government Archives, to do some research concerning Geography and History [...] at the Admiralty Office, where we went the same day, we saw a fairly large quantity of geographical plans, of which they promised Mr. de Lisle to communicate the list to him." Delisle, in an ethnographic approach, also observed the local customs. Their trip was marked by the death of the Empress Anna Ivanovna, and the ascendance, which would be of short duration, of the young prince Ivan to the throne:

> On the 8th, all that we were of our astronomical expedition, we went to the Cathedral Church, & we also lent the oath of fidelity to the Emperor Iwan III. The oath was first recited in Russian language in the presence of some clergymen; then each one kissed the Gospel & the Cross, & put his name at the bottom of the Writing which contained the oath.

They were in Moscow on August 23, and returned to St. Petersburg on December 29, 1740. This trip was exploited by his opponents to intensify their criticism of the French scholar (Jaquel 1976, p. 413). After his return, his relations with the Academy having become complicated, Delisle ceased purely and simply to frequent it.

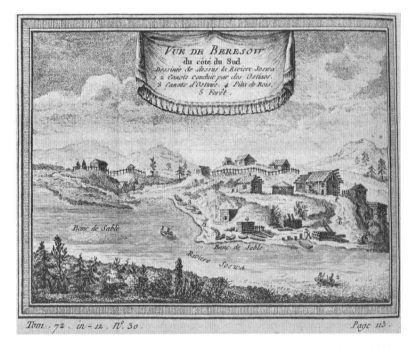

Figure 2.4. *View of Berezov from the south (Koenigsfeld 1768, p. 113)*

Figure 2.5. *Territory between the cities of Arcangel, St. Petersburg and Wologda. "This Map contains a part of the White Sea, Lake Onega, Lake Ladoga, Lake Bielosero & the Gulf of Finland; as well as the city of Wiburg, with Karelia, Olonetz, Kargapol, Ingria and a part of the Provinces of Novgorod & Wologda, together with the bordering territories" (Atlas Russien 1745)*

Delisle's delay in producing the map of Russia, which justified his eviction from managing the Department of Geography, was not attributed by his Russian detractors to an incompetence strictly speaking, as it was the case for the maps of the North Pacific established by Delisle on the occasion of the second unfortunate expedition to Bering, but rather it was Delisle's overly perfectionist character. He was someone who could not envisage a map of Russia that did not give rise to sufficient astronomical measurements of positions, which was precisely what he intended to do with his project of measuring the shape of the Earth in Russia. However, the Russians were in a hurry, and preferred an imperfect map in a reasonable time, rather than an accurate map in a distant future. The *Atlas Russicus*, or *Atlas Russien* finally appeared in 1745, even if Delisle, sidelined after his return from Siberia, ceased to participate in its elaboration from 1741. Moreover, he was at this time explicitly accused of espionage and also of hiding some of his observation results from the Imperial Academy. In order to move away from the Academy, with which he was in conflict, he asked in May 1741 the chancellor Andreï Ostermann to take over the direction of a Department of Geography that would no longer report to the Academy, but directly to the government, which Ostermann did not support (Appleby 2001, p. 199). He also applied at the same time to manage the Naval Academy, following the death of Farquharson two years earlier, but his application remained unfulfilled (ibid., p. 200), especially since Ostermann was arrested and deported at that time.

Thus, Delisle found himself without any possibility to continue his work outside the cenacle of the Imperial Academy of Sciences, in which life had become complicated. He wrote in a letter sent on June 7, 1742 to Celsius that his situation in St. Petersburg was painful, "because I am obliged to tell you that since my trip from Siberia, things have happened in such a way here, that I have been obliged to separate myself from the rest of the academy, & to forbid for nearly a year the entrance of the observatory to Mr Heinsius" (letter of June 5, 1742). (Heinsius was Delisle's assistant, who managed the Observatory during his trip to Siberia.) The year 1743 was moreover agitated at the Imperial Academy for other reasons, following the complaint filed in September 1742 by Andrey Nartov, one of the rare Russian members of the academy, with Mikhaïl Lomonosov against Schumacher for "dishonorable acts and embezzlement" (Leonov 2005). An inquiry commission was then created, and Schumacher was placed in detention. In February 1743, Lomonosov was banned from visiting the Academy Assembly because of his inappropriate and violent behavior towards some of his colleagues, and a series of serious interpersonal conflicts within the Academy ensued. In December, Schumacher was cleared and reinstated. We will return to this episode in the next chapter. Concerning Delisle, he wrote in a letter dated September 7, 1743 to Euler that he had not been paid his salary for more than two and a half years, in other words, since his return from Siberia and his dismissal from managing the Department of Geography. The reason, he said, was that Schumacher reproached him for communicating abroad observation results that he hid from the Russian Academy of Sciences. In a letter dated September 8, 1744, Euler, who had met in Berlin with German scientists who had obtained their leave from the Russian Imperial Academy, informed Delisle that these scientists had told him that he would be paid all his arrears and would obtain his leave when he had delivered all his observations to the Academy. And, indeed, at the same time, Delisle was busy putting in order all the observations he had made in St. Petersburg, to give them to the Academy before his departure, so as to be able to receive his salary (letter from Delisle to Hiorter of August 21, 1744). In a letter dated January 15, 1745 to Pehr Wilhelm Wargentin, he reiterated that he asked for his leave to return to France. But, as we learn it in a letter from February 26, 1746 to the astronomer Jean-Jacques

Marinoni of the Imperial Observatory of Vienna, "H.M.I. [His Imperial Majesty] wished me to stay a few more years until [the problem of] the academy, in which many disorders have occurred, is solved, this until the Observatory, which is still lacking a lot, is in the best possible condition, on which H.M.I. has already given his orders".

In a letter addressed long after his return to Paris, in 1758, to count Chouvalov, Minister of Education under Empress Elizabeth, a great Francophile and curator of the University of Moscow, Delisle similarly wrote that the Empress had "testified in July 1745 to His Excellency the Count of Vorontsov, then Vice-Chancellor of the Empire, that she would be pleased if I stayed even longer in Russia; after having made me the grace to promise, that she would remedy the abuses that I had represented to her, which had been introduced in the academy" (letter of July 6, 1758). Thus, the empress formalized with Vorontsov her wish to keep Delisle at the direction of the Observatory, and we learn in the same letter that she had then granted to Delisle a sum of 6,000 rubles (i.e. the equivalent of three years of his salary) for the purchase of the instruments which were still missing at the Observatory. Delisle changed his mind in the summer of 1745 and decided to stay in Russia, but he set out his conditions. These conditions are explained in a "Representation to the ruling senate" (letter from September 12, 1745). The first and most important of these conditions was that "the academy was regulated in such a way that the chancellery [headed by Schumacher] had no power over the professors or over all the things that should concern the sciences or even the economy of the academy". Delisle wrote that, according to the regulation of January 22, 1724, reflecting the intentions of Peter the Great, a professor should be named "President or permanent director". He informed that, considering his seniority, and his profound knowledge of the Academy, he suggested in his memoir of August 8 of the same year that he could be the first director so appointed. Thus, Delisle tried to substitute himself for Schumacher as the effective director of the Academy, the dispute between the two men having clearly become a power issue for Delisle. He asked to have nothing more to do with the chancellery, and that the new contract be made with the ruling Senate. On a financial level, he asked for additional means, and to have to account only to the Senate for the operating and equipment expenses realized for the Observatory. He thus asked that:

> I would like it to be stipulated in this contract that to the 1800 rubles of pledge per year that I have received up to now, 600 rubles should be added for my lodging, wood and candles, which were promised to me by my first contract, which has always been badly executed by the chancellery of the academy; having been badly lodged & having often been obliged to make advances and false expenses for the purchase of wood and candles. I also wish that the sum of 6 thousand rubles given by H.M.I. for the acquisition of the large instruments which are still

missing at the Observatory be given to me & that I give an account of it only to the directing Senate by providing the receipts of the Use which I will have made of it, that it is the same of a sum of 3 hundred rubles per annum necessary for the least expenses of the Observatory to know the repairs and the changes which there is often to be made to the instruments of which one makes use continuously & of which annual sum I will also give an account by the receipts which I will represent.

His second condition concerned access to personnel attached to the Academy, including student interpreters, which Delisle wanted:

As I have experienced the bad consequences of the authority of Counsellor Schumacher, who has led the Academy until now, has arrogated to himself to dispose of the student interpreters and all the other persons who are attached to the Academy and who can be useful to it; I asked in my memorandum of August 8 for the power to employ at my will the assistants, pupils or students of the nation, who have been given to me & those whom I can still find who will want to be instructed and help in the calculations and observations; without my being obliged to account for it to anyone other than the body of Professors.

He specified that he requested this not only for the Department of Astronomy but also for the Department of Geography. Finally, the third condition related to obtaining the post of Professor of the Académie de Marine, following the death of Farquharson to which he considered to be all the more entitled since it was him who, since the death of the latter, "instructed in practical and theoretical astronomy the principal masters and sub-masters of this Academy". This letter shows to what extent Delisle's ambitions in the Russian academic system were high, in the measure of the difficulties that he encountered there, from which he thus sought to extract himself by the top.

We understand, with the reading of his letter to Vorontsov already quoted (letter of July 6, 1758), that a new contract was ready to be signed when the president of the Academy, Count Razoumovski, recently appointed, suddenly served, on January 13, 1747, his order of leave to Delisle. At the time of leaving Russia, Delisle was prevented by Chancellor Bestoujev from seeing the Empress, certainly to prevent her, who was not aware of his expulsion (it was at least what Delisle thought), from ordering him to remain in St. Petersburg. The reasons that lead Bestoujev to expel Delisle, through his ally Razoumovski, were probably of a political nature, linked to the relations that Delisle, like other scholars expelled in the same conditions (such as Ribeiro Sanches, see Dulac 2008), had with the count Lestocq, doctor of the empress, schemer and author of a foiled conspiracy against Bestoujev, after having been in 1741, with the French ambassador La Chétardie, the instigator of the coup d'état which placed Elisabeth on the throne. Delisle was enraged by what he considered as a coup de force of the Russian Chancellery, led against him without the knowledge of the empress, but could only bend to the will of Razoumovski and Bestoujev. He committed, on the way back, an imprudence, by sending to Gerhard Friedrich Müller from Riga a letter dated May 30, 1747, which was intercepted and caused very serious problems to its addressee. This letter referred to manuscripts that Delisle had left with Müller before his departure, and that he finally preferred to take with him to France, asking him to give the bearer of the bill the portfolio in question. In this letter, he called the Academy of Peterbourg a *corps phantastique* (phantasmal body), a particularly pejorative expression at the time, which we learn in the *Dictionnaire de l'Académie Française* of 1694 to designate the incarnation of a demon. Müller was arrested in his house and interrogated several times by the professors of the Academy, who were thus exercising police functions (Mervaud 2009). Müller, in order not to have to leave Russia, later took Russian nationality, and became one of the main accusers of Delisle in the affair of Bering's tragic adventure during the second Northern Expedition. He nevertheless remained hated by Lomonosov until the end of his life, suspected of despising the Russian people and of looking for "stains on the clothes of Russia" (Mervaud 2009).

The following year, Delisle, charged to answer certain questions of the Imperial Academy which were put to him through Ribeiro Sanches, who had been sent back from St. Petersburg like him but who had become a correspondent in Paris of the Russian Imperial Academy (Dulac 2008), got angry. He declares that he did not want to have any more business with this chancellery, a contemptible institution which "unites a bad joy to the most lamentable ignorance", and that, if he has to write to the Academy of St. Petersburg, it would not be to those who declared themselves its leaders, or to those who shamefully submitted to them, but to the academicians whom he respected and whom he remembered with pleasure (Mervaud 2009, p. 465). Following this answer, Schumacher wrote to the Académie des Sciences de Paris and to all members of the Academy of St. Petersburg to demand that they cease all relations with Delisle (ibid.). And here is what the botanist Johann Friedric Gmelin said, former member of the Imperial Academy of St. Petersburg and participant in the second Northern Expedition, in a letter from Germany to his friend Delisle:

> We must go back to your connection with the Petersburg Academy. I have been told about it in a slightly different way than what you say. It is written that after a stay of more than half a year you have not written a single letter from Paris, neither to Mr. President, nor to any member of the Academy, that on this you were asked, if you do not act in the future in accordance with your contract, and that you

answered that you do not care about the Razoumovskys, the Winsheims, &c. & that you want to be free to print your observations as you wish and where you want. That on this the president has sent circular letters forbidding that no one of the honorary members have correspondence with you, because you had used unworthy expressions with the president. Allow me to tell you frankly that if the thing is as I was warned, it seems to me, however, that you could have been a little gentle towards such a person, who is put in such a high degree as the president of the Petersburg Academy. For the rest of us, we would hardly dare to use such expressions for fear that the Russian army, which is in Moravia, is brought back to force us to respect and obey (letter from December 22, 1748).

Schumacher's instruction had a negative impact on the functioning of Delisle's network, mainly in its German branch (Berlin Academy). We learn in the letter from Gmelin already quoted that Georg Wolfgang Krafft, a former student of Delisle's at the Observatory and a relative of Schumacher, who had returned to Germany in the meantime, did not reply to Delisle's letters out of consideration for the Academy. Gmelin himself wrote to Delisle that he would not pay his compliments to Krafft in order not to reveal to the latter his epistolary relationship with him. In a letter to the Marshall Count of Schmettau in December 1748, Euler indicated that he was "forbidden to have any trade with Mr Delisle". In Paris, the claims of the Russian embassy, demanding "1) that Delisle give back all the documents concerning geography and history that he had brought from Russia; 2) that he should not print anything about Russia without the knowledge of the Academy of Petersburg; and 3) that he should not be assigned anywhere until he had executed what he had promised in his contract", were, however, judged unenforceable by the Académie des Sciences (Mervaud 2009, p. 465). According to Euler, they made a bad impression in European scholarly circles. The correspondence of Delisle with Wargentin in Sweden, for example, was not affected at all. Beyond the immediate consequences described above, the repeated requests made by Delisle in 1755 and 1756 to the vice-chancellor Vorontsov, his former ally, to obtain the financing by the Imperial Court of the publication of his "great collection of all astronomical and physical observations" made in Russia, were not acted upon. He just received from Vorontsov a short and polite letter authorizing him to publish his observations, as long as they did not harm Russia:

> I have received your two letters of June 25 and August 14 last. In spite of all that has happened between you and the Imperial Academy of Sciences, there is no objection to the exchange of private letters that you intend to maintain with Professor Grischow, to whom you will be able to communicate the details of your observations and discoveries. As for the new observations made by the Academy after your

departure, I doubt that it will send them to you, but with regard to those that you have made previously yourself, it seems to me, Sir, that it will be a service to the public to have them printed insofar as they do not contain anything prejudicial to the interests of this empire (letter of October 27, 1756).

Thus, Delisle, who lacked means in Russia, and could not publish any other memoir than the first volume of 1738 because of the difficulties encountered with the Russian Imperial Academy in the 1740s, found himself after his return to Paris unable to publish his Russian observations, an operation that neither the Russian government, nor, visibly, the Académie Royale des Sciences, were willing to finance.

2.4. Brief synthesis of Delisle's scientific trajectory

Delisle's publication activity in the *Mémoires de l'Académie Royale des Sciences* reached between 1715 and 1720 a rate of three to four memoirs per year, on average. These publications are essentially about the events of 1715, namely the total eclipse of the sun and the various other eclipses, like those of Venus or of Aldebaran by the moon as well as on the questions of circumlunar luminous ring during the eclipses, of atmosphere of the moon, as well as on the use of the gnomon or the study of the diffraction of light. We have seen that the experiments on diffraction, whose results were read in a public assembly in 1717, were not included in the *Mémoires*, nor in the *Histoire*, of 1717. Delisle's publication activity decreased strongly from 1720, as we said, publishing between 1720 and 1725 only three memoirs, two on the transit of Mercury and one on the solar eclipse. Then, publishing stopped completely during his stay in Russia, to resume from 1748 on a basis of one to two publications per year on average between 1748 and 1754, then one publication per year until 1760. The subjects of the memoirs published by Delisle after his return from Russia were essentially about eclipses and transits, with the exception of an article on geography in 1751, and two articles on comets in 1759 and 1760. The decrease in Delisle's productivity from 1720 onwards corresponded to the period of preparation of his stay in Russia, the official invitation being made to him by Schumacher in 1721, but it undoubtedly also reflects, as we have shown, his lack of motivation following the setbacks he suffered concerning his work on the diffraction of light and his first project to measure the shape of the Earth.

We have seen that Delisle's material situation in France at the beginning of the 1720s was very satisfactory, both personally, thanks to his employment at the service of the States of Brittany, which brought him high emoluments, and professionally, as his recognition as one of the best astronomers of his time allowed him to successfully compete with the political authorities for prestigious positions at the Collège Royal and at the Académie Royale des Sciences, and to obtain the financing of powerful instruments, like his great quadrant. There is little doubt that Delisle would have found, by remaining in France, favorable ground for the realization of his scientific ambitions, in particular of his great project *Histoire Céleste*. There were various reasons for his departure and probably had to do, on the one hand, with the obstacles that his "activism" in favor of Newton's theories put in his way at the Observatory and at the Academy, on the other hand, with the attractiveness of a responsibility of astronomical observatory in the context of a nascent academy of sciences, created from scratch, which Delisle probably thought would offer him a much wider and less constrained framework for the realization of his scientific objectives. Delisle was indeed endowed with a solid ambition, in the best sense of the term, since he sought to put at the disposal of the community of European astronomers the tools most likely to support its development. This ambition was developed with energy through the epistolary network that Delisle patiently weaved from the middle of the 1710s. The portfolio containing the inventory of Delisle's correspondence from 1709 to 1768 (sold after Delisle's return to the Dépôt de la Marine in Paris) is rich in 2,606 letters, of which 1,514 are addressed to him, and 1,092 are copies of his replies. Of these letters and answers, 841 date from the time of Delisle's stay in Russia (Struve 1847, p. 98). The volume of this correspondence is thus exceptional, and Delisle's epistolary activity hardly slowed down during his stay in Russia, in spite of the obstacles he encountered there. There is no doubt that the attraction of the new structures set up in St. Petersburg, appearing to Delisle as virgin ground for the pursuit of his ambitions, especially as the conditions of life and exercise of his work which were promised to him there were excellent, and the rejection of the Parisian system in which the ebullient astronomer saw the projects that were dearest to his heart stifled and diverted, must have combined to push Delisle into exile, a decision whose long negotiations that led to the signing of the contract in 1725 show that it was not taken lightly.

But the scientific Eldorado that Delisle probably imagined in St. Petersburg quickly turned out to be a minefield. From the beginning, the secretary general of the Senate Kirilov, also a geographer, saw in Delisle a competitor and managed "to deprive him of the collaboration of some geodesists that he employed for his own works" (Isnard 1915, pp. 44–45), publishing himself a map of Russia in 1734. The authoritarian management of the chancellor Schumacher, without scientific coherence and out of any control by the academicians, being moreover inscribed in an unstable political context, generated significant turbulences in the institution, which culminated in the 1740s with the complaint lodged by Nartov against Schumacher. The significant means that Delisle enjoyed, which were out of all proportion to those of his fellow academicians in St. Petersburg, aroused jealousy, and Schumacher tried to limit and reduce them by all means in favor of other expenses that would allow him to enhance the library and other ceremonial services that he was responsible for. As much as Delisle's living conditions in St. Petersburg were excellent, as we have seen, his working conditions were not always up to his expectations, the means allocated to the astronomer being late in coming at the beginning of his stay, and the instrumental equipment reaching a satisfactory level only after 1735, when Delisle had already been in St. Petersburg for more than 10 years. Several times, Delisle wanted to return to France, because he considered that his cartography mission prevented him from devoting himself to his passion which was astronomy but under the pressure of the French political authorities, which used him as an informer, he needed to remain in Russia, where precisely his espionage activities began to be known and harm him. In the absence of sufficient human resources, in particular geodesists and draftsmen, the project of the map of Russia went nowhere, and the Russian authorities became impatient. The scientist Delisle wanted to treat the question as a good scientist, which required a great number of astronomical measurements of position, and thus a lot of time, much more than the policy could admit. In 1737, he tried everything with his measurement project of the shape of the Earth in Russia, and measured himself "on the ice, with all possible accuracy, a base of 11600 toises [22 km], which is the largest that one has yet employed in such a design" (Delisle 1737, p. 172). He wrote to an English journalist, with the obvious intention to give an international scope to his action. His project included a complete triangulation of all of Russia from the measured base, and the means to be implemented were considerable, although probably of at least an order of magnitude lower than the cost of the second Expedition of the North, then at the height of its realization. Delisle tried to kill two birds with one stone – to participate in the great scientific adventure of measuring the shape of the Earth, which he failed to do 17 years earlier in Paris, and to obtain more means to accelerate the realization of the map of Russia. He solicited the help of France, but obtained only an official letter of support sent by Mairan to the Russian authorities, and not the provision of a French astronomer to assist him in his task, as he requested (Chabin 1983, p. 176). The project definitely failed in 1739, and the measurement of the shape of the Earth escaped him once again. The delay taken by the project of a map of Russia combined with the accusations of espionage and of withholding scientific information from the Academy, resulted, upon his return from Siberia at the end of 1740, in his exclusion from the cartography project and the suspension of his salary. Delisle then broke off his relations with the Academy, and his situation did not improve until his return to Paris in 1747, in spite of the hope that arose in 1745 following the gesture of the Empress in his favor, which he tried in vain to take advantage of to become the director of the Academy. Delisle's break with the Academy made it impossible for him to publish under the seal of the Academy, and the first volume of *Mémoires* of 1738 was not followed by any other. The decade 1740 was, scientifically speaking, hollow for Delisle, who nevertheless continued his correspondence with the partners of his network. Back in Paris, as we have seen, he did not have the means to publish his observations made in Russia. He benefitted then, as Veteran to the Academy, of a pension of 3,000 francs a year and installed his observatory in the hôtel de Cluny, where Charles Messier and Jérôme Lalande took their first steps as astronomers. His full rights as an academician were not restored until 1761, and he taught at the Collège Royal until 1763. His last years were relatively miserable, and he died in 1768, at the age of 80.

2.5. Conclusion

Delisle's stay in Russia was therefore in many ways a failure. Can we say with Schaffer that it was "little more than a failed spy mission on behalf of the French government" (Schaffer 2014, p. 181)? The answer to this question seems to us to have to be no. The scientific feedback of the 22 years spent by Delisle in Russia may appear limited in terms of scientific production; however, important are the observations of aurora borealis collected in the *Mémoires* of 1738, and the activity of realization and dissemination of calibrated thermometers which was at the origin of a genuine European meteorological network, which fed the treaties on the subject of the second half of the 18th century, a question on which we will return in Chapter 4. In terms of cartography, the *Atlas Russicus* was published in 1745, and even if it was not as perfect as Delisle wished, the objective was reached, and the part taken by Delisle in its elaboration was not denied by anybody at the time in Russia, in spite of some reservations about his talents as cartographer, reservations that he himself shared. In astronomy itself, the number of observations made by Delisle in Russia was judged "prodigious" by Lalande (Struve 1847, p. 82), much more significant than only the measurements of eclipses of Jupiter's satellites for longitude estimation purposes published in the *Commentarii* during the first years. But these observations were not published, and the great *Histoire Céleste* that Delisle wanted to offer to the young generation did not see the light of day. Nevertheless, Delisle's observations were the subject of numerous exchanges between him and his correspondents in Europe during his stay in St. Petersburg, and undoubtedly contributed to the progress of astronomy in this crucial period of emergence of Newtonian celestial mechanics. The overall conclusion of Delisle's stay in Russia was therefore far from being negligible, even if the number of failures, in comparison with the considerable energy deployed, gave to the life of this exceptional scientist an undeniable tragic character.

Moreover, the foundation of the Imperial Observatory of St. Petersburg was not an adventure without a future, despite the departure of its two astronomers: Gottfried Heinsius, who returned to Leipzig in 1746, and Delisle, who left Russia the following year. Shortly after Delisle's departure, the Observatory was largely destroyed by fire. But the political will existed to repair it, and to equip it with better instruments, which was achieved in 1748, and two astronomers were appointed to replace Delisle and Heinsius. A third German-born astronomer, August Nathanael Grishow, was appointed the head of the Observatory, and the authorization was given to Delisle by Vorontsov to communicate with him. It was on Grishow that Delisle counted on to publish the whole of his Russian observations, but we saw that the project did not succeed for lack of means. Grishow died in 1760, after having drawn the plans of a new Observatory which was to be established away from the city, in a place more favorable to the observation. It was in Poulkovo, 19 km south of St. Petersburg, that the new observatory was built in the following century.

3

The Creation Ex-nihilo and the Beginnings of the Imperial Russian Academy of Sciences: The Influence of Christian Wolff

3.1. Introduction

As we have seen, Joseph-Nicolas Delisle's team at the Imperial Academy of Sciences of St. Petersburg provided de Mairan with a large number of aurora borealis observations in the first half of the 18th century. In addition, a member of this team, Friedrich Christoph Mayer, had imagined since 1726 a mathematical method enabling calculating, under certain simplifying assumptions, the height of the auroral arc. He also supported an aurora borealis system based on a mechanism of inflammable exhalations igniting in the atmosphere. Leonhard Euler, participating in the observations of aurorae in connection with Mayer, also worked out an aurora borealis system, very different, based on the principle of the formation of cometary tails imagined a century before him by Johannes Kepler, which he published only 20 years later, while he was posted at the Royal Prussian Academy of Sciences in Berlin. For Euler, the aurora was not an atmospheric phenomenon, but a cosmic one, comparable to the tails of comets and zodiacal light. Mayer, a follower of the philosopher Christian Wolff, experienced a major contradiction between his aurora system, very close to the one defended by Wolff, imposing an altitude not exceeding a few kilometers, and his mathematical method of determining the height of the aurora, giving it an altitude more than a hundred times higher. This contradiction, which one can qualify as a hiatus between experience and reason, rather than inciting him to build another system more in conformity with the observation, caused him a discouragement which resulted in a pure and simple abandonment of the aurora borealis as an object of study. Euler's reaction was very different, since he deduced from the great height of the aurora, which he did not doubt, that the mathematical method developed by Mayer led to the correct estimation of altitude and that the phenomenon occurred far above the atmosphere. His system led, however, to inconsistencies with the observation, which did not fail to be raised by Mairan in the second edition of his treatise on the aurora borealis. Euler's system, a main model in the first half of the 18th century, along with those of Halley and Mairan's, deviated from the Aristotelian explanation in terms of the inflammation of exhalations, did not resist criticism for long. Interestingly, Euler's approach, in order to escape the impasse into which the chemical hypothesis led Mayer, who could not seem to free himself from it, was on the contrary the deployment of the imagination, supported by physical concepts tested by his illustrious predecessors, to the detriment nevertheless of the attention paid to observation.

It is by starting from this particular problem, which must itself be placed in the specific framework of the development of astronomy and the setting up of the great European observatories of the time, that we will present the history of the creation of the Imperial Academy of St. Petersburg and of its Observatory, created ex-nihilo around a group of about 20 scientists, mostly called from Germany on the advice of Christian Wolff, follower of Gottfried Wilhelm Leibniz, the founder of the Berlin Academy and main interlocutor of Peter the Great for the realization of the project. Wolff happened to be the first, or one of the first, to organize a conference for the general public to provide his interpretation of the aurora borealis. The conference took place in Halle, in March 1716, one week after the great aurora observed throughout Europe, and Wolff, in response to a request for an explanation from the citizens, delivered a rational vision of the phenomenon, far from the superstitions and fears of apocalypse which agitated the Germanic world after the Lutheran reformation. The power of the Wolffian group at the academy, made up in large part of scientists with a theological background, in a context of the progressive emergence of Newtonian gravitation, supported in St. Petersburg by a minority of French and Swiss scientists, Delisle, Euler and Daniel Bernoulli, as well as Jakob Hermann, created an interesting context that we will analyze in more detail. The new academy, the center of animated discussions between Wolffians and Newtonians and subjected during its first 15 years of existence to a strict censorship of the orthodox religious authorities, refusing the heliocentrism of which Delisle and Daniel Bernoulli were the champions from the beginning, encountered serious difficulties, which led a certain number of scientists to leave it prematurely, like Daniel Bernoulli. The poor organization of the Academy, under the power of an authoritarian administrator, Johann Daniel Schumacher, who was subject to fluctuating and contradictory political directives, and who made the library and the museum of natural history the center of his preoccupations, led to a major crisis in the 1740s. This crisis, added to the political instability that reigned at the time of Anna Ivanovna's accession to the throne, led other scientists, such as Euler, and then Delisle, to leave St. Petersburg, depriving the Academy of its initial driving forces and plunging it into a profound crisis. This crisis, which affected astronomy and all other disciplines, would only cease in the second half of the 18th century, during the

20 years between 1766 and 1783, when Euler returned to St. Petersburg, at the request of Catherine the Great, to give a new impulse to the institution.

In section 3.2, we describe the Academy project imagined by Peter the Great, after having put it in its historical and cultural context, and by detailing the question of the birth of astronomy in Russia, a country which had then no past in the matter. Section 3.3 is devoted to the aurora borealis, describing first the historical context of Wolff's explanatory approach to the aurora borealis, and then the conference he gave in Halle on the subject. We then give biographical and scientific elements on the quartet of observers of the aurora borealis who officiated at the Observatory of St. Petersburg during the first years, after which we describe the hiatus between experience and reason experienced by Mayer in his approach to the aurora borealis, which discouraged him from continuing, and we provide elements on the physico-mathematical explanation proposed by Euler (we can also consult Chassefière 2021c on this question). Section 3.4 is devoted to the life of the Academy, since its beginning in 1726. It details, in particular, the opposition it encountered from the old nobility, and especially from the orthodox clergy, who censored works promoting heliocentrism, the oppositions between Wolffians and Newtonians, the serious problems of functioning encountered because of a conflict between the administrator of the Academy and the academicians. Then, we present the regulation instituted in 1748 to solve the difficulties which arose. In the conclusion, we briefly describe the evolution of the Academy in the second half of the 18th century, in particular the progressive creation, in astronomy, of a Russian community that took over from the few brilliant individualities from Western Europe who had initiated the movement.

3.2. The foundation of the Imperial Academy of Sciences in St. Petersburg

3.2.1. *Historical context*

Peter, son of Tsar Alexis Mikhailovich, who died in 1676, was proclaimed Tsar at the age of 10 in 1682, following the death of his brother Feodor. Ivan, the second son of Alexis Mikhailovich, should have succeeded Feodor, but Ivan's constitution was weak and it was Peter, the son of a second marriage, who took power. The reign of Peter I, called Peter the Great, was placed under the seal of a rapprochement of Russia with Europe and a strengthening of the power of the tsar, who claimed to be emperor of all the Russias. This approach corresponded to the desire to give Russia the appearance of a centralized and civilized state. However, the country remained governed according to a patriarchal principle, and reforms were imposed by force. The Great Northern War, which between 1700 and 1721 pitted a coalition led by Peter the Great, including Denmark, Norway, Saxony and Poland-Lithuania, against the Swedish Empire, ultimately saw the defeat of Sweden and the accession of Russia as the main power in the Baltic Sea. Sweden lost a large part of its influence, keeping only Finland and the northern part of Swedish Pomerania, a small piece of territory along the Baltic Sea, which was divided between Germany and Poland in the 20th century, among its possessions. St. Petersburg, the new capital of Russia, was founded in 1703 by Peter the Great, who logically wanted to bring the capital of his empire closer to the sea, to increase trade with Europe. He established a national educational system (see, for example, *Imago Mundi*, Russia in the 18th century), mainly for the sons of nobles and the sons of priests. Education was essentially practical, of direct use for the opening of the country to the ideas and sciences of the West, in particular in the fields of engineering, accountancy and navigation. He encouraged translation literature to promote the import of European ideas. Peter abandoned the Slavic alphabet, which was kept only for religious books, and created the Russian alphabet. He modernized printing tools. He was interested in geography, and in 1720, he instituted a school of cartography. In 1722, he ordered old correspondences contained in the archives of the monasteries to be collected and copied to enable the history of Russia to be written.

The rapid economic development of Russia attracted migrants from Europe. Following the revocation of the Edict of Nantes in 1685 and the repression of the Huguenots, it is estimated that 200,000 French Protestants went into exile abroad, especially in the Protestant states of the Holy Roman Empire. A number of Huguenots, coming from German states, appeared in Russia at the end of the 17th century. They generally mastered two or even three languages, French, German and sometimes Dutch, the latter two being more familiar to the Russian elites than French in the first quarter of the 18th century (Rjeoutski 2007). Around 1716–1717, French Catholics also emigrated to Russia, attracted by the economic boom of the country. They did not speak Russian, and most of the educated Russians did not know French, so they had difficulties integrating. Most of the time they were destined for menial jobs, while the Huguenots occupied relatively high positions in the army and the navy in particular, and also in medicine and industry. They were therefore, in fact, much closer to the country's elites. Many French Catholic migrants experienced poverty in Russia and returned to France mostly ruined and disappointed. The teaching of French in the Russian educational system began only after the death of Peter the Great, in the Imperial Academy of Sciences from 1725, and in the Corps of Noble Cadets of the Army from its creation in 1731. By the middle of the century, almost half of the cadets chose French, the others studying German. The first teachers of the French language in the Russian educational system were Huguenots, who had spent a long time in the German states and had acquired a relatively high level of education there, similar to that of the original Germans who emigrated to Russia. Many Frenchmen were introduced during the years 1730–1740 into the houses of the Russian nobility: cooks, wigmakers, confectioners, private tutors, etc., thus promoting the integration of the use of French in Russian high society.

Thus, at the turn of the 18th century, the foreigners employed in Russia at a certain level were all Protestants, either coming directly from Germany or French Huguenots who had passed through the Protestant states of the Holy Roman Empire, the foreign language used being German. It is in this context that Peter the Great made between March 1697 and September 1698 his trip called the Great Russian Embassy, during which he exchanged letters with Gottfried Wilhelm Leibniz, the father of the Royal Academy of Sciences of Prussia, founded in Berlin in 1700, on how to quickly make Russia a European state (Esteve 2018). The Great Embassy was a large-scale operation, mobilizing hundreds of people, and cost the imperial treasury 200,000 rubles, a quarter of its annual budget (Sigrist 2015). "During this trip", writes Sigrist, who compares it to a "gigantic enterprise of espionage and of technical, economic and cultural recruitment", "the tsar had the opportunity to become a carpenter in an Amsterdam shipyard while discovering Dutch mathematics, then to visit the Woolwich arsenal, the docks and an English cannon foundry, but also the University of Oxford, the Royal Society and finally the home of Isaac Newton" (Sigrist 2015, p. 92). In a letter to Pierre Lefort in July 1697, Leibniz confided that the Russian tsar wanted to "attract to his country the sciences, arts, and morals, especially of our Europe" (Esteve 2018, p. 263). In the same letter, Leibniz set out the seven points that he felt were necessary to implement in order to achieve this goal:

> 1. To create a general institution for the sciences and the arts. 2. To attract competent foreigners. 3. To import from abroad what is worthwhile. 4. To go over the various subjects with all due attention. 5. To educate the people on the spot. 6. To determine precisely the country's relations in order to know its needs. 7. To make up for what is missing (ibid., p. 264).

Leibniz and Peter the Great met in Torgau in 1711, Carlsbad in 1712 and finally in Bad-Pyrmont in 1716. On this occasion, Leibniz wrote for Peter the Great a project for the modernization of Russia, which showed his decisive influence on the creation of the Imperial Academy of Sciences in St. Petersburg, which was partly built in the image of the Academy of Berlin, of which it was in a way the heir. In addition, Peter the Great visited, as we have seen, the Académie Royale des Sciences in Paris in 1717, of which he became an honorary member, during the trip that brought him in contact with Guillaume Delisle about the great project of cartography of Russia, a project that led to the invitation made to Joseph-Nicolas Delisle to come and set up an astronomical observatory in St. Petersburg.

3.2.2. *Peter the Great's Imperial Academy of Sciences project*

Peter the Great presented a document describing the proposed academy to the Russian Senate in January 1724 (Esteve 2018). The academy consisted of three classes or departments: (i) mathematics and mathematical physics concerning the related sciences of astronomy, geography and navigation, (ii) the whole of the physical sciences, including experimental and theoretical physics, anatomy, botany and chemical sciences and finally, (iii) the humanities, encompassing rhetoric, the study of antiquities, ancient and modern history, law, economics and politics (ibid., p. 264). The first two classes were identical in their perimeters to the classes, or departments, of "mathematics" and "physics and medicine" of the Berlin Academy at its inauguration in 1711, the third more or less overlapping the perimeters of the Berlin Academy's classes of "philology" and "national history" (Bartholmess 1850, p. 55), even though the question of language and its roots was presented in St. Petersburg in a completely different way, since it was towards the study of foreign languages that the Academy needed to turn in order to move towards the desired Europeanization of Russia. As in Berlin, there was no department of pure philosophy, certainly under the influence of Leibniz, who reserved the benefits of the association between scholars to the exercise of experimental science, pure philosophy being the business, according to him, of "isolated geniuses". Contrary to the main academies that were created at the same time, or that were constituted in the second part of the 17th century, in Paris, London, Uppsala, Berlin or Bologna, in particular, the academic circle was not formed on the basis of pre-existing academies, or groups of scholars, but ex-nihilo. Leibniz's recommendation to bring in foreign scholars was followed to the letter, and it was essentially from Protestant Germany that the first class of scholars arrived. Among the subjects of interest for Peter the Great were all that concerned the application of mathematics, astronomy and chemistry to the construction of ships, canals and ports, and also to the improvement of navigation, artillery, mining and public health. This question of the utility of science for society was key, and revealed the influence on Peter the Great, through Leibniz and Christian Wolff in particular, of what was called German Cameralism, a Prussian-born school of economic and social thought that advocated the culture of utility, with the state expected to capture scientific and craft knowledge (and know-how) for the benefit of its own power, promoting the development of production and general well-being (Sigrist 2015). This cultural model, present in other developed European nations, was particularly prevalent in Germany, where it presided in particular, under the direct will of Frederick I, over the objectives of the constitution of the Berlin Academy.

Peter the Great's project, of very high ambition for the present and the future, stipulated the establishment of "a building that would serve not only to disseminate the Knowledge of today for the glory of the state, but also, through the teaching and propagation of it, to be useful to the nation in the future" (Esteve 2018, p. 264).

Given the almost non-existence, at the turn of the 18th century, of a national public education in Russia, the project included an affirmed university dimension. Without being anchored in a pre-existing university system, and following the wish of Leibniz, relayed at his death, in 1716, as adviser to the building of the Academy by his follower Christian Wolff, the project included the creation of a University attached to the Academy, composed of three faculties: law, medicine and philosophy, and of an Academic Lyceum intended to prepare young Russians for university studies. The building in which the Academy was located included several dedicated facilities: library, cabinet of curiosities and natural history, observatory, anatomical theater, physics cabinet, department of maps, workshop for construction of instruments and a printing works. The Academy was composed at the beginning of about 20 scholars (17 academicians, one auxiliary, one master of astronomical instruments, one master of elocution), about 20 technicians (seven engravers, two illustrators, six translators, three library assistants, seven printers), about 20 service employees, and about 20 students, for a total of a little more than 80 people (ibid., p. 264).

On the day of its official establishment, January 28, 1724, by a decree of the Senate which began with the words "His majesty the Emperor (Peter I) ordered the creation of an Academy where languages, as well as other sciences and fine arts, would be studied and works translated", "the Academy already had its prehistory" (Shafranovski 1967, p. 604). The books of the tsar's library and the anatomical preparations of the Department of Pharmacy had been put in order and transferred from Moscow to Peter the Great's Summer Palace as early as 1714. New books and natural history collections were acquired to complete the existing ones, under the coordination of the court physician Robert Areskin who died in 1718, and his assistant Johann Daniel Schumacher, who later became the Academy's Administrator and Librarian. In 1718, the books and collections of natural history were brought together in a house that served as a library and museum of natural history, where the books were available for reading and the collections were displayed to the public. When the Imperial Academy of Sciences was created, both entities were placed under its dependence, following the wish of Peter the Great, as written in his draft statute of the Academy: "It is necessary that the library and the natural history cabinet of the Academy be opened, so that the academicians do not lack the necessary instruments for their work. To direct this sector, it is necessary a particular librarian and it is necessary to acquire the books and the instruments which the Academy needs; it is necessary to order them abroad or to manufacture them at home" (Shafranovski 1967, p. 605). This point was essential because "it has often been said that in the years immediately following 1724 it was not the library and the museum that were attached to the Academy, but the Academy of Sciences that began its activity on the fringe of the library and the museum" (ibid.). The director of the library and the cabinet of curiosities and natural history was Schumacher, Areskin's deputy, in the first years. Schumacher, who was of Alsatian origin and had studied at the University of Strasbourg, was not a scientist, which generated, due to his difficult temperament, many disturbances within the academy during its first decades of existence, as we have already pointed out, but we will come back to it in the rest of this chapter.

3.2.3. *The birth of astronomy in Russia*

The first astronomer who worked in Russia was Jacob Daniel Bruce, a military engineer and cartographer of Scottish origin born in Moscow in 1670, who accompanied Peter the Great to England during the Great Embassy and developed Russian artillery during the Great Northern War (Sigrist 2015). He was the author, among other things, of a map of southern Russia (published in the Netherlands in 1705), a treatise on geometry and surveying, a calendar and a translation of Christian Huygens' *Cosmotheoros*, "a work that was a vindication for experimental natural philosophy as well as a vehicle for Copernican thought [and] made possible the development in the Empire of a worldly culture of astronomy" (ibid., p. 92). In 1702, Bruce installed in the upper part of the Sukharev Tower in Moscow what was probably the first observatory in Russia, equipped with specialized instruments (telescopes, quadrants, sextants, nocturnal – using the stars to measure time), and based on his observations, he produced a nautical almanac for the Russian Navy. He also made terrestrial and celestial globes that could be used by naval pilots. A school of mathematics and navigation was created in 1701 under the coordination of Bruce, intended to train naval architects and engineers, in which, as we have seen in the chapter devoted to Joseph-Nicolas Delisle, Henry Farquharson, a Scottish scholar he met during the Great Embassy, taught mathematics. In 1715, the school was transferred to St. Petersburg and renamed Naval Academy, and it became the first place for teaching astronomy in Russia. In 1716, Bruce transferred his observatory from Moscow to St. Petersburg, where it remained for 10 years, until the opening of the Academy's Observatory. At the end of this period, at the age of 56, he left St. Petersburg to settle in his estate near Moscow, where he occupied himself exclusively with astronomy, polishing himself the mirrors of his telescopes and making his instruments. He bequeathed his library of 1,500 works to the new Academy.

At the same time, some Russians were interested in astronomy (Sigrist 2015), like Alexander Menchikov, a military man belonging to the Neptune Society, who met at the Sukharev Tower Observatory created by Bruce in Moscow. He was elected to the Royal Society in London in 1714 and built a private observatory. Another member of the Neptune Society, the Ukrainian monk Theophane Prokopovich, advisor to Peter the Great on religious matters from 1716, also practiced astronomy as an amateur and had two observatories built. He was one of the founding fathers of the Imperial Academy of Sciences and its Observatory, from which he sometimes borrowed instruments and eventually bequeathed a total of 14 instruments. Among the enlightened amateur astronomers of the same period, we must also mention the ex-tsar of Georgia Vakhtang VI, and Antioch Dmitrievich Cantemir, son of the Lord of Moldavia and geographer Dmitri Cantemir. Refugee in St. Petersburg in 1724, Vakhtang VI developed research on oriental astronomy at the encouragement of Joseph-Nicolas Delisle. As for Dmitrievich Cantemir, also a refugee in Russia with his father, he served as an intermediary for the purchase of astronomical instruments as ambassador in London (1732–1738) and in Paris (1738–1744). We saw under which conditions Delisle negotiated with the Russians, and more particularly Laurentius Blumentrost, physician of Peter the Great, son of a German immigrant and the only Russian member of the Academy at the beginning, his conditions to accept the position of astronomer that was offered to him. Without going back over the details, we can summarize the program fixed by Delisle in a letter sent to Blumentrost, which closely associates astronomy, geodesy and geography in agreement with the intentions of Peter the Great:

1) The measurement of one degree along the meridian and parallel of St. Petersburg, a measure necessary to know the true figure of the Earth and therefore to establish more accurate land and sea maps.

2) Astronomical determination of the latitudes and longitudes of the main cities of Russia, with the implementation of triangulation as a necessary basis for drawing a map of the Empire.

3) The creation of an astronomical observatory in St. Petersburg and the organization of systematic observations coordinated with those conducted in other European observatories.

4) The determination of the exact distance of the Earth from the Sun, the Moon and other celestial bodies. Special attention was also given to the theory of the Moon's motion.

5) The study of atmospheric refraction. As a specialist in thermometry, Delisle proposed various experiments and physical observations intended to explain the differences in temperature between Russia and other European countries. It was thus an opening towards meteorology.

6) The training of Russian scientists, a point that Peter I was particularly keen on.

7) The writing of a comprehensive treatise on astronomy that would include the basics of this science, as well as its history (Sigrist 2015, p. 97).

Architecturally, the Academy consisted of two buildings located on Vasilievsky Island (Shafranovsky 1967). In the first one were grouped the observatory (the central tower), the museum of natural history and the library. This building was started in 1718, under Peter the Great, thus well before the foundation of the Academy, and its construction was spread over 10 years, under the responsibility of several architects. It was completed in 1727–1728, and Delisle installed his observatory on three floors of the tower. The second building, adjacent to the first one and built at the same time, contained a meeting room for the academicians (lecture hall), the administration of the Academy and the academic office under the direction of Schumacher, a geographical department which edited the maps of Russia, the offices of the translators involved in the publishing activities of the Academy, a printing workshop and a bookshop.

Figure 3.1. *Engraving extracted from Delisle's manuscripts preserved at the Library of the Observatoire de Paris representing constructions seen from the Observatory (Debarbat 1990)*

Before looking at the actual foundation of the Academy, which did not really start until 1726, and at its history over its first half-century of existence, we will look at the case of Christian Wolff, a great philosopher, follower of Leibniz, who was to be the inspiration of Peter the Great, and then of Catherine I after Peter's death in February 1725, in the choice of the 20 or so foreign scholars, mostly Germans, who joined the Academy for its start-up. Wolff himself was invited to become an active member and to preside over it, but he declined the offer and was made an honorary foreign member. However, Wolff was one of the first scientists, if not the first, to take the initiative of intervening directly with the public to explain and play down the occurrence of the famous aurora borealis of March 1716, which inspired Edmond Halley's aurora system. The two German astronomers of the Academy who assisted Delisle in the observation of the aurora borealis during the first 10 years of his stay in Russia, Friedrich Christoph Mayer and Georg Wolfgang Krafft, both of Wolffian persuasion, also offer an interesting example of the application of Wolff's thinking to the observation of the aurora borealis. The aurora borealis, as a natural phenomenon affecting the Earth's atmosphere, is therefore an appropriate theme to present the philosophical and scientific basis on which the Imperial Academy of St. Petersburg was created. This is an essential prerequisite to understand the spirit of the Academy, independently of the numerous organizational, relational and political difficulties that marked the life of the Academy during its first decades of existence.

3.3. Christian Wolff, the aurora borealis and their first observers at the Academy of Sciences in St. Petersburg

3.3.1. *Historical context*

At the end of the Middle Ages, the populations of Northern Germany were frightened by phenomena such as meteorites, solar or lunar halos, or rainbows (Schröder 2005). The sky, considered to be eternal and unchanging by nature, became a stage where frightening human and animal forms appeared, as diverse as they were inexplicable. Among the most important phenomena considered as harbingers of disaster were the aurora borealis. It was in the Holy Roman Empire that printed texts predicting the occurrence of terrible evils announcing an imminent end, often based on astrological predictions, were the most numerous. The apocalyptic fever that gripped Europe at the end of the 15th century, to which the invention of the printing press ensured a rapid diffusion, was the melting pot in which the Lutheran reformation was inscribed, splitting the Christian world. The schism, which refers to the texts of the Holy Scriptures, predicting a limited duration to humanity and describing an apocalypse of extreme violence, impressed the spirits, whose main concern was the struggle for daily bread. Luther's denunciation of the corruption of the Catholic Church, which equated the pope with the antichrist, and lent God such a wrath that he himself would precipitate the world towards its loss, further reinforced this feeling of the imminence of an end, to which the miraculous signs imprinted in the sky could not fail to resonate in the popular imagination. The announced arrival of the Antichrist pushed some Lutherans, entering in dissidence, to call the people to take up arms against the corrupted noble and religious elites, which reinforced the climate of violence inherent to the time, in a context of recurring misery. In the 17th century, the Thirty Years' War, which shook Europe until the middle of the century, contributed to the survival of the old superstitions. The penetration of scientific knowledge in the popular circles was then nonexistent, and no awareness, or questioning, on the nature of the worrying celestial phenomena, coming to disturb the purity of the celestial vault, came to temper the old ancestral fears. But, with the end of the Thirty Years' War, in the middle of the 17th century, life gradually became normal again, and the idea that the end of the world had been avoided took over. We have seen that, because of weak solar activity, almost no aurora borealis was observed during most of the 17th century, in any case since the aurora of 1621 observed by Gassendi. The phenomenon did not appear again until 1707, and in a sustained and repeated way only from 1716.

It has been mentioned that Maria Winkelmann-Kirch, in a letter to Leibniz at the end of 1707, shortly after her observation of the aurora borealis of that year, expressed a questioning as to "what nature has tried to tell us" (Schiebinger 1987, p. 183). It must be understood that the aurora borealis may have been a message from God, the meaning of which the astronomer wondered about. The Kirches were pietists (Mommertz 2005, p. 164), belonging to a movement founded in the 1670s, which encouraged the faithful of the Lutheran Church to a more personal and intense exercise of piety, promoting individual religious feeling over purely doctrinal knowledge of religion. The pietistic movement was born in the aftermath of the 30-year war, which saw a third of the population dead, during a period of deep economic and spiritual crisis, following a century of bloody confessional confrontations between Catholics and Protestants. Its founder, Philipp Jacob Spener, wanted to reinstill in the population a true and sincere faith, far from the dominant orthodoxies which he judged to distort it. This return to the purity of faith, exercised far from worldly circles and within the framework of a strict personal moral conduct, on which the Kirches met and united their efforts in their common practice of astronomy, could not fail to arouse in Maria Winkelmann, who believed in the signs and wrote astrological treatises, a fearful reaction to unusual atmospheric phenomena such as the aurora. God's judgment on the distortion of true faith was for her likely to manifest itself in the celestial sphere. Although mediated by the scientific practice of astronomical observation, which mitigated against the panic fears of the end of the world that these phenomena provoked in the populations of previous centuries, the perceptible anxiety that Winkelmann aroused by the aurora of 1707, almost the first seen in Europe for almost a century, was indicative of the still strong hold of irrationality in the Germanic world of the beginning of the 18th century.

3.3.2. *Christian Wolff's conference*

The aurora borealis of March 17, 1716, observed all over Europe, aroused in Halle, a city close to Leipzig, a wish for a rational explanation, as a reaction against the ancestral fears caused by these phenomena. It was towards Christian Wolff, a renowned mathematician and philosopher of the time, follower of Gottfried Wilhelm Leibniz, that the population of the city turned to. Here is how Wolff expressed himself on this subject:

> When last Tuesday after darkness on the 17 of March of the present year 1716 in the evening an unusual light appeared to us in Halle towards the North in the sky (which/as we learned soon afterwards/was at many other places/which lie on parallels not far away from that of Halle/also observed) and many who are inexperienced in their knowledge of nature felt great consternation; therefore many and diverse asked my opinion about this strange phenomenon and especially wanted to know if it can be explained anyhow with some basis. As I felt that in this case the people had some confidence in me and considered me as appointed to explain the book of Nature; I decided to expound my ideas about it at a Lectione publica on the 24th March, where a large audience participated… (Schröder 2001, p. 109).

Christian Wolff was born in 1679 in Breslau, a city in Silesia, now Wroclaw in modern-day Poland. As the son of a brewer, he was destined to study theology, but he was particularly precocious and interested in the sciences during his primary and secondary studies at the Breslau Lyceum. Witnessing the endless theological disputes between Catholics and Lutherans in his school, Wolff had an early desire to replace theology with mathematical truth as a medium for unquestionable deductive reasoning:

> But because I lived among the Catholics here, and from childhood observed the disputes of the Lutherans with the Catholics, I noticed in this connection that each believed to be right; I could not get out of my head the thought that it was not possible to show in theology the truth so distinctly, as that which is removed from contradiction. As now I heard after this that mathematicians proved their propositions so certainly, that every one of them should be recognized as true, I desired to learn mathematics for the method, in order to endeavor to bring theology to an incontestable certainty (Neveu 2014, p. 23).

It was above all as a first step in the acquisition of a rigorous method of exercising thought, far from dogmatic theological discourse, and not as an end in itself, that Wolff envisaged the study of mathematics. For him, the authority of scientific discourse, of a causal and demonstrative nature ("philosophical" knowledge), had to be placed above that of religious discourse, which approaches its object in a factual and descriptive manner ("historical" knowledge). Once this pre-eminence of philosophy over theology had been established, with Wolff centering his approach on the "knowledge of the world", where his Germanic predecessors placed at the center the "revealed knowledge of God", the whole of Wolff's work was marked by the desire to articulate, in the philosophical approach to the world, the two great axes of reason and experience (ibid., p. 17). Such an articulation was at the time a subject of discord between the two rival schools of empiricism, embodied by Locke in England, favoring concrete sensitive experience, and of rationalism, incarnated by Descartes in France, favoring deductive demonstration. This synthesis, this "marriage of experience and reason", as intimate as possible, was for Wolff the most important operation of philosophy, being useful to establish the greatest number of certainties, and to advance "prodigiously" in science. He saw in the current observation of nature the source of all possible knowledge, and in experience the first positive stage of rigorous science, the human mind showing itself active and directive in this matter. The rational experience of the sensitive was thus the intermediary between pure sensation and pure demonstration, which realized the marriage of experimentation and reason. However, experience and reason do not play symmetrical roles, the prevalence coming back in fine to the exercise of reasoning, the results of experience constituting a passive material, which reason must work on to make it intelligible. For Wolff, sensitive experience concerns individual objects, and cannot explain the why of things, which is due to the relations between objects, inaccessible to experience alone: "A man who is ignorant of Philosophy can indeed learn by experience about many possible things, but he cannot explain their possibility. Experience tells us that it can rain but does not tell us why it rains nor how it rains" (ibid., p. 71).

Wolff's *Elements of Aerometry*, published in 1709 in Latin under the name *Aeris vires ac proprietates juxta methodum Geometrarum demonstrantur*, offers an interesting example of Wolff's approach, and constitutes one of the first treatises on meteorology in the sense of a branch of physics based on observations of atmospheric quantities (Wolf 1938, pp. 274–276). Meteorological observations were just beginning, at that time, to be systematized by astronomers, and it would take more than half a century for all the material thus collected to be gathered and analyzed in a scientific way, in spite of the difficulties inherent to the physical formalization of meteorology, for example, by Louis Cotte (1774). Wolff's treatise, in accordance with the philosopher's doctrine, is mathematical in its form, based on definitions of terms, such as aerometry, scholia commenting on the meaning of words, axioms, such as the fact that heavy bodies press down perpendicularly on the bodies below them, theorems, such as Boyle–Mariotte's law, or problems, such as the construction of an air pump. The treatise returns to some famous experiments carried out in the previous century by Galileo, Torricelli and Boyle, takes stock of the question of the height of the atmosphere determined by the duration of twilight, or by the pressure of the air using the barometric law, and presents some purely meteorological points. Concerning the winds, Wolff attributed them to sudden expansions or compressions of air caused by the fluctuations of the heat deposited locally by the sun. He briefly describes the measuring instruments, with emphasis on the hygrometer and the anemometer, of which he gives a more detailed description. In the first case, he used the variation of length with the degree of humidity of a hemp rope passed over a pulley, and to which a weight was suspended, measured by a pointer attached to the pulley. In the second case, the wind turned a small windmill driving a wheel by a worm screw. The wheel had a radial arm carrying a weight at its end, oriented vertically downwards at rest. The torque exerted by the wind on the wheel caused it to rotate until the weight exerted exactly the opposite torque, the equilibrium angle of rotation of the wheel then providing the wind speed. The author also describes an air pump system, giving the mathematical expression for the residual amount of air in the receiver, namely, at the nth pump stroke, the nth power of the receiver's capacity compared to the nth power of the capacity of the receiver and the cylinder (where the air is pumped) combined. This treatise constitutes a condensed version of Wolff's philosophy, in which are found all the elements that, according to him, constituted the strength of the scientific approach: mathematical reasoning, physical experimentation, both in terms of instrument design and of the results of observation, and attempts at a general explanation of the causes of phenomena, such as the wind.

It is precisely this approach that guided him in the conference he gave in Halle on March 24, 1716 on the subject of the aurora borealis. He was then professor of mathematics and physics at the University of Halle, a chair he had obtained thanks to Leibniz, one of his teachers along with Galileo and Descartes, 10 years earlier. In the meantime, Wolff had extended the scope of his teaching to philosophy, and met with great success among his students. He extended his deductive rationalist approach to the truths of faith, a claim which in 1723 earned him an injunction to leave Prussia on pain of death. He spent 17 years in Marburg, before being recalled to his professorship in Halle by Frederick the Great in 1740, and becoming vice-chancellor of the University. In the public note accompanying the announcement of the conference, Wolff wrote that he wanted to discuss the following points:

> 1) Whether our phenomenon is something special, or whether it has been previously seen at other places and whether it has been observed by diligent observers;
>
> 2) Whether this belongs to the group of meteor(ological events)s as they are called by naturalists, and if this question is answered in the affirmative, in which class of meteors should it be included;
>
> 3) Thirdly, I promise, to investigate briefly the causes by which these curious phenomena in the air originate and
>
> 4) I want to briefly touch on the consequences, and also the significance of such phenomena (Schröder 2001, p. 110).

Wolff, in his lecture, addressed all these questions and affirmed that this optical phenomenon did not express a judgment of God, but was a natural behavior of the Earth's atmosphere. This event was a turning point. A professor, fulfilling what he considered his duty, opened the "book of nature" to deliver to an audience of interested common people a rational explanation free of theological bias. The conference participants had all witnessed the aurora borealis the previous week, and the answers Wolff offered to explain it opened their minds away from the eternal truths professed by the Church. The sky was no longer the immutable and perfect sphere disturbed by worrying phenomena, because they broke this immutability, but the atmosphere of the Earth, the center of natural phenomena that had to be understood. Contrary to the cases of the requests for clarification addressed by the English and French governments to the Academies of the two countries, charging, respectively, Edmond Halley and de Mairan, to provide a rational explanation to the phenomenon, it was here from the people themselves that, locally, the request for clarification came, the great recognition which Wolff enjoyed in Halle designating him as an enlightened and legitimate informant.

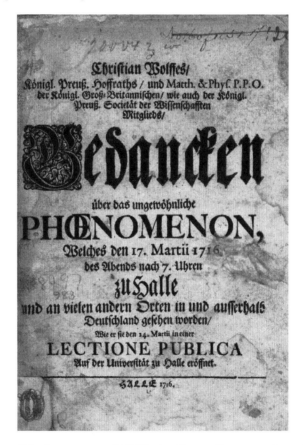

Figure 3.2. *Cover page of Wolff's treatise on the aurora borealis of March 17, 1716, which led to the public lecture of March 24 (Wolff 1716)*

3.3.3. *The quartet of aurora observers at the Academy of Sciences of St. Petersburg*

Joseph-Nicolas Delisle's *Mémoires pour servir à l'histoire et au progrès de l'astronomie, de la géographie et de la physique*, which he printed in St. Petersburg in 1738, placed his own contribution to the observation of the aurora borealis in the period which extended from September 1734 to the end of 1736. Before him, two academicians took turns in this observation task: first Friedrich Christoph Mayer from October 1726 to the end of 1729, date of his death, then Georg Wolfgang Krafft took over after Mayer's death, as well as for the daily meteorological observations. Mayer, as well as Krafft, were students of Georg Bernhard Bilfinger, a German philosopher and mathematician, a follower of Leibniz and Wolff.

Bilfinger, born in 1693, came from a family of academics, his father and his maternal grandfather both being theologians, and exercising ecclesiastical responsibilities. He studied at the monastery of Tübingen from the age of sixteen, and in 1711 obtained a doctorate in arts, followed by a degree in theology. He was already interested in mathematics, which he learned by reading Wolff and Leibniz. After his exams, he served as a curate in the cloister of Blaubeuren, then in Bebenhausen. In 1715, he became a lecturer at a theological school. In 1719, he obtained a loan which allowed him to study in Halle under Wolff for two years, after which he returned to Tübingen where he became associate professor of philosophy at the University in 1721. Wolff's philosophy was still little known in Tübingen, and Bilfinger was to become its spokesman. His treatise linking philosophy and theology (Bilfinger 1725), which essentially took up Wolffian doctrine, was a great success. Bilfinger's range of study was encyclopedic, like that of his teachers, covering physics, astronomy, botany, theology, as well as practical skills in the field of fortifications, which he acquired from a professor of mathematics specializing in these matters. In 1724, he was also appointed professor of moral philosophy and mathematics in a college for the nobility near the University. But, as we have seen, Wolff attracted the wrath of the church by his rationalization of religious discourse, earning him an accusation of atheism which led to his excommunication in 1723. The opprobrium fell on Bilfinger, who was rejected by his peers and whose courts were deserted. He was dismissed from his teaching position in 1724. It was this event that led him to apply, with Wolff's help, for a position as professor of physics at the new Imperial Academy of Sciences in St. Petersburg. He joined the Academy as soon as it was inaugurated, at the end of 1725.

Friedrich Christoph Mayer, born in 1697, arrived in St. Petersburg from Tübingen, where he was Bilfinger's student in the disciplines of mathematics, theology and philosophy, and served three years as a curate after his doctorate (Bayuk 2018, p. 7). Very little biographical information is available about Mayer. He probably stayed in St. Petersburg as a visitor in 1725, before being appointed the following year as an extraordinary professor, thus without an explicitly defined discipline. He was recognized in the new academy as a specialist in "mathesis", that is, "the foundation of knowledge", a rationalist project inherited from Descartes and Leibniz based on the fact that it should be possible to understand the universe from a small number of simple laws, the universal order thus proving accessible to reason (Rabouin 2009). During the four years he spent at the Academy, Mayer, who worked with Delisle at the Observatory, composed a calendar, the first and only one for a long time not to include astrological elements (Sigrist 2015, p. 96). It is also known that from the arrival of Leonhard Euler in 1727, he helped the Basel mathematician on various problems of celestial mechanics applied to the determination of the orbit of the sun, the movement of the planets and the calculation of the lunar eclipses. In addition, he worked with two other Basel scientists, Jakob Hermann and Daniel Bernoulli, on a theory of the moon based on Delisle's extensive astronomical observation program, which includes occultations of stars and planets by the moon, as well as lunar and solar eclipses (ibid., p. 102). These works give place to several publications of Mayer in the first volumes of the Acts of the Academy, the *Commentarii Academiae scientiarum imperialis Petropolitanae*. One of Mayer's most outstanding subjects of study is the observation and theory of the aurora borealis. Delisle indicated that Mayer observed, within the framework of his activity of collecting meteorological observations, many aurorae during the first year of his stay in Russia, and that he published in the first volume of the *Commentarii* as early as October 1726 a collection of his observations, accompanied by "his thoughts on the cause of this phenomenon, & the first foundations of his system which he tried to perfect until his death three years later" (Delisle 1738, pp. 77–78). He was rather negative about Mayer's talents as an observer, judging that the latter did not take the trouble to put in writing the details of his observations of aurorae, often indicating only the day. We will return to Mayer's system, as well as on a mathematical method, that Delisle did not mention, although it was published in the same article, that he imagined to estimate the height of the auroral arc without measuring parallax, thus from an observation in a single place.

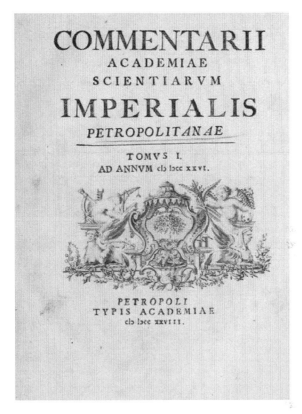

Figure 3.3. *Cover page of the first volume of the* Commentarii Academiae scientiarum imperialis Petropolitanae *published in 1728 (Commentarii de Petropolitanae 1728)*

It was Georg Wolfgang Krafft who took over the meteorological monitoring after Mayer's death, in 1729, monitoring to which was attached the observation of the aurora borealis. He was born in July 1701 in Duttlingen, a town located about 100 km south of Tübingen. His mother was the daughter of a prominent person in the town. From the age of 16, he entered the Blaubeuren monastery and studied theology, mathematics and natural history under Bilfinger and other teachers, and was in charge of a cabinet of natural curiosities. In 1720, following a similar path to Bilfinger, he joined the cloister of Bebenhausen, where he completed a two-year training course designed to give him access to the university. At the end of his training, in 1722, he joined the University of Tübingen, where he took courses in mathematics and physics taught by Bilfinger, a professor since the previous year. The two men became friends. In 1728, Krafft received the degree of Master of Arts. Bilfinger, who was already in St. Petersburg at the time, obtained for him a position at the new Academy of Sciences, which he accepted. He immediately set off, accompanied by the anatomist Johann George Duvernoy, who had been appointed to a professorship at the University of Tübingen after completing his doctorate in Paris, and for whom Bilfinger had also obtained a professorship in St. Petersburg. After an eventful journey, the two men arrived at their destination in late 1728. Krafft was given the task of teaching mathematics at the newly founded Imperial Academy, while waiting for a position as professor of mathematics, which would come five years later. In the meantime, Krafft, in contact with his fellow academicians, improved himself in different disciplines, ensuring the follow-up of meteorological observations, a task in which he seemed to give all satisfaction since he was promised the direction of the Observatory when it became vacant. This did not happen since Delisle stayed in Russia even longer than Krafft, the latter returning to Tübingen in 1744 to take a position as professor of mathematics and physics. Krafft, like Mayer, belonged to the circle of mathematicians of the Academy, namely, Euler, Bernoulli and Hermann, and Christian Goldbach from Königsberg, whom Delisle asked to exploit his numerous astronomical observations. Concerning the observation of the aurora borealis, Delisle seemed to be more satisfied with the work of Krafft than with that of Mayer. He indicated that Krafft continued Mayer's meteorological observations and that he "also marked most of the days on which these lights appeared in the sky, & did not omit the principal circumstances of those that were most remarkable" (Delisle 1738, p. 78).

To the trio constituted by Delisle, Mayer and Krafft, we must add Louis Delisle de La Croyère, who came with his brother to Russia and observed many aurora borealis during his trip to Siberia, in the years 1727–1729, during which Mayer made his observations at the Academy's Observatory. A section of the *Mémoires* of 1738 is devoted to these observations, the first of which is dated September 3, 1727, and the last one on September 15, 1729 (Delisle 1738, pp. 21–76). Most of them were made at very high latitudes, near or beyond the Arctic Circle. These observations, representing more than a quarter of those reported by Joseph-Nicolas Delisle in the *Mémoires*, are reported by de Mairan in the second edition of his treatise (1754, p. 509), but he did not include them in his statistics. The reason for this is that Mairan was concerned not to bias his sample by including regions with a particular climate, either because the weather was always overcast, or because of the proximity of the pole, the sun being above the horizon for entire months. In both cases, retaining these observations would bias the sample towards the very large, very bright aurorae, the only ones likely to be seen through the clouds, or in the permanent brightness of the Arctic summer. Moreover, it is necessary to note the doubt which existed on the real capacity of Louis Delisle to describe scientifically what he saw. Thus, in a letter of April 12, 1728, Joseph-Nicolas Delisle, after having indicated to his brother that he wished him to measure the aurora borealis in degrees and not in feet, specified that he could use the compass for this purpose: "you orient it", he told him, "towards each of the points (of the aurora borealis) and one can in this manner measure many angles in a very short time" (Klein 2001, p. 21). When examining the third volume of Louis Delisle de La Croyère's diary from his Siberian voyage, which is entirely devoted to the description of the aurora, the descriptions of the aurora become more succinct, perhaps reflecting an effect of weariness (Klein 2001). It is known that, in addition, the competence of Louis Delisle in astronomical measurement was questioned by his companions, in particular during the Northern Expedition of 1733–1743 (Anonymous 1753a, p. 49).

3.3.4. *The rejection of aurora observations by Mayer*

Mayer and Krafft came directly from the Wolffian school through Bilfinger's teaching in Tübingen. Both studied theology, and also turned to mathematics to give all possible rigor to their work as scientists. In St. Petersburg, they had, one after the other, the responsibility of meteorological monitoring, which included the observation of the aurora borealis, a phenomenon considered to be of meteorological nature (a "meteor"), in accordance with the Aristotelian thought on the matter. Mayer presented his theory of the aurora borealis in an article written in Latin published in the first volume of the *Commentarii* published in 1728 (Mayer 1726), a full English version of which is given in the Appendix. He first described what he considered to be the two most commonly observed types of aurora borealis:

> The first kind is common and more frequent in our regions: shaped as an arc, it is located in the north, sends towards the top rays that look like light beams and shines with a quiet light. The second kind, which is rarer, turned towards the sky of which it occupies every regions, is fragmented into different parts where are most commonly fixed beams that are either very short or non-existent. The splendor of this kind of polar lights, as it is often thought, is that it waves or trembles (Mayer 1726, p. 351).

Concerning the first type, he stated its characteristics, in terms of geometry of the arcs, of their variability (displacement, division/merging), of properties of the region of the sky located under the arc (black, but transparent to the light of the stars, except at the level of the dark interior margin of the arc), of the existence of parallel and intermittent vertical beams, animated by horizontal displacements, rising towards the top of the arc, and sometimes exceeding it, of the appearance of clouds higher than the ordinary ones. About the second type, he described it as a general blazing of the sky, with small bright clouds scattered all over the sky, sometimes extinguishing, sometimes lighting up, sometimes projecting beams of light, which he brought closer to the phenomenon of lightning, and noted that it was often accompanied, or preceded, or followed, by aurorae of the first type. He then identified the properties common to both types of aurorae, including an appearance between the autumnal equinox and the spring equinox. He concluded "that these two kinds differ in appearance only. Most of their characteristics are indeed common [...] and, regarding those of which one might think that they are peculiar to one kind of polar lights, I shall later explain that it is only because of the diversity of places, diversity which is itself observed".

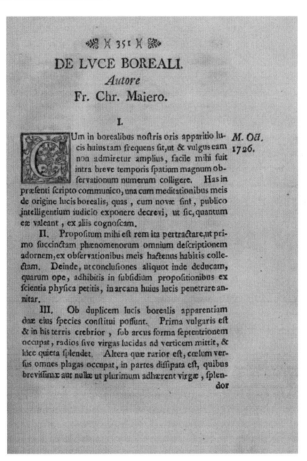

Figure 3.4. *First page of Mayer's 1726 treatise on the aurora borealis (Mayer 1726, p. 351)*

He then explained that (i) either the light comes to us in a direct line from the bright matter of the aurora, (ii) or it is due to the refraction of light from a source located behind the region where the auroral matter is present, (iii) or it is produced "by the reflection of the rays cast by a light matter which is not present at the place where we see the beams". He rejected the first hypothesis because, in this case, due to the rotundity of the Earth, the plane of a distant aurora deviates from the vertical local to the observer, inducing by an effect of perspective a geometrical deformation of the auroral structures, which means that distant observers of the same aurora located in different places, or an observer of different aurorae occurring in the course of time in different places, should never see the same thing, which he considered contradicted by observation, without further precision. He rejected, for not very explicit reasons, the refractive hypothesis, and came to the conclusion that only the hypothesis by the reflection of the light was tenable, an hypothesis that he then endeavored to prove. First, he assimilated the auroral matter to the matter of the clouds, based on the fact that the dark lower edge of the arc "has to be regarded as a cloud", especially since the aurora often shows small luminous pulsating clouds, in conclusion of which: "it is obvious that this matter has all the characteristics of the clouds and thus occupies the same position and is equidistant from the Earth". And he invoked another reason: the arc could be explained according to him by a ring of auroral matter located at a certain altitude, at a certain latitude, and going around the Earth along the corresponding parallel (a representation which is the basis of his method of estimating the height of the aurora). In this case, the auroral matter, like that of extended clouds seen on the horizon, has a sharp lower edge, showing again, according to Mayer, the resemblance between auroral and cloudy matter. This analogy with clouds allowed him to suggest that the two types of aurorae constitute the manifestation of the same phenomenon, the distinction coming only from the place from which it is observed: "the matter of this light is scattered and disjointed in the small luminous clouds; but the first kind of light appears continuous [...] and more radiant [...] because in this case the matter is seen from the distance and in an oblique way". Just as the clouds, seen from near, appeared individualized, in their diversity of forms, and seen from afar, in the vicinity of the horizon, as on the contrary a uniform layer of vapor with a well-defined lower edge, the aurora of the second type (the general blaze of the sky) was an aurora seen from near, that of the first type (the arc and the pulsating beams), an aurora seen from afar.

It remained to be explained where the beams came from, these vertical jets and columns rising towards the top of the arc. Here Mayer referred in particular to his observation of a distant fire from his window on a snowy evening, when he saw a column of light rising into the sky as the light from the fire was reflected back to him by the snowflakes. He also evoked the reflection of a star in calm water, which gave rise to a column because of the irregularities generated by the agitation of the liquid surface: "the moon or any luminous star irradiating the surface of a quiet water sends back to the eye a beam similar to the one I described". And he hypothesized that this was the origin of the jets: "I assert that the polar lights' beams are issued from the reflection of the light that is present in the small luminous clouds [...], while these small clouds project towards the top beams on the flat surface of the very tenuous steams that overlook them, by which light is after reflected in the form of beams". For, according to Mayer, "the clouds' flat and inferior part, is wavy, flat and it results from the polished water or ice particles". It was this lower surface which reflected the light of the small clouds towards the observer, generating the columns of reflection which constitute the observed beams. It remained to define the physical cause of the light emitted by the auroral matter constituting the small clouds, which was far from being obvious, or in any case was much less obvious, according to Mayer, than the optical theory previously developed: "The explanations I gave up to now have, I suppose, a degree of certitude; but there are still other phenomena whose explanations I do not promise to be equally certain". Mayer stated the principle, according to him not totally certain but which he presented "for posterity's profit", according to which during the summer, "the flammable exhalations [sulfur, nitre, salts], warmth helping, mingle and merge with the watery steams and are afterwards split up from them when cold comes, when the watery steams gather faster than the flammable exhalations". He proposed a flashover-extinction cycle related to the fact that the heat generated by the ignition caused the exhalations and vapors to mix, which upon subsequent cooling of the air separated again, reconcentrating the exhalations and creating a new ignition:

> The flammable exhalations little by little leave these small clouds and gather on top of them […]; once a sufficient amount of these exhalations has gathered up, it produces a flame and catches fire […]; a new gathering is formed by this ignition, it catches fire and the history repeats itself as many times as possible. I think that suits admirably to them what the very famous professor Wolff said about them, when he called matter of lightning the matter of the polar lights (Mayer 1726, p. 364).

He explicitly referred to Wolff's theory of the "imperfect storm". Note that the same theory was found in the work of another follower of Wolff, Johann Christian Heuson, who provided a similar interpretation of the aurora borealis of February and March 1721, and November 1729 (Schegel and Silvermann 2011). The explanation proposed by Mayer accounts for the fact that the aurora borealis, according to him, only occurred between the fall and spring equinoxes, thus during the fall and winter. In this period, the lower air cools down, making the warmer air inherited from the summer rise, and it is at the top of the cold layer that the aurora borealis are born by the mechanism described. The hypothesis of the ignition of exhalations, inspired by that of Aristotle, was quite frequently made at that time. Halley exposed, as we have seen, an explanation of the same type in his article of 1717 on the aurora borealis of the previous year, to immediately judge it unrealistic and propose rather the effect of magnetic matter (Halley 1717). Generally speaking, there was then a whole current of scientists, including, for example, Pieter van Musschenbroek, who explained all the meteors (atmospheric lightning, flying lights and shooting stars, aurora borealis) by the same chemical origin, linked to the ignition of sulfurous matter released by the Earth in the atmosphere (van Musschenbroek 1739, pp. 813–847). In France, Jacques Philippe Maraldi and Bernard le Bovier de Fontenelle, throughout the decade that followed the aurora of 1716 and saw the regular appearance of aurorae, progressively elaborated a hypothesis by ignition (Fontenelle 1717), which was dethroned in 1731 at the Académie des Sciences by the system of de Mairan invoking the precipitation of solar matter at great height in the atmosphere (de Mairan 1733).

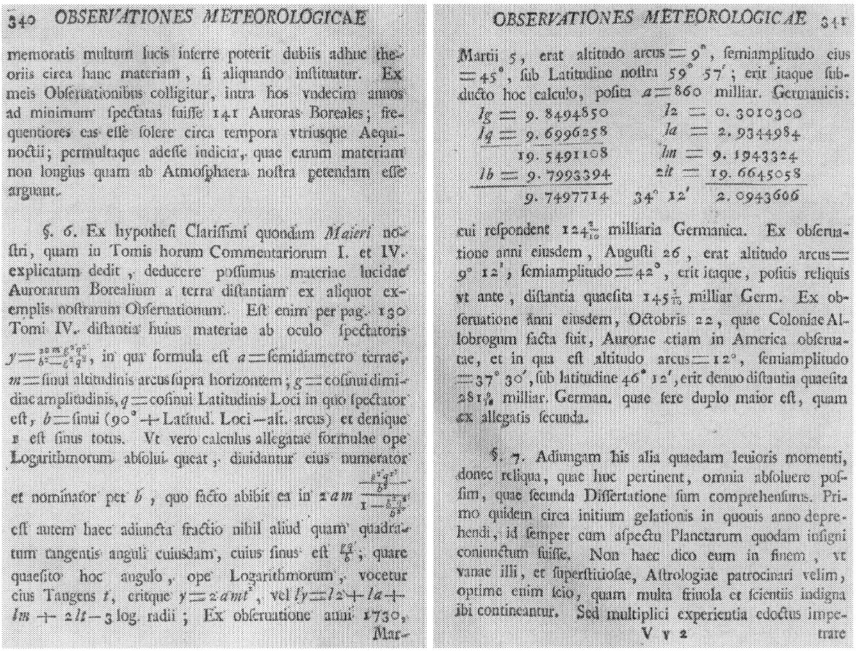

340 OBSERVATIONES METEOROLOGICAE

memoratis multum lucis inferre poterit dubiis adhuc theoriis circa hanc materiam, ſi aliquando inſtituatur. Ex meis Obſeruationibus colligitur, intra hos vndecim annos ad minimum ſpectatas fuiſſe 141 Auroras Boreales; frequentiores eas eſſe ſolere circa tempora vtriusque Aequinoctii; permultaque adeſſe indicia, quae earum materiam non longius quam ab Atmoſphaera noſtra petendam eſſe arguant.

§. 6. Ex hypotheſi Clariſſimi quondam *Maieri* noſtri, quam in Tomis horum Commentariorum I. et IV. explicatam dedit, deducere poſſumus materiae lucidae Aurorarum Borealium a terra diſtantiam ex aliquot exemplis noſtrarum Obſeruationum. Eſt enim per pag. 130 Tomi IV. diſtantia huius materiae ab oculo ſpectatoris $y = \frac{2am\,g^2q^2}{b^2 - g^2q^2}$, in qua formula eſt a = ſemidiametro terrae, m = ſinui altitudinis arcus ſupra horizontem; g = coſinui dimidiae amplitudinis, q = coſinui Latitudinis Loci in quo ſpectator eſt, b = ſinui (90° + Latitud. Loci − alt. arcus) et denique 1 eſt ſinus totus. Vt vero calculus allegatae formulae ope Logarithmorum abſolui queat, diuidantur eius numerator et nominator per b, quo facto abibit ea in $2am \frac{\frac{g^2q^2}{b^2}}{1 - \frac{g^2q^2}{b^2}}$ eſt autem haec adiuncta fractio nihil aliud quam quadratum tangentis anguli cuiusdam, cuius ſinus eſt $\frac{gq}{b}$; quare quaeſito hoc angulo, ope Logarithmorum, vocetur eius Tangens t, eritque $y = 2amt^2$, vel $ly = l2 + la + lm + 2lt - 3$ log. radii; Ex obſeruatione anni 1730, Mar-

OBSERVATIONES METEOROLOGICAE 341

Martii 5, erat altitudo arcus = 9°, ſemiamplitudo eius = 45°, ſub Latitudine noſtra 59° 57'; erit itaque ſubducto hoc calculo, poſita a = 860 milliar. Germanicis:

lg = 9.8494850		$l2$ = 0.3010300	
lq = 9.6996258		la = 2.9344984	
19.5491108		lm = 9.1943324	
lb = 9.7993394		$2lt$ = 19.6645058	
9.7497714	34° 12'	2.0943606	

cui reſpondent $124\frac{2}{10}$ milliaria Germanica. Ex obſeruatione anni eiusdem, Auguſti 26, erat altitudo arcus = 9° 12', ſemiamplitudo = 42°, erit itaque, poſitis reliquis vt ante, diſtantia quaeſita $145\frac{1}{10}$ milliar Germ. Ex obſeruatione anni eiusdem, Octobris 22, quae Coloniae Allobrogum facta fuit, Aurorae etiam in America obſeruatae, et in qua eſt altitudo arcus = 12°, ſemiamplitudo = 37° 30', ſub latitudine 46° 12', erit denuo diſtantia quaeſita $281\frac{8}{10}$ milliar. German. quae fere duplo maior eſt, quam ex allegatis ſecunda.

§. 7. Adiungam his alia quaedam leuioris momenti, donec reliqua, quae huc pertinent, omnia abſoluere poſſim, quae ſecunda Diſſertatione ſum comprehenſurus. Primo quidem circa initium gelationis in quouis anno deprehendi, id ſemper cum aſpectu Planetarum quodam inſigni coniunctum fuiſſe. Non haec dico eum in finem, vt vanae illi, et ſuperſtitioſae, Aſtrologiae patrocinari velim, optime enim ſcio, quam multa friuola et ſcientiis indigna ibi contineantur. Sed multiplici experientia edoctus impe-

V v 2 trare

Figure 3.5. *Simplified version of the trigonometric formula established by Mayer, as presented by Krafft, with examples of applications (Krafft 1737, pp. 340–341)*

Moreover, with his knowledge in mathematics, Mayer proposed at the end of his article a trigonometric method to estimate the height of the auroral arc from its observation at a single point. He represented, as we have seen, the auroral matter by a circular ring centered on the axis of rotation of the Earth, which he placed at a certain altitude, and at the vertical of a parallel placed at a certain latitude. The latitude and the altitude of the ring were the two unknowns of the problem. He established a trigonometric formula which enabled, from the angular height of the top of the arc, and its angular width on the horizon, two observable quantities, to determine the height of the ring above the ground, and its latitude. This method encountered certain success in the community of the astronomers and mathematicians of the time. The demonstration of the result was made in Volume 5 of the *Commentarii* of the Imperial Academy of 1729 published in 1735, five years after Mayer's death (Mayer 1729). Pierre Louis Moreau de Maupertuis, an ardent defender of Newton's theory of universal attraction, who a few years later led an expedition to Lapland to measure the shape of the Earth, gave a demonstration in the *Mémoires de l'Académie Royale des Sciences* in 1731 (de Maupertuis 1731). Krafft presented a simplified version of it using logarithms in Volume 9 of the *Commentarii* of 1737 published in 1744 (Krafft 1737, pp. 340–341). The Reverend Father Roger Joseph Boscovich, in his treatise on the aurora borealis published in 1738, while he was studying theology in Rome, similarly took up Mayer's problem, for which he gave his own demonstration (Lisac et al. 2011). He did not share Mayer's opinion on the origin of the phenomenon, siding with the conception expressed by Mairan in his treatise published in 1733.

As Mairan noted in his clarification XIII of the second edition of his treatise (de Mairan 1733, pp. 404–436), and as Mayer himself noted in his second memoir (Mayer 1729), the hypothesis of the circular shape of the ring, and of its concentricity at the pole (or at the axis of the Earth), was in practice rarely verified, a reason why Mairan preferred the method of parallaxes which required two distant observers observing simultaneously the same structure. Here is what he said about it:

> His first Memoir, "de Luce Boreali", read in 1726 at the Imperial Academy of Petersburg, [...] contains, as we have said, only the simple exposition of his Problem or his Method, & the analytical Formula which results from it. But as early as 1728, he gave to the same Academy & the geometrical construction & the demonstration, which however did not appear until 1735, with the remarks that a great number of Aurora Borealis that he had observed since then had been able to provide him. It is in this second Memoir that he announces the corrections that must be made to the first one as a consequence of these remarks, & the restrictions that must be brought to all his theory. [...] However, in spite of these simple limitations, which are reduced to considering as ordinary and more frequent what he had treated as constant and absolute, in spite of these appearances of the Aurora Borealis, which have become so common in Petersburg that the people were no longer astonished by them as they were in the past, in spite of, I say, such a favorable position and time, he ends his Memoir by admitting that he had not yet been able to find any of these phenomena whose observations he could apply to his Rule [...] (de Mairan 1754, p. 410).

56. *TABLE des différentes hauteurs de l'Aurore Boréale au deſſus de la ſurface de la Terre, déterminées dans cet Éclairciſſement.*

Numero.	ANNÉES, MOIS & JOURS.	VILLES & LIEUX d'Obſervation.	OBSERVATEURS.	LIEUES de Hauteur.
20	1621. Septembre, 12	Peynier,	*Gaſſendi,*	160.
2	1726. Octobre, 19	Paris, Rome,	*Godin,* *Bianchini,*	266.
39	....	Rome, Coppenhague,	*Bianchini,* *Horrebow,*	187.
2	1730. Février, 15	Genève, Montpellier,	*Cramer,*	160.
23	1730. Mars, 16	Péterſbourg,	*Krafft,*	47.
23	..... Septembre, 6	Péterſbourg,	*Krafft,*	58.
23	..... Novembre, 2	Genève,	....	170.
2	1731. Octobre, 2	Coppenhague, Breuillepont,	*Horrebow,* *De Mairan,*	250.
41	1732. Septembre, 1	Paris, Coppenhague,	*Buache,* *Horrebow,*	214.
42	..... Novembre, 12	Paris, Coppenhague,	*Godin,* *Horrebow,*	174.
43	1734. Février, 22	Paris, Coppenhague,	*Godin,* *Horrebow,*	211.
44	1735. Février, 22	Paris, Coppenhague,	*De Mairan,* *Horrebow,*	165.
45	1736. Décembre, 22	Paris, Torno,	*De Fouchy,* *Celſius,*	194.
46	1737. Janvier, 21	Paris, Torno,	*De Mairan,* *Celſius,*	155.

Iii

Numero.	ANNÉES, MOIS & JOURS.	VILLES & LIEUX d'Obſervation.	OBSERVATEURS	LIEUES de Hauteur.
47	1737. Décembre, 16	En Angleterre, Padoue,	.... *Poleni,*	275.
48	....	Paris, Montpellier,	*De Fouchy,* *Plantade,*	200.
50	1740. Novembre, 3	Upſal, Sain-port,	*Celſius,* *De Mairan,*	157.
21	1750. Février, 3	Paris,	*De Fouchy,*	169.
22	....	Genève,	*Abauzit,*	134.
52	....	Paris, Toulouſe,	*De Fouchy,* *d'Arquier,*	173.
53	....	Paris, Genève,	*De Fouchy,* *Abauzit,*	154.
54	....	Genève, Toulouſe.	*Abauzit,* *d'Arquier,*	175.
Sup. p. 393	..... Février, 27	Paris, La Haie,	*De Mairan,* *Gabry,*	168.

Figure 3.6. *Table giving the height of different aurora borealis in leagues (1 league is 3.9 km) (de Mairan 1754, pp. 433–434)*

Mairan estimated the height of the auroral arc by applying Mayer's method in the cases where the conditions of application of this method seem best verified. He thus found the arc of the aurora of 1621 (observed by Gassendi) to be well centered on the geographical north and presenting a half amplitude of 59.5° and a height of 40.5°, and by using Krafft's logarithmic formula, a height of 160 leagues (770 km). In the case of the aurora of February 3, 1750, particularly clear and sharp, and observed from different places in Europe, the height deduced by Mayer's method was 134 leagues (650 km), almost identical to that obtained by the parallax method. He also cited three observations of aurorae made in 1730, two in St. Petersburg and one in Geneva, to which Krafft applied Mayer's method, giving for one of them a height of 170 leagues (820 km), but only about 50 leagues (250 km) for the two others. Mairan finished his clarification by a table in which he reported the heights of about fifty aurorae estimated, either by the method of parallaxes or by that of Mayer when the conditions made it possible to apply it, deducing an average altitude of the aurora of 175 leagues (850 km).

Mairan then tried to understand why Mayer himself applied his method so little to his own aurora observations. The hypothesis he made to explain his lack of conviction to use his own method is interesting:

> He believed with an illustrious Philosopher [most probably Wolff], that the matter of the Aurora Borealis was hardly anything else than an indigestible mass of that which produces the Lightning [...] & consequently he did not make its luminous Arc rise above the region of the clouds [...]. What could he think of such a principle, when his rule sent him back a hundredfold of this distance, and sometimes beyond! However, his rule was good, it was demonstrated, he was a skilful calculator; it was therefore necessary to attack the observations which he could not fail to find faulty, and infinitely more faulty than they were, since they resulted in such an enormous alleged error. This, if I am not mistaken, is the outcome of the difficulty, & why Mr. Mayer undoubtedly neglected to report to us on attempts which must have seemed so defective. He gave the same year or the following one some other Memoirs in which there is no more question of the Boreal Light, & he died on December 5, 1729 (de Mairan 1754, p. 411).

It probably explains Delisle's surprising finding about Mayer's aurora borealis observations:

> But with regard to the particular observations that he made on each Aurora borealis, it seems that he was content to fill his imagination with them, without putting them in writing, since one has found in the

collection of his meteorological observations, only very few circumstances of these observations (Delisle 1738, p. 78).

This lack of eagerness to describe the observed phenomenon of the aurora in Mayer's work, noted by Delisle, corroborates Mairan's explanation that Mayer did not have confidence in auroral observations. There is no doubt that he was a good enough mathematician to judge that his method of estimating the height was accurate, but the result of the observation was in contradiction with the theory he had developed. The height of the aurora, according to this theory, could not exceed the height of the highest layers of clouds, supposed to reflect towards the observer the light of the small clouds of auroral matter, height hundred times lower than the value found from Mayer's mathematical model. One is struck, when reading Mayer's article, by the theoretical and deductive character of the way of thinking that led him to his system, relying only on some general observational characteristics of the aurora: the presence of an arc materializing the ring of auroral matter, whose sliced lower edge evokes that of a distant cloudy layer seen on the horizon, the existence of vertical columns reminiscent of the moon's reflection on a body of water, calling for the idea of a discrete light source (a small cloud) reflecting on a slightly agitated surface (the lower surface of a layer of vapors). Mayer did not seem to see the contradictions inherent in his approach, the first and principal one of which ruined his system, namely, the great height of the aurora. His argument to reject the hypothesis that the light of the aurora is seen in a direct line (without the interposition of a reflecting obstacle) was based on the (correct) evidence of geometric distortions of the figure of the aurora due to the rotundity of the Earth which can exist only if, precisely, the aurora is located at very high altitude (as logically shown by the result of its mathematical model), far above the clouds.

It is instructive, at this point, to compare the case of Mayer with that of Euler, another mathematician, who also developed an aurora borealis system. This system was precisely based on the observation of the very great height of the aurora.

3.3.5. *Euler's physical–mathematical explanation*

Euler was noticed in Basel, where he wrote his thesis on the comparison between the natural philosophies of Descartes and Newton, as an exceptional mathematician by his teacher Jean Bernoulli, father of Daniel Bernoulli. The latter, who was among the first scholars to join the Academy of St. Petersburg, wrote a letter to Euler in 1726 in the following terms: "Sir, it is some months since I wrote to you by order of our President, Mr. Blumentrost, and invited you in his name to come and take the place of a Student in our Academy… You are awaited with great impatience; come therefore as soon as possible…" (Esteve 2018, p. 266). Euler left Basel on April 5, 1727, went to Marburg to see Christian Wolff and arrived in St. Petersburg on

May 24, 1727. From the beginning of his time at the Academy, he was closely associated by Delisle with the work of the Observatory, in astronomy, and also in geography. It was probably at the Observatory of the Academy, where he himself was able to observe many aurorae, that his interest in this phenomenon was born, which would lead him, some 20 years later, to publish his own theory (Euler 1746). It is known that in 1729, he studied different conjectures on the nature of the aurora borealis (Esteve 2018, p. 268), and it was undoubtedly this thought, shared with his colleagues at the Observatory, Mayer and Krafft, who interacted with each other only one year between Krafft's arrival, at the end of 1728, and Mayer's death, at the end of 1729, that led him to his theory of the aurora, published much later, after his departure from St. Petersburg, in 1746. Euler did not believe that the solar atmosphere could extend far enough from the sun to precipitate in the atmosphere, as Mairan thought. However, he noted a resemblance between the aurora and the tail of comets in terms of texture and transparency. As soon as he arrived in St. Petersburg, he became aware of Mayer's theory, and could not but share his doubts, especially since he was bound to interact with him on the subject during 1729. For Euler, who followed the opinion expressed by Johannes Kepler a century earlier on the basis of his model of atmospheric refraction, the height of the Earth's atmosphere not exceeding one German mile (Euler 1746, p. 131) or about 6 km. According to him, the atmosphere was thus a very thin film of air covering the Earth, whose height gave an idea of the typical maximum altitude up to which sulfurous exhalations could be raised, in the theory defended by his colleague Mayer. This certainty that the atmosphere was fine resonated in Euler and the mathematical plausibility of the method developed by Mayer to estimate the height of the aurora, 100 times greater than that of the atmosphere, logically led Euler to reject Mayer's theory.

Euler developed his own theory, which he tried to make compatible with the considerable height of the aurora borealis, which he did not doubt. This theory is of an astronomical nature, Euler placing the aurora very high above the thin Earth's atmosphere. Mairan, to solve the difficulty represented by the great height of the aurora borealis, imagined the presence at high altitude of a subtle air, extremely light, mixing with the solar matter. Not believing in the extension of the solar atmosphere, Euler had no reason to retain Mairan's hypothesis, and he preferred to imagine the auroral matter as a tail of the terrestrial atmosphere, in the manner of the cometary tails, the atmospheric particles "pushed" by solar light (according to a mechanism that he had difficulty in imagining, because he did not believe in Newton's corpuscular description of light) moving away at very great distance (tens of thousands of kilometers) and being definitively lost in the interplanetary space. In his explanation of the tail of comets, constituted according to him of cometary matter pushed by the solar rays, Euler followed in the footsteps of Kepler, who had the same idea. Euler attributed the great value of the height of the atmosphere deduced from the duration of twilight (according to him, 180 km, in reality 70 km) to the solar light scattered by the air particles escaping under solar thrust, thus by the same mechanism according to him responsible for the aurora borealis. In his memoir of 1746, he explained in a similar way zodiacal light ("tail" of the sun), comet tails and the aurora borealis ("tail" of the Earth). For him, all three phenomena were the manifestation of the same physical effect, that is, the thrust exerted by the sun's rays on the most subtle part of the matter surrounding the bodies subjected to its radiation.

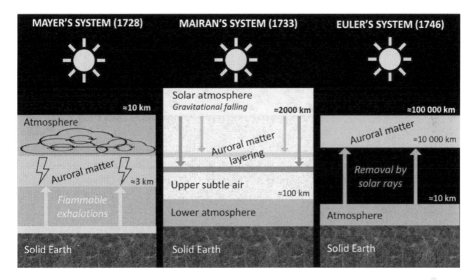

Figure 3.7. *Schematic views of the aurora borealis systems proposed by Mayer, Mairan and Euler (Chassefière 2021c, p. 181)*[1]

This theory, which has the merit of allowing the presence of atmospheric matter at a very high height above the atmosphere, was nevertheless quickly perceived as untenable, in particular because it did not correctly account for the structure and the observed dynamics of the aurora. In particular, it predicted a too high altitude of the phenomenon, and it is moreover difficult to understand why the matter torn continuously from the atmosphere by the solar rays forms clusters of matter detached from the atmosphere, whereas the auroral light should rather constitute a continuum gradually fading upwards. Euler's theory, which was quickly challenged by Mairan himself in the second version of his treatise on the aurora borealis (1754), was considered, for example, by Biot (1820), as the creation of a pure mathematician, little concerned with the observed realities of nature. It shows here two opposite approaches, between Mayer, on the one hand, who, convinced of the validity of his system of the aurora borealis, rejected the observation and continued his

---

1 See: https://www.tandfonline.com/.

investigations, and Euler, on the other hand, who, to satisfy the observation of the great height of the aurora, developed a mathematically and physically sophisticated theory (although not taking into account the detail of the observation).

3.3.6. *Mayer's philosophical position and possible reasons for his abandonment of aurora observation*

The lasting reappearance of the aurora borealis in 1716, in a period marked by the spirit of the Enlightenment and the development of experimental physics, generated intense observation activity for half a century throughout Europe and the Russian world. The recurrence of the aurorae, their seasonality, their variability of shapes and colors, indeed called for patient observation and the progressive accumulation of observation data, so as to elucidate the causes. We have seen that in his Foreword to the *Histoire de l'Académie Royale des Sciences* of 1699, Bernard le Bovier de Fontenelle, Permanent Secretary of the Académie des Sciences, advocated starting by collecting observation data before building systems. The aurora borealis, like meteorological phenomena, shows such complexity that any development of a system that would not be based on a search for coherence between a large number of observations would be reduced to conjecture. It is in this spirit, as we have seen, that after six years of regular observation of aurorae, Jacques Philippe Maraldi believed he noticed in 1721 a correspondence between the occurrence of aurorae and the dryness of the climate (Maraldi 1721), which made him attribute the aurora borealis to the ignition of exhalations rising from the Earth, in connection with weather conditions. For him, as for Aristotle, then Descartes and Wolff, the aurora borealis was a meteorological phenomenon that occurred in the lower atmosphere.

The great spatial extension of the phenomenon of the aurora borealis obliged, to draw all the scientific benefit from it, treating simultaneously observations acquired from one end to the other of the European continent. This imperative encouraged certain scientists, like Joseph-Nicolas Delisle in St. Petersburg, or Anders Celsius in Uppsala, and during his great European voyage of the 1730s (Viik 2012), to set up a European-wide network for the exchange of the results of these observations, as we will return to in Chapter 4. The observation of the aurora generated collaborations that transcended the philosophical divide between Cartesians and Newtonians in a tense period of penetration of Newtonianism on the European continent (Crépel and Schmit 2017), which did not prevent, as we have seen, certain scholars of the Académie Royale des Sciences from making Mairan's system an element of affirmation of Cartesian cosmology. Great observers of the aurora, such as Delisle or Celsius, who were also convinced Newtonians refrained from proposing an explanation for the phenomenon, remaining within the logic traced by Fontenelle of a first purely observational approach preparing a future system. The aurora borealis thus constituted a particularly unifying theme around which many scholars of all philosophical persuasions gathered. The work initiated in St. Petersburg by Mayer around the explanation of the aurora borealis, and the estimation of their height, taken up again first by Krafft, then by Gottfried Heinsius, a geographer and astronomer from Leipzig recruited in 1736 to assist Delisle at the direction of the Observatory, constituted the first study of this type in Russia (Eather 1980). Mayer's system is mixed, borrowing at the same time from Descartes, proposing that the light of the dawn is reflected by the particles of the clouds (Descartes 1824, pp. 263–264), and from Aristotle, then Wolff or Maraldi, invoking as a source of light the ignition of exhalations raised from the ground into the atmosphere. The dispute between Wolffians and Newtonians at the Imperial Academy of Sciences of St. Petersburg did not seem to cross as such the small community of aurora observers of the Observatory, made up of scientists of the two sides (Delisle and his brother, on the one side, Mayer and Krafft, on the other side), the observational approach, in St. Petersburg as elsewhere, transcending the groups which were divided only on the causes.

We know very little about Mayer. He was certainly, as a former student of Bilfinger, of Leibnizian obedience. We know that he was a specialist in mathesis and recognized as such in the Academy. Mathesis is a concept from Greek antiquity, taken up by Descartes, then Leibniz. It constitutes the hypothesis of a universal science, based on the model of mathematics, allowing the world to be described in a rigorous way (Rabouin 2009). The fact that Mayer did not question his system, despite the flagrant contradiction with the great heights of auroral matter predicted by the application of his mathematical method to his observations, and the lack of interest he showed in the reports of auroral observations that he had to write from day to day, seem to denote an approach of an aprioristic nature, granting primacy to deductive reasoning and neglecting observations. Was such an approach the consequence of Mayer's philosophical conceptions, notably his commitment to mathematics, or was it the result of a mind that is naturally more oriented towards the imagination than towards the real world? We do not know. As a practitioner of mathesis, Mayer can be considered most likely Leibnizian. He seemed to be further away from Wolff, whose philosophy of nature was closer to that of the French scientists of the Académie des Sciences, such as Fontenelle, whom we have quoted, advocating a systematic physics, in which observation served as the basis for the elaboration of systems. But, whereas for Fontenelle, the systems could only come in a second stage, after collecting many observations, for Wolff, data collection and theory building were largely simultaneous and interdependent (Vanzo 2015), a view that it should be noted is very similar to that expressed by Mairan in his foreword to his *Dissertation sur la Glace* (de Mairan 1749). The conceptualization in hypothesis permanently guided the observer in his choice of observations to make, and in return, the new observations led to revising the hypothesis in case of disagreement between theory and observation, Wolff using the example of astronomy to illustrate his point. Thus, Wolff's natural philosophy was close to that developed by the

French mechanists of the Académie Royale des Sciences, such as Mairan, in the first half of the 18th century. Wolff, who himself advanced a system of the aurora borealis from 1716, found in Mayer a successor, who took back the same idea according to which the matter of the aurora was that of the clouds and did not hesitate to publish this idea quickly, which was enabled by Wolff's doctrine, leading to the introduction of hypotheses at all stages of the study of a phenomenon. It is in this that we can qualify him as Wolffian. But he did not draw all the lessons from Wolff's philosophy that "if a hypothesis is incompatible with one single observed phenomenon, it must be abandoned" (Vanzo 2015), since instead of modifying his hypothesis, he gave up the game, either because he lost interest in the problem or because he decided to grant reason pre-eminence over observation (Chassefière 2021c).

In a way, it was Leonhard Euler, whom he met at the Observatory at that time, who, 20 years later, proposed a system compatible with the great height of the aurora. In the dispute which opposed the Imperial Academy of Sciences of St. Petersburg, Newtonians and Wolffians, Euler, at first reserved, ended up siding, from 1736, with the Newtonians, Delisle and Bernoulli, against Leibniz's monads theory (Calinger 1996). In the early 1730s, he helped Delisle by recording astronomical observations in the journal of the St. Petersburg Observatory. From the end of 1734, he made independent observations for the construction of meridian tables, which he published the following year (ibid.). From this time on, he also participated in an increasing number of technical projects on magnetism, machines, shipbuilding and education. Although he carried out few experiments and observations himself, he attached great importance to the validation of the results of mathematical calculations by experiment, which was obviously necessary for the success of application projects. In his "Letters to a German Princess", concerning the test of Newton's theory of tides against Descartes' (Euler 1802, p. 251), we find a very clear expression of the necessity of using experiments to choose the right theory. As we have seen, Euler was aware of Mayer's work on the mathematical determination of the height of auroral matter (a few hundred kilometers), which was much greater than the height of the atmosphere. He thought, following Kepler, that it was weak (a few kilometers), which led him to believe that the aurora borealis was a phenomenon outside the atmosphere. It is on this postulate that he built his mathematical model, which he published in 1746, faithful to his principle of using mathematics and experience together. Of course, his work was sometimes approximate, as in the estimates he made of the geometric characteristics of the atmosphere, leading to an excessive height of the aurorae, and not explaining their polar location, but he had the audacity to question the atmospheric character of the aurorae, which only Halley before him had done 30 years earlier by attributing a magnetic origin to them (Halley 1717).

Meanwhile, in the early 1740s, Anders Celsius and Olof Peter Hiorter, as we will return to in the next chapter, demonstrated the link between the aurora borealis and magnetism through a long series of observations of the irregular agitation of the magnetic needle during the aurora (Biot 1820), a discovery that should have given the advantage to Halley's system. But, from the beginning of the 1750s, with the discovery and exploration of atmospheric electricity, the electrical theory, developed independently by John Canton in England (Canton 1753) and Mikhail Vasilevich Lomonosov at the Imperial Academy of Sciences of St. Petersburg (see, for example, Yevlashin et al. 1986), came to overshadow all the previous hypotheses. It was not until the beginning of the 20th century that the phenomenon was perfectly understood (Birkeland 1908), borrowing at the same time from the magnetic, solar and electric theories.

3.4. The Imperial Academy of Sciences of St. Petersburg

The Imperial Academy of Sciences of St. Petersburg was officially established in January 1724. Peter the Great died a year later, on February 8, 1725, and thus did not have time to see the effective opening of the Academy, which took place on January 7, 1726. The death of the emperor raised doubts for a while about the establishment of the institution, but Catherine I, who succeeded her husband, took over the entire project, which had already been in the works, both architecturally and intellectually, for almost 10 years. It is very likely that the empress attended the opening ceremony.

3.4.1. *The setting up of the Academy*

Twelve members, including President Blumentrost, Russian, but son of a German immigrant, and Secretary Schumacher, who emigrated from Germany in 1714, were present at the ceremony (Bayuk 2018). Of these 12 members, eleven were German or Swiss. The Swiss were all three mathematicians: Jakob Hermann, as well as the Bernoulli brothers Daniel and Nicolas, sent by their father Jean Bernoulli who, having declined the invitation to join the Academy, was made an honorary foreign member. Nicolas died of appendicitis a few months after his arrival. Besides Schumacher, the Germans were the physicists Bilfinger and Mayer, and the philosopher Christian Friedrich Gross, all three from Tübingen; the physicist Christian Martini and the historian Gerard Friedrich Müller, both from Leipzig; the mathematician Christian Goldbach, from Königsberg and the philologist Johann Peter Kohl, from Kiel. Among the eight German scholars attending the opening ceremony, all of them Protestants; seven had studied theology or even held a religious position before coming to St. Petersburg. Seven other scientists joined the academy the same year: the physicist Krafft and the anatomists Johann Georg

Duvernoy and Josias Weitbrecht, all three from Tübingen, belonging to the Bilfinger movement; the art historian Gottlieb Teophil Siegfried Bayer, from Königsberg; the physician and botanist Johann Christian Buxbaum, in Russia since 1721 as head of the botanical garden of the Medical College of St. Petersburg, originally from the Leipzig region and the astronomer Joseph-Nicolas Delisle, accompanied by his brother Louis de la Croyère, both from Paris, the only two Catholics in the Academy. It was not until 1739 that another Catholic, the Jesuit Antoine Gaubil, missionary astronomer in Peking, was elected to the Academy. The total number of academicians in 1727 was thus 20, adding to those already appointed Leonhard Euler, who arrived from Basel that year, followed by the botanist Johann Georg Gmelin, who came from Tübingen, and the theologian and mechanic Johann Georg Leutmann, from Wittenberg, recruited at Hermann's suggestion. Most of them were chosen on the recommendation of Wolff, and his follower Bilfinger, whose movement from the University of Tübingen alone represented one-third of the membership of the nascent academy. The group of Wolffians, led by Bilfinger, who had a large majority in the Academy, was part of the Leibnizian opposition to Newton. The Basel group, Daniel Bernoulli and Euler, together with Delisle, formed the Newtonian group of the new Academy, even though Euler was far from adhering to all of Newton's opinions, and the dividing lines between the different schools were anything but precisely defined. Jakob Hermann, for example, belonged to both schools. Most of the members of the new academy were under 30, and the nascent institution was not yet as hierarchical as the older academies in London and Paris.

The academy began to function effectively from the moment of its creation. In accordance with the commitment of Peter the Great, it was allocated an annual funding of 25,000 rubles, which was comfortable for the time. Peter the Great wanted a state-funded Academy, as was the case in Paris, but, as we have seen, not in London. The logistical means at the disposal of the academicians, in terms of buildings, equipment and technical personnel, were significant. Concerning the Observatory, its installation and the realization of its stock of instruments, however, were delayed for several years, as we have already noted. But the system could nevertheless function. The academicians met weekly to present and discuss scientific topics, meeting at 4 p.m. on Thursdays and Fridays (Esteve 2018). They were required to publish a few dissertations per year, and present them for discussion at these biweekly conferences. They dealt with topics of current interest, such as the question of the shape of the Earth, or that of intelligent life on the moon. They were a place for debate on the physical theories of Leibniz and Wolff, which could not fail to provoke discussion between members of the different schools of thought. The results of the research were published, under the direction of the Secretary of the Academy, Christian Goldbach, in the *Commentarii Academiae scientiarum imperialis Petropolitanae*, printed at an annual or biennial rhythm on the Dutch-made Presses de l'Académie, acquired in 1728. A total of 14 volumes were published over the first 20 years, covering all the work achieved between 1726 and 1746. Each volume is composed of three sections: "mathematics category", "physics category" and "history category", with a special section at the end of the volume reserved for astronomical observations made by Delisle and his brother.

3.4.2. *The clerical and noble opposition*

Catherine I died on May 17, 1727. Peter II, grandson of Peter the Great, became emperor at the age of 12, and remained so until his death in 1730 from smallpox. Too young to govern, he left power to the Dolgoruki family, an old Russian princely house that took the opposite direction of the progressive policy initiated by Peter the Great and Catherine I. This was a period of instability for the Academy (Calinger 1996). The old Russian nobility, which controlled the young emperor, viewed the Academy negatively, considering it an intrusion of foreigners into Russian culture. Through its ministers in the government, it postponed the granting of recurrent financing, froze the credits necessary for the opening of professorships, and fomented quarrels with the foreign members. It blocked the entrance of students to the Academy's Lyceum and University. Most of the recruited students came from foreign families, very few from Russian aristocracy. From 112 in 1727, the Lyceum's enrolment dropped to 74 in 1729. At the University, there were only eight students from Vienna, as the Russian schools did not offer enough education to prepare the students for admission. The court was repatriated to Moscow. Two members of the Academy, Blumentrost, as physician of the young emperor, and Goldbach, as tutor, moved to Moscow, Blumentrost ceding the presidency of the Academy to Schumacher.

Besides the old nobility, the Russian Orthodox Church was also hostile to the scholars of the Academy, but for a dogmatic reason. Like a part of the Catholic Church of the time, it still rejected the Copernican doctrine, which Delisle and Daniel Bernoulli ostensibly introduced in the Academy as early as 1728. The speech Delisle gave on March 2 of that year, accompanied by a reply from Daniel Bernoulli, published by the printing house of the Académie des Sciences, was interpreted by the church as a declaration of war. In this text, Delisle first asked the question to know whether it is possible to demonstrate, by the only astronomical facts, that the Earth turned. He began his speech by mentioning the two types of prejudices: the religious one, against the movement of the Earth, conforming to the literal meaning of the Holy Scripture, establishing the rest of the Earth; the scientific one, in favor of the movement of the Earth, as thought by many mathematicians and astronomers. In addition to answering the mathematical question posed, he intended to specify a system of the world, in relation to which to define the rest or movement of the Earth:

By the phrase System of the World, I mean the arrangement & situation of the celestial bodies among themselves, & with respect to the earth. Thus to know the true system of the world; it is to know at what distance, the celestial bodies, & the earth, are really the ones of the others; it is to determine, which are in movement, & which in rest. It is finally to know exactly what are the rules of these movements (Delisle 1728, pp. 2–3).

After which he states that "one cannot demonstrate by mathematical rules alone, without making suppositions; or without admitting physical hypotheses [...] whether there are some heavenly bodies at rest, and what they are", taking up the equivalence point of view already expressed by Galileo in the second day of his "Dialogue on the two great world systems": "Moving the Earth is the same as moving everything else in the world, since this movement acts only on the relation between the celestial bodies and the Earth and only this relation changes". All the progress made in the knowledge of celestial motions did not yet allow for "deciding which of the two hypotheses of the rest or motion of the Earth is true" (Delisle 1728, p. 3), and even though we had the knowledge to predict all the celestial phenomena, such as eclipses, that would not be enough to decide the motion or rest of the Earth. He then detailed his point of view by comparing the Earth to a ship on the sea, the other planets to other ships and the fixed stars to distant coasts, then by declining in different ways the principle of relativity of the motions of the different objects (see, for example, Gapaillard 1993). He began his explanation with the stars, revolving around us, and asked the question, taking up his image, whether it was the ship that turned on itself, or the distant lands that revolved around the ship. We cannot mathematically conclude from the observations. He continued his reasoning by dealing with the relative motions between the vessels (the Earth and the planets) to arrive at a comparison between the systems of Copernicus and Tycho Brahe. In Copernicus' system, assuming two motions for the Earth, the annual and the diurnal, it was not necessary to give the sun and the fixed stars a movement, each planet having only one simple motion around the sun. The moon, the satellites of Jupiter and Saturn, with only two motions, one around their main planet and the other around the sun. In Tycho Brahe's system, if we give the Earth only the diurnal motion, each of the planets must have two motions, and each of the satellites of Jupiter and Saturn three motions, the moon and the sun one motion. If we give no motion to the Earth, we must give the diurnal motion to the sun and the stars, three motions to each planet, four motions to the satellites of Jupiter and Saturn, two motions to the sun and the moon. Thus, Delisle explained, an astronomer, if he is given the assumption of the motion or immobility of one of the bodies, can deduce the motion or immobility of all the others. But he could not decide mathematically, in the absolute, the immobility or the motion of a body. To decide, he had to have recourse to "plausibilities or physical hypotheses". He mentioned that if one could measure a parallax to the fixed stars, which had not been done so far, one would demonstrate "the local motion of the Earth with respect to the fixed stars".

Delisle concluded by the fact that it was necessary to continue the observations, these observations making the movement of the Earth more and more probable. He invoked the argument of simplicity, already advanced by Copernicus, then Galileo in his *Dialogue sur les deux grands systèmes du monde*:

> Is it conceivable that the whole expanse of the heavens makes a revolution around the earth every day, rather than giving the earth a small movement on its axis, a movement, moreover, of which there are examples in almost all the celestial bodies, since it is certainly known that most of them rotate on themselves, in addition to the local movement that I have marked above that each one must have in order to explain the appearances in the different systems.
>
> The same is true of all the rest of Copernicus' system; the movements of the celestial bodies are reduced to such great simplicity and harmony that most Astronomers and Philosophers, charmed by the beauty of this System, do not doubt its truth, even though they have no purely Geometrical demonstration independent of any supposition.
>
> That if one adds, in the future, new observations, to assure oneself of the Astronomical facts which are not yet entirely constant, one must not doubt that one draws from it new proofs or stronger likelihoods for the true system of the world (Delisle 1728, pp. 15–16).

He declared that it was in this perspective that Peter the Great laid the first foundations of the Observatory's astronomical tower, "which like a lantern was to illuminate the whole Empire". In his answer, Daniel Bernoulli took up again Delisle's arguments, by insisting on the force of the deductive step, against the irrational beliefs, because one arrives at the conclusion of the heliocentrism "only by dint of reasoning and stripping oneself of prejudices" (Delisle 1728, p. 17). He developed more particularly the question of the research of the parallax of nearby stars, adding that, even though the motion of the Earth was demonstrated by this method, there would be supporters of Ptolemy who would object that it was the star, and not the Earth, that moves in the zodiac. Taking up the principle of simplicity advanced by Delisle and his illustrious predecessors, he asserted that no observation could mathematically prove the movement of the Earth, and invoked the wisdom of the Creator in support of the Copernican hypothesis: "I ask first if a Phenomenon being able to be explained in several ways, one should not choose the simplest one, or at least reject those which are the most composed? Do we not see nature everywhere acting by the shortest way? & how could she do otherwise being directed by a sovereignly wise Being?" (Delisle 1728, pp. 21–22). He did not fail to evoke Kepler's law which linked the distance of a planet to the sun and its time of revolution, a relation later demonstrated by Newton. Now the Earth, he said, followed exactly the same law as the planets. It was therefore necessarily a planet and, as such, revolved around the sun. In December 1728, the month of the speech's publication, Delisle added a warning that observations of parallaxes of stars from Denmark seemed to allow their distance to the Earth to be determined. But he did not mention the discovery by James Bradley of the aberration of stars and the interpretation he gave in his article in the *Philosophical Transactions*, published on January 1, 1728, in terms of the composition between the speed of light and the speed of the Earth on its orbit (Bradley 1728). This discovery was considered as a proof of the motion of the Earth relative to the fixed stars. It is possible, if not probable, given the slow flow of mail at the time, that Delisle and Bernoulli had not yet been informed of Bradley's discovery at the time of the speech's printing.

This speech was obviously very badly perceived by the religious authorities of St. Petersburg. Two years later, in 1730, Antiokh Cantemir, who we saw was interested in astronomy, and who had been a student at the university of the academy in 1726–1727, carried out a translation into Russian of *Entretiens sur la Pluralité des Mondes* by Bernard de Fontenelle, published in Paris in 1686. During the first evening, the talks dealt with Copernican astronomy, judging it as very probably correct. In the second and third evenings, they stated the discoveries of Galileo and Kepler, then, in the fourth and fifth, presented the cosmology of Cartesian vortices. The orthodox authority considered that the heliocentric doctrine, which did not place any more humanity in the center of the world, and the mechanism of Descartes, constituted a desacralization of the Earth. For 10 years, it was opposed to the publication of Cantemir's translation on the Academy's presses, which had not been freed from censorship by the imperial authorities. Euler used all his influence so that the translation appeared. The authorization to print was finally obtained only in 1740, but Cantemir, meanwhile converted to Newtonianism, decided to print his translation of *Newtonianisme pour les dames*, published the year before by Francesco Algarotti, a project which did not succeed. The translation of Fontenelle's *Entretiens* appeared in 1740, published by the Academy, with Cantemir inscribing on the cover page of the book the date of 1730, when the manuscript was actually completed.

The conflict with the clergy was responsible for the departure, as early as 1730, of two prominent Wolffians from the Academy, Bilfinger and Hermann; the first moving to Tübingen, where his home university was located, and the second to Basel to take up a chair in ethics at the university that had been offered to him three years earlier (Calinger 1996, p. 128). It was Daniel Bernoulli who succeeded Hermann as the first professor of mathematics. Euler who, shortly after his arrival in 1727, had accepted a position as physician at the Naval Academy, based on both his work on ship's masts and his medical knowledge, returned in 1731 to the Academy, where he succeeded Bilfinger as professor of physics. As mentioned in the previous chapter, a dispute broke out between Euler and Schumacher, as Euler did not feel that he was sufficiently paid for his abilities. His salary was 400 rubles per year, as was the case for many of his colleagues, but it did not compare to the salary of Jakob Hermann (2,000 rubles) or Joseph-Nicolas Delisle (1,800 rubles), "special" guests of Peter the Great. Schumacher considered that the academicians were globally overpaid, which gave them an additional "arrogance" (ibid.), to the detriment of the power he sought to exercise over them. When Peter II died in January 1730, he was replaced by the niece of Peter the Great, Anna Ivanovna. The new empress surrounded herself with German officials and moved the capital back to St. Petersburg in 1732. The relations of the Academy with the power were arranged. The conflict that broke out between Euler and Schumacher certainly resulted in Euler's proposal to have Jean Bernoulli, Daniel's brother and friend, appointed to the Academy, who visited them in 1732, failing, the process blocked by Schumacher on the pretext of a balance to be respected between the representations of the different countries. In this climate, Daniel Bernoulli, tired of the censorship exercised by the Orthodox Church, of the hostility towards the Germans and of Schumacher's behavior, which he judged to be tyrannical, left St. Petersburg in 1733, accompanied by his brother Jean, to take up a post as professor of botany and anatomy in Basel. Euler became the first professor of mathematics, and his salary was increased to 600 rubles per year, which gave him enough money to get married in early 1734.

Thus, in only a few years, the academy lost three of its most remarkable members: Bilfinger, Hermann and Daniel Bernoulli, mainly, but not only, because of religious censorship. Other early members soon left St. Petersburg: the philologist Kohl in 1728, the physicist Martini in 1729, not to mention the death of Nicolas Bernoulli in 1726, and that of Mayer in 1729, and the resignation of Gross, who left his post in 1731 to pursue a diplomatic career. Thus, almost half of the original members of the Academy were no longer present by the beginning of the 1730s.

3.4.3. *Wolffians versus Newtonians*

Wolff's philosophy of nature was globally of Cartesian inspiration by its dualism and its mechanism, but it differed from it by the addition of a dynamism of Leibnizian inspiration involving an "active force" (vis activa), principle of movement and change and a "passive force" (vis inertiae), principle of resistance to movement, which explain entirely, according to Leibniz, the diversity of composed bodies and their physical changes (see, for example, the summary of the general principles of Wolffian cosmology in Neveu 2014, pp. 95–96). Leibniz was, in fact, opposed to the reductive conception of Descartes, who saw in the cause of the movement of a body only the mechanical action of another body coming into contact with it, namely, its "efficient" cause. Leibniz subordinated this to a "final" cause, a final reason founding the necessity of movement, the proper faculty of matter to which he gave the name of "force", the active force having rather to be assimilated to an energy (what we now call kinetic energy) than a force in the Newtonian sense (Rey 2004). This force, active or passive, is contained in the invisible elementary substances constituting the physical bricks of any sensible natural body, simple units, indivisible, without extent and non-material. While for Leibniz, these elementary substances were "metaphysical points ... [which] have something vital and a kind of perception" (the "monads"), Wolff recognized them only a physical existence without pronouncing himself on their possible "vital" character (École 1964). According to the principles of Leibnizian mechanics, the active force present in a body caused it to change its place and move obstacles in its path, the passive force causing it, on the contrary, to resist movement and prevent the advance of another body that came into contact with it (see, for example, Stan 2012). The action of a body thus consists of the effective exercise of its active force, so that it changes the state of passivity, or inertia, of another body. But, Wolff added, a body only activates its active force if there is a reason for it to act on another body, in this case if it is prevented by this other body from following its path. As long as there is a clear distance between two bodies, and thus they can, according to Wolff (as he thought daily experience dictated), move freely, neither of them hindering the path of the other, there is no reason why one should act on the other. This conception was inherited from Leibniz's principle of sufficient reason which, in its most general form, says that nothing is without reason, in other words that there is no effect without a cause. In Wolff's mechanistic conception, a cause is needed for the activation of the living force, and this cause is the contact with a resisting body that impedes it. In the absence of contact, there can be no force, hence the unintelligibility, according to Wolff, of the notion of gravity. It is on this philosophical background, inherited from Leibniz, that the criticism came of Newtonianism by the Wolffians after Leibniz's death in 1716.

The notion of intelligibility is radically different in Newton's work. For him, centripetal forces were intelligible from the moment they are deduced from observed phenomena, proving to be the necessary and sufficient causes of Keplerian orbits. From the moment that these forces are "universally established by astronomical experiments and observations", the provisional ignorance of their causes does not, according to Newton, prevent us from admitting these forces themselves as intelligible causes of motion, if they are deduced from the motions in a correct and coherent way. Newton accepted that he did not understand the causes, as long as the forces made it possible to explain the movements. For Wolff, on the contrary, as for Descartes, the laws governing motions had to be founded by a mechanism of deduction from an identified cause, or by the principle of sufficient reason. But on the precise reason that led him to posit contact between bodies as the exclusive reason for physical action, as he set out in his General Cosmology of 1737, he did not provide an indisputable argument, as Stan (2012) analyzed. There he asserted that all physical change always occurs "by motion", and deduced without transition that "no body can produce motion in another in the absence of mutual contact", and immediately added that "experience too confirms [that all change occurs by contact], for we do not observe any mode being induced in a body except by a contiguous one" (excerpts from Wolff's *Ontologia* cited by Stan 2012, p. 473). As for the question of what was the cause of the attraction mechanisms that occur in magnetism and electricity, Wolff did not give a clear answer. Stan analyzes the four levels of Wolff's arguments in his critique of Newtonian action at a distance, and questions the relevance of his arguments. First, Wolff argues that the cause of such attractions at a distance "is no evident to the senses", but should the same conclusion not be drawn for Leibniz's vortices of ether, which he claimed was the cause of the orbits of the planets? On the contrary, according to Wolff, it is on the basis of the physical experimentation that we make of the tendency of bodies to continue their movement when we try to stop them, that we can attribute to bodies the active force, but, in this case, the attraction by the magnet of a piece of metal, which similarly imposes itself on our senses, does it not legitimize the existence of an attraction at a distance, even though the cause remains obscure? Secondly, Wolff said that the attraction at a distance experienced by a body would be a "sheer indeterminacy", not responding to any sufficient reason, whether it is proper or external to this body. But nothing in the Wolffian doctrine, Stan objects, implies that an external reason must be contiguous to the body, which, in fact, invalidates the criticism made of Newton's theory. Third, Wolff objected to this theory that we are unable to prove that a real attraction can exist without contact, meaning that attraction cannot be proved by deduction from dynamic laws known a priori. But Newton, having shown from observation that gravitation is the necessary and sufficient condition for the planets to follow a Keplerian motion, deduced the mutual character of the attraction between two bodies from the law of action and reaction, which Wolff precisely mentioned and affirms a priori in his *Cosmologia Generalis*. Thus, Newton's theory does not contradict the conditions set by Wolff for its acceptability. Finally, Wolff asserted that we cannot understand how two distant bodies can approach each other without another contiguous matter pushing them towards each other. But the way in which two bodies coming into contact exchange force and motion remains mysterious, such transfers, operating on quantities supposed to be intrinsic properties of substances, were for Leibniz and Wolff metaphysically inconceivable. Wolff acknowledged, moreover, that he did not have a clear and distinct vision of the notions of living force and passive force, the two forces which, in his metaphysical dynamics, are exerted by any two bodies that collide. Thus, impact is unintelligible in Wolffian metaphysics, which should not allow Wolff to reject Newtonian attraction at a distance precisely because of its unintelligibility. In the end, Wolff's philosophical argument against Newton brings few new elements to the controversy, returning to the Cartesian idea that two distant bodies can only act on each other if they are pressed by another matter, in this case the ether.

The quarrel between Bilfinger and Daniel Bernoulli on this question was lively at the Imperial Academy of Sciences in St. Petersburg. As much as Jean Bernoulli, Daniel's father, had fought Newtonianism under the banner of Leibniz, his son became the main defender of Newtonian science at the Academy against Bilfinger and the group of Wolffians, with Euler, although intellectually close to Bernoulli, staying out of the hostilities (Calinger 1996, pp. 148–149). Bilfinger, like many opponents of attraction at a distance, considered Newton's hypothesis, which appealed to occult forces, to be "vulgar" and therefore regressive, and opposed Newton's refutation of the Cartesian vortices. Taking up the theory of the ether, he proposed to the Académie Royale des Sciences a dissertation on the cause and mechanism of gravity for the 1728 prize. To solve the famous objection that the theory of vortices should lead to the fact that bodies fall towards the axis of rotation of the Earth, and not towards its center, an objection pointed out by Huygens (1690, pp. 134–135) and to which Saurin had provided an answer within the strict framework of the Cartesian theory (1703), he imagined a vortex turning at the same time around two perpendicular axes, hoping to deduce from it the direct fall of the heavy bodies towards the center (Encyclopédie de Diderot et d'Alembert 1757, GRAVITÉ article). He described in his memoir a machine intended to test experimentally his idea, in which a sphere filled with water was animated by these two movements. He assumed, from this thought experiment, that the water particles, because of this double motion, described large circles concentric to the sphere. However, the author of the *Encyclopédie* article objected that such a supposition was inaccurate because the curves described were rather in the shape of a figure of eight, "as can be ascertained by experience & by analysis". Nevertheless, the proposed explanation earned its author the prize of the Academy, giving him a certain legitimacy to lead the fight against Newtonianism at the Imperial Academy of St. Petersburg. The same year, Daniel Bernoulli replied in the *Commentarii* of the Imperial Academy that the Cartesian hypothesis was "insufficient". He reproached Bilfinger for his lack of knowledge in mathematics. A dispute arose between the two men over the theory of capillarity developed by Bernoulli on the basis of the theorems in Newton's *Principles* (Calinger 1996, pp. 148–149). Bilfinger and Hermann set up a think tank at the Academy around the natural philosophy of Leibniz and Wolff, defining as primary substance, not the Cartesian corpuscles or Newton's atoms, but the animate monads, which constituted another form of opposition. The quarrel nevertheless subsided before Bilfinger left for Tübingen in 1730, and Euler only took a position against Leibniz's theory of monads later, in 1736.

The 1730s were marked by French expeditions to Lapland and Peru, on the orders of the King of France, to measure the shape of the Earth, and to decide between the conclusion drawn by Jacques Cassini from his geodesic measurements of an elongated Earth along its axis of rotation (Cassini 1718), around which the defenders of Cartesian vortices quickly gathered at the Académie Royale des Sciences, and the Newtonian prediction of an Earth flattened on its equator. We have already described the strong implication on this question of Joseph-Nicolas Delisle, who, as early as 1720, proposed without success a measurement of the degree of the parallel of Paris, and made a new attempt in 1737 by submitting to the Russian authorities a measurement of the shape of the Earth in Russia, associated with a complete cartography of the country by triangulation. This project did not succeed, probably because the expedition led by Pierre Louis Moreau de Maupertuis in Sweden gave from 1738 a result favorable to the Newtonian hypothesis of the flattening of the Earth, perhaps also because the cost was considered too high by the Russians, and the delay of realization too great, perhaps finally by concern of the French authorities, which did not seem to support strongly the project with respect to the Russian State, to keep in France all the profit of its two expeditions. Nevertheless, Maupertuis' results were not unanimously accepted in St. Petersburg. Euler, in particular, although a Newtonian, was cautious. He insisted on the need to obtain experimental confirmations, not only from the study of the shape of the Earth, and also from the study of the tides, the movement of the moon and the trajectories of comets, all subjects on which Delisle, through his observation program, played a major role. Euler was very involved in the cartography of Russia, and as such on the geodesic methods required for the measurement of the shape of the Earth. As early as 1726, he worked as vice-director of the geography office created by Delisle at the Academy, in order to realize the map of Russia which motivated his recruitment as an astronomer by Peter the Great. This office became in 1735 a department, endowed with a conservative secretary (Pierre-Louis Leroy), a translator and six geodesists. In 1736–1737, Euler followed closely the expedition in Lapland, and wrote articles on this subject for the general public in the St. Petersburg newspapers. In a letter from March 29, 1738, he rejected Daniel Bernoulli's assertion according to which the results of the Maupertuis expedition definitively settled the question of the shape of the Earth (Calinger 1996), the precision of the measurements being for him insufficient. Jacques Cassini in Paris and Jean Bernoulli in Basel also rejected Maupertuis' conclusions. In a letter dated May 20, 1738, Maupertuis began a correspondence with Euler on this question, telling him of his admiration for his *Mécanique*, published in 1736 as a supplement to the *Commentarii*, and at the same time communicating his book on the journey to the Arctic Circle to determine the figure of the Earth (Maupertuis 1738). In his reply to Maupertuis, Euler judged that the measurements had certainly been made with the best instruments, but were still insufficient to determine the shape of the Earth. Euler said to retain his judgment while waiting for the results of the expedition in Peru led by Bouguer and La Condamine. Assuming that the density of the Earth increased towards the center, he calculated a shape for the Earth significantly different from that found by Newton (less flattened). In the decade that followed, while in Berlin, Euler worked, as did

Daniel Bernoulli and Alexis Claude Clairaut, who had taken part in the expedition led by Maupertuis in Lapland, and later Jean le Rond D'Alembert, from 1745 onwards, on the development of differential equations dedicated to a precise description of the shape of the Earth, of the tides, of the lunar movements and of the trajectories of comets in order to verify Newton's law of universal attraction. He announced in 1747 at the Académie Royale in Paris that the result of his differential equations, describing the motion of the moon in the three-body system that it composed with the Earth and the sun, suggests the necessity of a correction to the law in $1/r^2$. But Clairaut showed the inaccuracy of Euler's equations, and confirmed Newton's law in 1749.

In an article published in 1746, after the question of the shape of the Earth had been decided in favor of the Newtonian hypothesis following the results of the Peruvian expedition, Daniel Bernoulli, placing himself within the framework of the Newtonian alternative to the vortex theory, mechanically modeled the magnetism on a Cartesian principle according to which "all in the world arrives via matter and movement", therefore by the action of contact. He showed the insufficiency of the mechanism and argued in favor of "the mutual attraction of all matter". According to him, without an attractive force, mass would continuously flow out of any finite volume, and the density of matter would constantly decrease, as matter could not form a stable structure. Here is how Bernoulli expressed himself on this subject:

> If there were only matter and movement in the world, it seems that the world could not exist, whatever movement one wanted to conceive in matter, either circular, as would be that of vortices, or rectilinear, as of central torrents, or a movement of agitation, in which the parts of matter colliding, go back and forth reciprocally, or finally such other movement as it is possible to imagine: it is certain that the matter which composes this Universe, should always move away, & consequently become more & more rarefied, & finally dissipate. This consequence is certain, according to the universally recognized laws of mechanics; & the difficulty is not removed by saying that the world is of infinite extent, since the defect of permanence will always remain (Bernoulli 1752, pp. 119–120).

But, wrote Bernoulli, matter is stable and "its state of permanence consists in an equilibrium between the tendency of matter to spread (because of its motion, of whatever nature) and mutual attraction". Thus, according to Bernoulli, gravity at a distance must be a real and essential property of matter, since it is required to counterbalance the force responsible for rebound in elastic shocks. In this, affirming gravity as an essential property, he passed a milestone that Newton had not taken, considering that it "is universally established by astronomical experiments and observations", but not asserting that it is "essential to bodies", as he specified in a remark added to the third edition of his *Principles* published in 1726 (Newton 1999, p. 795).

3.4.4. The problems of the functioning of the Academy in the decades 1730–1740

As we have seen, the library and the museum of natural history were in themselves important institutions, pre-dating the foundation of the Academy, the latter starting its activities alongside them or even on the fringes. They were directed by the emperor's librarian, Johann Gabriel Schumacher, an authoritarian man who made many enemies among the scholars. Mikhaïl Vassilievitch Lomonosov, one of the first Russian members of the Academy, a polymath with a wide range of scientific and literary interests, called him "the enemy of Russian sciences" (Shafranovskij 1967, p. 605). Schumacher was not a scientist, but he had a master's degree in philosophy from the University of Strasbourg. Hired by Peter the Great as a librarian, he became in 1714 secretary of the Chancellery of Medicine and responsible for the Imperial Library founded in 1703. In 1721, Peter the Great sent him to various European countries to invite scientists to join the future academy. In 1724, he was appointed Secretary of the new Academy, combining this role with his responsibilities as director of the library and museum. This double responsibility, given Schumacher's personality and the difficulty inherent in his role as intermediary between the political power and the scientists in a politically agitated period, had negative consequences on the functioning of the Academy. Schumacher, because of his personal interests, gave less importance to the work of the academicians than to the organization of the various establishments working for the Academy: printing, bookshop, engraving workshop, workshop for the manufacture of physical instruments, bookbinders' workshop and turners' workshop. These establishments were coordinated by Schumacher, and absorbed a significant part of the budget allocated to the Academy. He also placed great importance on pomp and circumstance, incurring lavish expenses at the expense of the budget devoted to science (ibid., p. 606). In the library, well-bound books systematically took first place, independently of their contents. He always strived to show the premises of the Academy, the Library and the Museum to guests: notables, foreigners curious to be informed, amateurs of sciences, systematically putting himself in front and presenting himself as the head of the Academy. In 1728, the president of the Academy, Blumentrost, as we have seen, had to return to Moscow after the transfer of the government to this city after the death of Catherine I. He then delegated all his powers to "Mr. Librarian" (Leonov 2005, p. 81). A struggle for primacy ensued between the Assembly of Professors and Assistants and the Academic Secretariat held by Schumacher. Delisle, in particular, struggled to obtain from Schumacher the credits which he needed to develop the Observatory. At the return of the government to St. Petersburg, in 1732, the power of Schumacher was already well established.

The second president of the Academy (from 1733), Baron Johann Albrecht von Korff, disregarded Delisle's complaints and took Schumacher on as an advisor, officially giving him responsibility for the Academy's accounts. Schumacher, who regarded some academicians, especially those from Germanic states, as "lazy" careerists who had come to Russia to earn money and acquire a high social status, showed ostensible contempt for a large part of the Assembly of Professors.

On September 22, 1742, Andrey Nartov, a mechanical physicist and one of the few Russian members of the young Academy, filed a complaint against Schumacher for "dishonorable acts and misappropriation of funds" (Leonov 2005, p. 82). It is likely that Lomonosov, who had been recruited to the Academy the previous year, was involved in the drafting of the complaint, which thus emanated from the Russian component of the Academy, believing, on a global level, that the Academy did not sufficiently serve the interests of Russia. The complaint began with a panegyric of Peter the Great, founding father of the Academy, emphasizing the promising beginnings of the institution:

> And that is why after the beginning, the state of things was in order and effective. But when, obeying the all-powerful God's will, the lifetime of Their Majesties ceased, then this Academy, having during certain time been under the directorship of the same definite members (among whom the then and present Counselor Schumacher had power over all of the board), did reduce to such a poor state as to bear no fruit to Russia except harm only through the actions that run counter to the former establishment of His Majesty Peter, and unprofitable to government, that I, a faithfully subjected servant of Her Imperial Majesty, looking at all this and preserving the highest interest of Her Imperial Majesty according to my post, as a member of that government and a son of the fatherland, could not keep silent and not inform the governing Senate about the following (Leonov 2005, pp. 84–85).

Nartov then listed all of the things that Schumacher was accused of. First, he mentions the absence of clear and publicly displayed rules of the Academy, which allowed Schumacher to do whatever he wanted. In particular, he blamed the departure of a number of professors who returned prematurely to their countries on the lack of understanding among scholars of the Academy's objectives. The departure of these scholars was detrimental to the interests of Russia. Second, Schumacher was accused of having, on his own initiative and without consultation with scientists, identified and contacted scientists abroad to replace the departing ones, promising them considerable salaries and pensions. Third, he assigned too much money to certain tasks, which could have been reduced if more care had been taken in the way the money was spent. Fourth, Schumacher managed his budget in an opaque manner, without anyone knowing, and never presented the Faculty Assembly with a detailed and up-to-date statement of accounts. Fifth, Schumacher was accused of presenting himself externally as the head of the Academy and of diverting money originally earmarked for the Academy of Arts to the Academy of Sciences:

> To show himself off to the former government he invented to present the description of academic establishment as if it occurred due to his efforts, and he carried this out and presented it to the former Princess Anna and then handed it over to print on behalf of himself only; by superficial and false evidence he closed down affairs of the High Monarch well-known all over the world (1) having written that the intention of the Sovereign Emperor Peter the Great was to subject the Academy of Arts and Crafts to the Academy of Sciences; such a thing has never happened, and as accusatory evidence I have the autograph order made with His Imperial Majesty's own hand that there was assigned a special sum for the purpose, but he, Schumacher took it without order for other use in the Academy of Sciences; (2) in the same book he included honorary members to the number of 23 persons, and in the staff he presented he wrote in 14; (3) he claimed as though civil typography, casting and other crafts had appeared owing to his efforts, while these have already been in reality during the Sovereign Emperor Peter the Great lifetime through His inventions to which He devoted a lot of energy; (4) according to the highest kindness of His Imperial Majesty I am a Member of this Academy, but because of his malice, Schumacher didn't include me among the others (Leonov 2005, p. 85).

The sixth criticism of Schumacher is that the discussions that came out of the Academy were printed in foreign languages, but not in Russian, which did a disservice to Russia, depriving Russian youth of the fruits of the Academy.

The year 1743 was particularly turbulent (ibid., pp. 88–89). The individual behavior of Lomonosov, who drank and insulted the whole faculty, further complicated the situation. Everyone fought against everyone else, the academicians against Schumacher and Lomonosov, Nartov and Lomonosov against the inquiry commission appointed following Nartov's complaint, the inquiry commission against Lomonosov. In February 1743, Lomonosov was forbidden from attending the Assembly of Professors, which further strengthened his resentment towards his colleagues. Schumacher was detained at that time during the work of the inquiry commission. In December 1743, when the investigation was completed, an imperial decree reinstated Schumacher. A month later, Lomonosov was exempted from punishment because of his scientific value, but he was found guilty of impertinence and obscenity and his salary was halved for one year. Thus, the investigation of the complaint against Schumacher was quickly closed, and Schumacher, who had the support of high officials in the imperial administration, escaped without significant damage. From then on, the library became an object of increased attention from the professors, who wanted it to be put back to the service of science. On October 3, 1745, a long report by the professors of the academy was transmitted to the senate, signed by nine academicians, including Delisle, Gmelin, Müller and Lomonosov (ibid., pp. 91–92). This report concluded that the management of the library had to be subordinated to the faculty, contrary to the practice instituted by Schumacher, which resulted in many dysfunctions: non-ordered books, badly maintained books and covered with dust, insufficient follow-up of the works and loss of some, bad classifications, inadequate opening hours, etc. Following this report, an inspection of the library was carried out, and the catalogs were supplemented according to the wishes of the scientists. Schumacher, as compensation for the moral offence suffered, was promoted to State Councillor. On May 21, 1746, the empress Elisabeth Petrovna appointed the young count Razoumovski, the younger brother of her favorite, as president of the Imperial Academy of Sciences. Succeeding Schumacher, was Grigory Teplov, a philosopher and historian of modest origin, translator into Russian of Wolff's writings. He took over the direction of the Register, and ensured de facto the direction of the Academy. Schumacher, a man of power at heart, wrote in a letter to Teplov in February 1748: "It is not I, Schumacher, who is repulsive to them but my rank. They want to be masters, to have distinguished ranks with huge salaries without caring at all for anything else!" (ibid., p. 92).

3.4.5. *The regulation of 1748 refounding the Academy*

Following these difficulties of functioning, a new regulation was instituted in June 1748, of which we can find the detail in the *Journal des Savants* of the same year. The Academy, divided into an Academy of Arts and an Academy of Sciences, was placed under the immediate authority of the emperor, who gave his orders directly to the president. The latter had all the powers, in particular to reward, dismiss and call to order the members of the Academy (Journal des Savants 1748, p. 500). The Academy of Sciences consisted of the Academy itself and its University. There were now four categories, not three: (i) astronomers and geographers, (ii) physicists (botany, natural history, anatomy, chemistry), (iii) those who applied themselves to experimental physics and mixed mathematics and (iv) geometers (ibid., p. 501). Thus, there was no longer a category of humanities, and the department of geography, created in 1735, was given a specific category, linking astronomy and geography. A permanent secretary was chosen among the academicians. The number of members remained extremely small: 10 academicians, 10 deputies and 10 honorary members ensuring the liaison with the European academies. The deputy, rather than an interpreter who was not familiar with the subject matter, was responsible for translating into Russian the works of the foreign academician whose deputy he was, as well as the books related to the science of this academician. At the beginning of each year, each academician submitted to the president his written program and gave a progress report after four months. For every important experiment that was known to have been conducted elsewhere, the academician specializing in the field was responsible for setting it up and carrying it out before the academy. The academicians also evaluated works published in other academies. On the contrary, "each academician will compose works in his science, which tend to the glory & to the advantage of Russia; they will be translated into Russian, & they will be printed with the approval of the President, after having been read in the Assemblies by the Authors themselves, or by those whom the President will have committed to this examination" (ibid., p. 502). He also had to read the new works which were provided to him by the library and communicate his remarks to the company. The academicians now met three times a week. The pieces intended to be read were written in Latin or Russian. "They will be presented to the President who will order the Secretary what to do with them. For without the approval of the President, nothing should be undertaken at the Academy. All the experiments made by any Academician at home, must be repeated at the Academy. Finally, it is there that all that concerns the sciences will be proposed". The president, or failing that the secretary, kept a journal and a register of documents. A volume of the documents approved at the assemblies by the president was printed every year. There were three public assemblies per year. The Academy also had an annual prize to reward the resolution of a problem that it proposed at the beginning of the year, as it was done in the great European academies.

The University was divided into the University proper and the Lyceum (ibid., pp. 503–504). The University was directed by a rector and a historiographer, and there was a professor for each of the five subjects: (i) poetry and eloquence, (ii) logic, metaphysics and morals, (iii) civil and natural law, (iv) mathematics and physics and (v) antiquities and literary history. The teaching was given in Latin or in Russian. There were 30 students at the Academy's expense. They were drawn first from the schools of the Empire, and then from the Lyceum. The courses were also open to other people, who were not obliged to serve the Academy. The Lyceum was headed by a rector. Latin, Greek, German, French and Italian, elements of mathematics and physics, the principles of history, geography, genealogy, blazon and drawing were taught. The academy maintained 20 students at its expense, and they became student members at the University, if they had the ability. Others could attend the Lyceum at their own expense. The Academy maintained at its expense a priest (of Greek Orthodox religion) to "catechize every Saturday the students and the schoolboys". As for the Academy of Arts, it was divided into three departments: (i) artists (architects, painters, sculptors, engravers, medalists, stone polishers), (ii) craftsmen (mechanics, manufacturers of mathematical and physical instruments, turners, clockmakers, locksmiths, carpenters, etc.) and (iii) printers (printers, letter founders, bookbinders, clerks in charge of the sale). All the members of this academy were paid on the credits of the Academy and the incomes of the bookshop. Thus, the Academy of Arts, which included among the arts the trades at the service of knowledge, was directly put at the service of the Academy of Sciences and the University. In a coherent way, the whole, to which one must add the Library and the Cabinet of Curiosities and Natural History, was placed under the authority of a unique president, who also directed the Observatory, the Department of Geography and History, the Theater of Physics and Anatomy, the whole being financed by the credits of the Academy. The president, in order to direct effectively, was assisted by a council of two members, this within a chancellery "from which the president will give all the orders suitable to direct this Body" (ibid., p. 505). For any need concerning science or otherwise, whether it emanated from a professor of the Academy or from an outside institution, the applicant had to apply to the Chancellery, not to the Colleges. The Chancellery arbitrated all activities, and had to ensure that expenses were kept within the allocation for the academy, which was 53,000 rubles (Sigrist 2015, p. 98), more than double the allocation initially given by Peter the Great for the first phase of the institution's operation.

Thus, the new Academy, which instituted the presence of foreign honorary members, in equal number to the resident members, introduced the Russian language next to Latin as a language of scientific communication, integrated the technical disciplines coming in support of the scientific activity, introduced training as a key element of the development of knowledge, was placed under the authority of a President, taking orders directly from the Emperor, and of a Council intended to support the President, aimed at rationalizing the system, and at providing it with new dynamics, after the difficulties encountered. The text notes what had not worked in the necessary synergy between research and training, namely, that (i) some scholars failed in their scientific duties under the pretext of teaching and vice versa; (ii) literature and science being mixed, some did not pay sufficient attention to the pieces of the others, physicists and mathematicians hearing nothing of literature and vice versa, and "the assemblies often ended in bitter & ungenerous disputes" (Journal des Savants 1748, p. 506). It is for this reason that in the new arrangement, the literature category (the humanities) was cut off from that of the sciences, in order to form the university, in which, however, a professor of mathematics and physics was added. In addition, to avoid any hole and detour of funds from one to the other, a separate fund was allocated to the Academy of Arts. To avoid any internal authoritarian downward spiral, the Emperor took the Academy under his immediate protection, and made the President dependent only on his orders, thus confirming the Chancellery by putting it "on a sure and firm footing, in order to preserve good order". The instructions towards the academicians were very clear:

> Since it is well demonstrated by experience that the sciences do not suffer little, when those who must work in them, meddle with

something else, His Imperial Majesty has been kind enough to forbid them to meddle any longer either with the direction of the affairs of the Academy, or to address themselves directly to the Colleges, either in relation to the sciences or in relation to their other affairs, but to address themselves to the Chancellery, which will assist them in whatever may be convenient (ibid., p. 507).

However, the judgment of the Academy's scholars on Schumacher's management must be qualified (Werrett 2010). The Academy was not exclusively a scientific institution, but incorporated, as we have seen, a wide range of competences, also concerning the practice of the arts. Schumacher had to constantly adapt the directions and productions of the Academy to the changing demands of expertise at the Imperial Court, in a period of great political instability, during which the Academy was constantly threatened with closure. In particular, Schumacher oriented the Academy more firmly towards the arts in the 1730s, which generated tensions, and led to a clear separation of budgets in the 1748 regulations. Although the new charter granted the Academy more autonomy from the court on paper, court-appointed presidents and outside patrons continued to shape academic affairs throughout the 18th century. The success of the Academy, and even its early existence, depended on the goodwill of the Court, which not only provided patronage for the arts and sciences but also shaped, from a distance, many of the factions and rivalries that emerged within the institution. By the time of the 1742 survey, Schumacher had achieved his goals. The debates were no longer about whether or not the Academy should continue to exist, but rather about how to resolve conflicts and what form academic expertise should take (ibid., pp. 124–125).

3.5. Conclusion

Euler's departure in 1741, in the climate of hostility to foreigners that followed the death of Empress Anna Ivanovna, marked a first turning point in the evolution of the young Academy. Euler was indeed an extremely prolific author, who was involved in numerous questions of mechanics and mathematics, and produced a large number of articles for the *Commentarii* which he continued to do throughout his time at the Berlin Academy before returning to St. Petersburg in 1766. The 1740s were marked by a poisonous climate, which resulted in the impeachment of Schumacher by the academicians, as we have seen. On his side, Delisle, accused as early as 1738 of espionage, no longer participated in the meetings of the Assembly of Academicians, withdrawing to his observatory and seeking an escape route to pursue his activities outside the Academy, either as director of the Naval Academy to replace Farquharson, who had recently died, or in a Department of Geography freed from the supervision of the Academy of Sciences, which he did not succeed in obtaining. He encountered serious problems with the administration of the

Academy, his salary ceasing to be paid after his return from his trip to Siberia at the end of 1740, as mentioned in the previous chapter. More generally, Schumacher considered, because of his objective position as an intermediary between a political power with high expertise requirements and a foreign scientific community that was by nature volatile and not very sensitive to Russian national interests, that the work carried out by the academicians was the property of the Academy, which financed it, and could not be used as they wished by the scholars. Beyond the accusation of espionage against Delisle concerning his sending of maps from Russia to France, it was thus the very heart of his network mode of operation in the young Republic of Letters that was affected by Schumacher's policy of secrecy in scientific matters. Others than Delisle, like Johann Georg Gmelin, suffer the same attack on their open philosophy of the practice of science. Delisle objectively did a lot for the recognition of the Imperial Academy of St. Petersburg in the 1730s. He not only developed astronomy, geography and meteorology in a perspective of coordinated observations at the European level thanks to the network that he did not cease to develop during all his stay, a question on which we will return to in the following chapter, and also took part in the diffusion of this knowledge to the public via the *Journal of St. Petersburg*, and the organization of public courses of astronomy. His departure in 1747, under pressure from the Russian authorities, was a major loss for the Academy, after that of Daniel Bernoulli in 1733 and of Euler in 1741. The years 1746–1748 saw the number of publications in astronomy in the *Commentarii* fall to almost zero, as a consequence of both the fire that destroyed the Observatory in 1746 and the departure of Delisle.

The Academy, however, was saved from the risk, latent during its first 20 years of existence, of its premature disappearance. The period which opened in 1747, coinciding with the replacement of the *Commentarii* (1726–1746) by the *Novi Commentarii* (1747–1775), saw a diversification of astronomical investigations, with at the same time a slowing down of research and also the beginning of a Russian academic succession (Sigrist 2015, pp. 106–109). Under the impulse of Lomonosov, who was hired as an assistant member in 1742, Russian scholars were integrated into the Academy. There had already been in 1731 the mathematician Vassili Adodurov, who was hired as a translator, then as an assistant member. Lomonosov favored the engagement of Nikita Popov and Nikolai G. Kurganov, the latter taking over from Delisle for the teaching of astronomy at the Naval Academy. The Observatory was rebuilt and re-equipped in 1748, and its direction entrusted to Christian Nicolas von Winsheim, 54 years old, a pupil of Delisle already present in St. Petersburg before the foundation of the Academy. The observations started again in July 1748 thanks to Josias Adam Braun and Nikita Popov. Following the death of Winsheim, the German mathematician Augustin Nathanael Grishow, who was called from Berlin, took over as professor of astronomy and director of the Observatory. After Grishow's death in 1760, the direction of the Observatory passed to the German physicist Franz Ulrich Æpinus, who had little interest in observation.

Thus, the key positions in astronomy, as in the other disciplines, remained in the hands of foreign scientists during the entirety of this period, which was globally a period of decline for the Academy. In 1765, following the death of Lomonosov, only 15 members remained, of which only seven were scientists (Sten 2014, p. 11). President Razumovsky, whose term of office ran from 1746 to 1758, lost effective influence in 1766. Catherine II of Russia, in power since July 1762, entrusted Count Vladimir Orlov with the effective direction of the Academy. From the beginning of her reign, the "Great Catherine" was impatient to restore the past splendor of the Academy, and negotiated with Euler, then in Berlin, his return to St. Petersburg. Euler returned in 1766 and embarked on a vigorous reform of the institution. This marked the beginning of an extremely prolific period in the annals of the Academy, which saw, among other things, a threefold increase in the amount of space devoted to astronomy in the *Novi Commentarii* (Sigrist 2015, p. 109), ending in 1783 with Euler's death. This marked the end of a second glorious period of the development of astronomy in Russia, the place of astronomy in the memoirs of the Academy falling until the end of the century to a very low level, equivalent to what it was before 1766. This gap corresponded to a phase of change, during which the researchers with a broad spectrum of fields that Delisle had gathered around astronomy and cartography gave way to specialists trained in Russia. Friedrich Theodor Schubert, of German origin but born in Russia, became at the beginning of the 19th century the major figure of Russian astronomy, who paved the way to the creation of the Observatory of Pulkovo in 1839. It took almost a century for the creation ex-nihilo of an Academy and an Observatory in St. Petersburg, under the leadership of high-level European scientists, such as Delisle or Euler, to lead to the formation of an autonomous Russian astronomical community and hold its rank internationally. The Observatory of St. Petersburg, under the leadership of Delisle, held its place in front of the equivalent institutions of London and Paris, and in the 18th century of Berlin, Bologna, Uppsala and Stockholm, by preparing the ground for a new age of Russian astronomy.

4

Anders Celsius and the European Observation Networks, Setting Up a Science Society and an Astronomical Observatory in Uppsala

4.1. Introduction

Anders Celsius, who left his name mainly for his temperature scale, played an important role in many fields of nature observation, which go far beyond the sole field of thermometry. In addition to the numerous observations of aurora borealis which he carried out, he was the first, helped by his assistant Olof Peter Hiorter, to have highlighted, at the beginning of the 1740s, the link between the aurora borealis and magnetism through a long series of observations of the irregular agitation of the magnetic needle during the aurora. This major discovery, which should have confirmed Halley's system, did not, however, have a considerable echo, in a period where the hypothesis of an electric origin of the aurorae was quickly imposed, and at most the measurements of Celsius and Hiorter were used by some, like the abbot Louis Cotte in his treatise of meteorology (1774), to deduce from the link between magnetism and aurorae a hypothetical relation between magnetism and electricity. Celsius also devoted himself with constancy to meteorological observation, which required the development of reliable and correctly calibrated thermometers. Thus, in addition to his activity as an astronomer, devoted in particular to the astronomical determination of longitudes, and to the observation and Newtonian reconstruction of the trajectories of comets, Celsius was an essential representative of the European observation networks that were being set up around the aurora borealis, the variation of the magnetic needle, or meteorology, networks of which Joseph-Nicolas Delisle, through his abundant correspondence, was one of the most active members.

The European dimension of Celsius is perfectly illustrated by the five-year study trip he made throughout Europe, passing through Germany, Italy, France and England, during which he met some of the greatest astronomers of his time, who taught him the techniques of astronomical observation. This trip was notably the occasion of meetings that accelerated the diffusion of the results of observation, and promoted the establishment of long-lasting collaborations. It ended with the participation of Celsius, both as an astronomer and an expert of the northern regions, in the expedition of Pierre-Louis Moreau de Maupertuis and his companions in Lapland for the measurement of the meridian degree, in order to determine the shape of the Earth. Thus, Celsius was also an actor of the major debate of his time between supporters of the Earth elongated around the axis of its poles, essentially the Cartesians around Jacques Cassini, director of the Observatoire de Paris, and the supporters of the Earth flattened on its equator, notably the Newtonians Maupertuis and Louis Godin, the leader of the expedition to measure the degree of parallel in Peru. It is undoubtedly this European dimension of Celsius which gave him in Sweden the necessary influence to set up the Royal Society of Uppsala, of which he became the Permanent Secretary, at the origin of the Royal Academy of Sweden and the astronomical observatories created as well in Uppsala, as in Stockholm. His premature death did not allow him to see the full achievement of his projects, under the aegis notably of his pupil Pehr Wilhelm Wargentin, who himself became Secretary of the Royal Academy of Sciences of Sweden. The two men, pioneers in the field of astronomy, encountered enormous difficulties in organizing astronomy at the national level in Sweden, especially due to a chronic lack of means, and to the fragility of the institutions set up, whose dynamics of development rested essentially on the goodwill and the genius of some rare determined individuals.

The objective of this chapter is to show how astronomy in Sweden developed under Celsius' influence, and how this evolution was part of the European dynamic of the constitution of informal networks of observations, and of the establishment of academic societies intended to reinforce and amplify the scientific work undertaken, and to relay the needs at the political level. In section 4.2, we present a summarized biography of Celsius, centered on the trip he made between 1732 and 1737 in various European countries. Section 4.3 is devoted to the three great networks of observation of nature, apart from astronomical observations proper, in which Celsius participated, around (i) the aurora borealis, (ii) the variation of the magnetic needle and (iii) meteorology. Section 4.4 deals with the Royal Society of Uppsala and the legacy that Celsius left to his successors. We first place the creation of the Royal Society of Uppsala in its historical and cultural context. Then, we trace the establishment and the first years of the Royal Society, before examining the relations between this Society and the University of Uppsala, relations that presided over the later creation of the Uppsala Observatory. Finally, using exclusively the correspondence between Joseph-Nicolas Delisle and Hiorter, and Delisle and Wargentin, for the most part extracted from the *Bibliothèque Numérique* –

Observatoire de Paris (in the absence of indication to the contrary, the letters quoted come, as in the preceding chapters, from this source), we follow the principal events that punctuated the life of the Observatory of Uppsala, bequeathed by Celsius to his successors, during the decade which followed the death of the master in 1744.

4.2. The life of Celsius

4.2.1. *The first years*

Anders Celsius was born in 1701 in Uppsala. His grandfather, Anders Spole, was a renowned astronomer who traveled to all the major astronomical centers of Europe, in Holland, England, France and Italy, where he met some of the most renowned physicists and astronomers of the time such as Frans van Schooten, Christian Huygens, Robert Hooke, Nicolaus Mercator, Giovanni Battista Riccioli, Jean-Dominique Cassini and Athanasius Kircher (Stempels 2011). Called to Lund, where a new university had just been created, he became professor there in 1666 and built an observatory on the roof of his house. He traveled several times to Uranienburg on the then Danish island of Ven, a mecca of European astronomy with its observatory founded by Tycho Brahe in the previous century, where he met Jean Picard, who was sent by the Académie Royale des Sciences in France to measure the precise longitude of the Uranienburg observatory. Lund was devastated by fire during the war with the Danes, and Spole moved to Uppsala, where he obtained a position as professor of astronomy in 1679, a position he would hold until his death 20 years later. One of his best students was Niels Celsius, son of Magnus Celsius, professor of mathematics at the University. Niels Celsius, a follower of heliocentrism in a university in Sweden still dominated by the religious orthodoxy, married one of Spole's daughters, a union from which Anders Celsius was born. The Observatory that Spole had rebuilt in Uppsala was destroyed by a fire that ravaged the city shortly after his death in 1699. It was Pehr Elvius who succeeded him at the chair of astronomy, which he occupied until his death in 1718, followed by Niels Celsius, then Erik Burman, who would only serve for a few years due to their early deaths. The young Anders Celsius, who first studied law, quickly turned to astronomy under the guidance of Burman. In 1719, at the age of 18, he joined a scientific society called *Bokvettsgillet*, literally "the society of book wisdom" (Viik 2012). He became its secretary in 1725, and at the same time taught mathematics in a private school. He spent some time in Stockholm, then returned to Uppsala. Samuel Formey (1750), summarizing the eulogy of Celsius pronounced at the end of 1745 by Baron André de Höpken, mentioned that Celsius published a history of arithmetic in 1726. Celsius was appointed Assistant at the University of Uppsala in October 1728, working very quickly as Professor to replace Samuel Klingenstierna, a renowned Swedish scholar, who had left for a few years abroad, and thus combined his roles as a teacher, both at the university and at the secondary level, with those of Secretary of the Society. On November 11, 1728, the Society, after receiving a royal charter, was transformed into the Societas Regia Litteraria et Scientiarum, or more simply into the Royal Society of Sciences (Viik 2012). When Burman died in 1729, Celsius succeeded him as chair of astronomy and gave up his position as a high school teacher. He discovered the same year a new method to estimate the distance of the sun from the Earth.

Samuel Klingenstierna was a prominent figure in the Society and University of Uppsala (L'Esprit des Journaux 1783). Interested in his youth by the mathematical works of Christian Wolff, as well as by the works of Leibniz, Newton and Bernoulli, he composed at the age of 25, in 1723, two dissertations, one on the height of the atmosphere, and the other on the way to improve thermometers that the Society published in its memoirs. In 1727, he obtained from the chancellor of the University of Uppsala a grant allowing him to travel in Europe, which led him first to Marburg where he met Wolff, on whom he made a strong impression. Following the death of the professor of mathematics at the University of Uppsala, he obtained the post on Wolff's recommendation to the king of Sweden, but temporarily gave it to Celsius for the duration of his trip. He went to Basel, where he met Jean Bernoulli, then Paris, where he took part in the theoretical work on the determination of the shape of the Earth by Maupertuis, and later published on this question in the *Memoirs* of the Academy of Sweden in 1744. His trip then led him to London, where he met Halley, and tackled the squaring of rational functions, work reported in the *Philosophical Transactions*, and published in 1740 in the *Memoirs* of the Royal Society of Uppsala. He returned to Uppsala in 1731 to take up the professorship that Celsius had replaced him in, as Celsius was then beginning his term of office in the chair of astronomy.

Klingenstierna, like Celsius, was a follower of Christian Wolff's mathematical method, which we mentioned in the previous chapter in connection with his *Éléments d'aérométrie*, and was the main introducer of Wolff's thinking in Sweden. In a textbook on arithmetic published in 1727, Celsius mentions the existence of Wolff's mathematical method (Frängsmyr 1990, pp. 38–43). The following year, Celsius used this approach in a philosophical treatise on the existence of the soul, explicitly embracing Wolff's philosophical system. Before his departure for his European trip, Celsius published another essay detailing the Wolffian system, and showing the major importance of mathematics and physics for all other sciences. We have seen that Klingenstierna returned to Uppsala in 1731 to take up his post as professor of mathematics. From then on, his lectures were full of references to Wolff, as well as to his student Georg Bernhard Bilfinger. Mathematics thus served as a gateway to philosophy. Klingenstierna gave a series of seminars *in philosophiam naturalem*, but he was opposed by the chancellor of the university, Gustav Cronhielm, who in 1732 warned the professors against an ill-considered diffusion of the new philosophy. Two years later, he forbade professors to preside over theses outside their discipline, targeting mathematicians such as Celsius and Klingenstierna who ventured into the field of philosophy. A year later, students enrolling at the university were notified that they had to make a declaration of faith committing them to reject the new philosophy. As in Halle 10 years earlier, the rationalization of the religious discourse that was the foundation of Wolffian philosophy was opposed by an accusation of atheism from the religious institution, which felt its power threatened. We have seen that Wolff had been excommunicated and had taken refuge in Marburg, and that Bilfinger, dismissed from his teaching position, had decided to go to St. Petersburg. The reaction was not as severe in Uppsala, but the Wolffian scholars nevertheless came up against the wall of censorship. The next chancellor, Gustaf Bonde, appointed at the end of 1737, was much more conciliatory, recommending, as a bulwark against atheism, Wolffian philosophy, which his predecessor had proscribed precisely on the pretext that it encouraged atheism. The chair of physics created in 1755, entrusted to Klingenstierna, is clearly a chair of experimental physics, marking the definitive victory of natural philosophy in a system that had until then been dominated by scholastic thought. The scientist then became tutor to the king of Sweden and devoted the last years of his life to the study of optics. He dedicated his last years to the study of optics, in particular through the realization of optical systems which earned him a prize from the Imperial Academy of Saint Petersburg.

Newtonianism spread very rapidly in Sweden from the end of the 17th century (Kragh 2012). Andreas Spole owned a copy of the *Principles* as early as 1692, which was later purchased by Niels Celsius, Anders' father. Harald Vallerius, professor of mathematics at the University of Uppsala between 1690 and 1712, referred to Newton's work as early as 1694. A thesis defended in Uppsala in 1693 mentions the question of the shape of the Earth as it was discussed in *Principles*. In the 1690s, Sven Dimberga, professor at the University of Tartu in Estonia (then occupied by Sweden), introduced elements of Newtonian natural philosophy into his curriculum. Elvius, Spole's successor at the chair of astronomy at the University of Uppsala, possessed a copy of *Principles*, but did not accept, any more than Spole, nor Niels Celsius, the attraction at a distance and adhered to Descartes' theory of vortices, like the majority of the Swedish scientists of the beginning of the 18th century. But from 1720, the influence of Cartesianism began to diminish in Sweden, as elsewhere in Europe. Erik Burman, successor of Niels Celsius and predecessor of Anders Celsius at the chair of astronomy, introduced Newton's theory into his curriculum, teaching in 1725 the two great systems of the world, the Cartesian and the Newtonian, and between 1726 and 1729 Newtonian elementary astronomy. But it was Anders Celsius, Burman's successor, who was the first true Newtonian of any significance in Sweden, Newtonianism that he adopted as an astronomer being strongly influenced, in its philosophical dimension, with Wolffianism. Other followers of Wolff's mathematical method, such as Martin Knutzen, appointed professor of logic and metaphysics in Königsberg in 1735, and of whom Immanuel

Kant was a pupil, similarly abandoned Wolffian mechanism in favor of Newtonian mechanics and optics (Frängsmyr 1975, p. 663), while retaining Wolff's "natural theology", applying the mathematical approach to the knowledge of God. Anne-Lise Rey underlined the complexity of the elaboration of natural philosophy in the first half of the 18th century (Rey 2013), the philosophy of Leibniz (and Wolff) not being reducible to a metaphysics founding physics, since experience played an essential role, that of Newton not being symmetrically limited to a pure empiricism, since he did not assume the property of attraction essential to matter, as we have said, requiring a metaphysical hypothesis on divine intervention. A certain number of scholars of the time, among whom we can include Celsius and Knutzen, and also the Marquise du Châtelet (ibid.) thus tried to articulate the two thoughts. Concerning the mechanism, as much as Niels Celsius had shown his preference for Cartesian vortices, his son Anders denounced them as a chimera in the 1740s, notably in his treatise *Vortices Cartesiani, ut non-Entia* of 1743. Celsius' European journey – including elements such as his participation in the agitated sessions of the Académie Royale des Sciences on the question of the shape of the Earth, his participation in Maupertuis' Swedish expedition and his epistolary exchanges with Joseph-Nicolas Delisle on the determination of comet trajectories in order to validate Newtonian mechanics – quickly make the young Wolffian evolve towards an adherence to the Newtonian world system.

4.2.2. The European journey

Samuel Formey mentioned that "for lack of Observations & instruments, the Science of the Stars had not yet reached in Sweden the degree of perfection that it had in the other Kingdoms. This defect excited the zeal of our Scholar, & it was in these dispositions that he undertook his voyages with the permission of the Court" (Formey 1750, p. 223). Thus, following in the footsteps of his grandfather and his colleague and friend Klingenstierna, Celsius started a European tour that kept him away from Uppsala for five years. In 1731, he asked the King for permission to travel while retaining his salary, as well as for a two-year scholarship. His request was accepted, and he left Uppsala in August 1732, the year after the return from Klingenstierna. He began his journey by passing through Gothenburg and Ystad. As Viik (2012) recounts, "a quasi-prophetic event occurred when he left Ystad: the sky exploded into an aurora borealis as if to announce that it was important to study this phenomenon". He spent the first year of his journey in Germany (Gros 2010), where he stayed in Berlin and met Christfried Kirch. He took courses with him and trained in astronomical observation at the observatory that Kirch directed. In the spring of 1733, he observed a partial eclipse of the sun. At the end of May, he went to Nuremberg, passing through Leipzig and Wittenberg. In Wittenberg, Celsius met the mathematician and astronomer Johann Friedrich Weidler, an observer of the aurora borealis who published a catalog of them. Celsius had his own aurora observations printed in Nuremberg, as well as those of some Swedish colleagues, 316 observations in total which, together with those of Weidler, would constitute an important part of the aurora catalog used by de Mairan in the statistical study that he delivered in the second edition of his treatise on the aurora borealis (1754). He wanted to make astronomical observations there, but noticed that the instruments of the founder of the Observatory, Johann-Philipp von Wurzelbau, who had died eight years earlier, were in bad shape and unusable (von Zach 1822). He therefore used a small portable quadrant which he had purchased in Berlin from the instrument maker Johan Ernst Esling, and limited himself to measuring the meridian height of the sun, from which he deduced the latitude of the city. In Nuremberg, Celsius tried to persuade Michael Adelbulner, the printer of a journal of physics and medicine, in which he published the results of Carl Linnaeus' trip to Lapland, to publish a similar bulletin in astronomy. He discussed with the astronomer and mathematician Johann Gabriel Doppelmayr the possibility that the latter could become the editor-in-chief. Celsius even wrote an introduction for the bulletin, as well as an appeal to the readers, but the project did not come to fruition, due to a dispute between Adelbuner and Doppelmayr. Adelbulner, who was not only a publisher, but later a professor of mathematics and physics in Altdorf, founded his own astronomical journal shortly afterwards.

In August 1733, Celsius went to Italy, where he visited Venice, Padua, Bologna, where he spent seven months, and finally Rome. In Venice, he met Martin Folkes, vice-president of the Royal Society. During a short visit to Padua, he met the astronomer and mathematician Giovanni Poleni, with whom he discussed the determination of the local meridian. Poleni, who had been informed by Mairan of his theory of the aurora borealis, and of the verifications that the latter wanted to make from the aurorae which would be observed later, put in the form of tables the collection of observations that Celsius had printed in Germany, and which did not circulate yet in France, and communicated these tables at once to Mairan. But it was in Bologna that Celsius decided to settle the longest, to work there with the astronomer Eustachio Manfredi, and his assistant Eustachio Zanotti, and to familiarize himself with the astronomical instruments then used at the Bolognese Observatory, in particular "a semicircle and two quarter-circle walls of six, four and three feet of radius", which he judged of excellent quality (von Zach 1806, p. 106). The observations took place at the Observatory, or in the Cathedral of San Petrone, where a large meridian line was drawn on the pavement. He made measurements of the intensity of the light, and others relative to the obliquity of the axis of the Earth on the ecliptic. The results of these observations were exposed by Mairan to the Académie Royale des Sciences (Fontenelle 1735), and also published in Bologna. In this city, Celsius gave a lecture on the supposed variation of the level of the Baltic Sea, which he attributed to the evaporation of sea water. He felt perfectly in his element, since he said in a letter to friends that he had found in Bologna a second "Berlin", and in Manfredi a second "Kirch" (von Zach 1806, p. 106). He then went to Rome, where Cardinal Da Via lent him his quadrant and an English pendulum clock, and Pope Clement XII authorized him to observe from his palace on the Quirinal, on Monte Cavallo. He continued to measure the intensity of light by trying to calculate the brightness ratios between the crescent moon, the full moon and the solar disk. He verified the position of the meridian established by Francesco Bianchini and Jacques Philippe Maraldi in the church of Santa Maria degli Angeli, highlighting an error of 2 minutes of arc, and observed at the Quirinal the solar eclipse of May 3, 1734.

After having obtained the extension of his scholarship, Celsius headed in August 1734 to Paris, accompanied by Francesco Algarotti, a writer and philosopher who published in 1737 a book on Newtonian theory, *Le Newtonianisme pour les dames*, translated into English, which Cantemir translated into Russian a few years later in St. Petersburg. In Paris, Celsius was hosted by Delisle's mother and sister. Here is what Formey said about Celsius' arrival in Paris, in the middle of the preparations for the two expeditions to measure the shape of the Earth:

> He hurried to Paris before the end of that year. He found himself in the time that the Geometers & Astronomers were busy fixing the figure of our Globe. Opinions were divided, and no better expedient was found to verify them, than to measure one degree under the Equator, and another towards the Pole. The Count of Maurepas easily felt how useful Mr. Celsius could be in a similar operation & he engaged him to accompany Mrs. de Maupertuis, Clairaut, Camus, le Monnier & Ourbier in the expedition of the North (Formey 1750, p. 224).

Celsius entered quickly the circle of the Newtonians and attended stormy sessions at the Académie Royale des Sciences around the question of the shape of the Earth. It seems that Maupertuis, who had convinced Maurepas to finance an expedition to northern Europe to measure the length of an arc of one degree of the meridian, had initially planned to go to Iceland, but that Celsius recommended a place in northern Sweden, on the Bay of Bothnia, enabling taking advantage of large expanses of frozen water to precisely size a network of triangles along the arc of the meridian studied (Viik 2012). Celsius also had extensive experience in the astronomical measurement of angles and meridians. Invited to participate in the expedition under French financing, in the double title of astronomer-geodesist and connoisseur of the Swedish great north, he was also charged to buy in England, a country which he had to visit after his passage in France, the instruments of precision necessary to the measurements. Celsius spent nearly a year in Paris. He tried during this period to convince the University of Uppsala's chancellor to finance the purchase in France of astronomical instruments to equip the Observatory of Uppsala. He obtained 3,000 thalers, which allowed him to buy a "Langlois quadrant of three feet with telescopic diopter and micrometer which was sent to Uppsala by sea" (Viik 2012).

Celsius asked for, and again obtained, an extension of his salary and scholarship. In the summer of 1735, he went to London. In a letter addressed to Erik Benzelius, the original creator of the Royal Society of Uppsala, Celsius reported that in London he stayed with the Secretary of the Royal Society, Cromwell Mortimer, and that he got to know its president, Hans Sloane. He also met there the two great English astronomers of the time, Edmond Halley and James Bradley, as well as several London instrument makers: George Graham, John Hadley and John Ellicot. For the Lapland expedition, he acquired a 10-foot zenith sector and a Graham pendulum clock, and purchased a one-foot reflector telescope made in Edinburgh on his own dime. He made astronomical observations and trained himself in the use of instruments new to him. He published in *Philosophical Transactions* observations of the aurora borealis (Celsius 1735) and of a lunar eclipse made from Graham's house in London (Graham et al. 1738). During his stay, he was elected member of the Royal Society.

4.2.3. *Maupertuis' expedition in Lapland*

At the end of April 1736, Celsius prepared to return to France in order to be in Dunkirk at the end of May to meet the other members of Maupertuis' expedition. The expedition first reached Stockholm, where the crew arrived on May 20 (Viik 2012). Then the group split up, one part continuing its journey by land, the other taking the sea for Tornea, in the Bay of Bothnia. The exact location of the operations was not yet defined. The idea of Celsius to trace the base line on the frozen water of the bay required waiting for the winter, but Maupertuis did not want to lose time. The Torne Valley, oriented north-south and bordered by hills lending themselves naturally to the establishment of the markers (conical structures formed of tree trunks stripped of their bark, standing out well against the dark background of the forest), was chosen to build the triangulation. The construction of the wooden markers was a big job for the team, which continued throughout the winter. The measurements of angles of the entire network of triangles were completed on September 6. The astronomical observations were carried out in parallel by Lemonnier and Celsius. Using the zenith sector acquired in London by Celsius, the two astronomers measured the difference in latitude between the northern and southern extremities of the network of triangles. It remained to trace and measure with precision the base line of the triangulation, and it was a length of 14.4 km of the ice covering the Torne which was chosen. A metallic reference bar of one toise (1.949 m) was brought from Paris. It was used as a standard for the manufacture of additional wooden bars. The question of the expansion of the fir wood with the temperature arose, and Maupertuis judged that it was negligible, what Celsius showed by precise measurements after the expedition. From St. Petersburg, Joseph-Nicolas Delisle, with all the interest that he carried to the measurement of the shape of the Earth, wrote in May 1737 to Celsius (letter of May 10, 1737, Clauraut.com 2022), asking him for news of the measurement of the baseline, and the conclusions of the expedition as for the size of the meridian degree at Tornea. The members of the expedition returned to Paris at the end of the summer of 1737, and Maupertuis presented to the Academy the result obtained, namely, that the arc of the meridian of one degree being longer than the one measured on the meridian of Paris, which seemed to confirm the Newtonian hypothesis of a flattening by the poles. It was in a report to the Academy of Sciences of Sweden entitled "The Form and Actual Size of the Earth" that Celsius gave his own account in December 1741 (Viik 2012). The leading role played by Celsius during this expedition, appreciated as much by Maupertuis as by King Louis XV, earned Celsius an annual pension of 1,000 pounds from the French state. Maupertuis offered Celsius the small portable quadrant used for the measurements, as well as the expedition's vehicle. The success of the expedition, in which Celsius was a major actor, contributed to reinforcing the international fame that he had acquired during his European journey.

The project also stimulated the establishment of regular correspondence between Celsius and Lemonnier, on the one hand, and Celsius and Delisle, on the other hand (see Clauraut.com 2022). Delisle wrote to Celsius in November 1737 to inform him that he would receive a copy of his project of measuring the shape of the Earth in Russia, presented to the Academy of St. Petersburg at the beginning of the same year, as soon as he returned to Stockholm (letter from November 8, 1737). Delisle wanted to know what the members of Maupertuis' expedition thought of his project. He asked Celsius to send him the details of the observations of Tornea, which would be very useful to him to refine his own project. He was aware that Celsius, an official member of the expedition, was bound to confidentiality until the publication of the results by the team, and promised him not to reveal anything. In a letter from March 25, 1738, already mentioned in Chapter 2, Celsius answered to Delisle, saying to him that the French team showed displeasure to learn of its project of measuring the shape of the Earth in Russia. The measurements made in Sweden were disputed at the Observatoire de Paris by Jacques Cassini, who "wanted to make our observations doubtful, because we did not turn our sector: and one published his objection in all the periodicals". Celsius attached to his letter several copies of an "Observation on the figure of the earth by Mr. Cassini" which he wrote in reaction to the criticisms of the director of the Observatoire de Paris, a text on which he asked Delisle his opinion. Delisle acknowledged receipt in his letter of August 1, 1738. He told Celsius he had not yet received from Paris the detail of the observations made in Sweden (Celsius thus did not violate his duty of confidentiality). Concerning Celsius' text written in reaction to the criticisms of Cassini, Delisle judged that Celsius drew at the end of his dissertation a "very rough" consequence, "namely that the observations of the southern part of France are so uncertain that one cannot conclude the figure of the Earth from them. Could we not", he said, "hope for a softening of this proposition? that by making some new particular observations, or even just by learning the details of all the operations of these gentlemen that they did not think to publish, we will be able to rectify most of what seems defective". Celsius, visibly infuriated by the aggressiveness of Cassini's attacks, answered him in a letter on August 16, 1738:

> But it is necessary to draw such a consequence, supposing that our figure of the Earth is true, because I speak of his [Cassini's] published observations, from which he concluded the figure of the Earth, and I cannot know what corrections could be made to them to make them agree with ours. I thought I would be able to clarify this in Mr. Cassini's reply, but you will judge better than anyone whether he has answered all my objections perfectly. He has answered a few, which are less essential. I am unhappy that my dissertation has stung Mr. Cassini, who is trying to make me obnoxious in France [...] Mr. Cassini is indeed the aggressor because he attacked our observations at the Academy, and then a rumor spread in all the gazettes about the uncertainty of our operations. Mr. de Maupertuis took the trouble to answer his objections in Paris, and I, as a foreigner, tried to defend us in other countries. Supposing our measure is correct, there must be some hidden faults in his operations. That is why I wanted to show on which side one could probably suspect the errors, and all this simply, in a way that does not seem piquant. [...] Mr. Cassini [...] does not want to show us where the errors are, on the contrary he always claims against me, that all his observations are exact. He admits very small errors, but I have marked in my dissertation that the errors are big enough to make the Earth elongated along the axis of the poles.

Thus, Celsius, after being involved in the realization of the project, was equally involved in the defense of its results in the face of strong criticism from Jacques Cassini, the official architect of French cartographic operations.

4.2.4. *The last few years*

Celsius returned to Uppsala in the fall of 1737, five years after leaving. He resumed his activity as professor, in which he had been replaced during his absence by his assistant Olof Peter Hiorter, as well as the animation of the Royal Society of Uppsala and the publication of his *Acta literaria* (Viik 2012). He taught, until his death in 1744, not only astronomy, but also mathematics, geography, chronology, gnomonics (the science of sundials) and the art of navigation. He directed a large number of theses covering both scientific and technical fields. The theses of astronomy carried out under his direction related to subjects as varied as the habitability of the moon, the satellites of Jupiter, astronomical refraction, the shape of the Earth, the influence of comets on natural terrestrial phenomena or constellations. Celsius was particularly interested in the constellations, assigning to the future Observatory the task of establishing a detailed catalog of the stars of these constellations with their magnitudes:

> The method of determining stellar magnitudes in this catalog was perhaps the most important result. It was carried out with the help of the one-legged telescope and glass plates. According to Celsius, by placing two glass plates in front of the telescope objective, the stellar magnitude was reduced by one point. The magnitude 12 was determined by the faintest star still visible through the telescope. Sirius became invisible when 24 glass plates were placed in front of the objective. The results obtained by this method differed from those determined by Flamsteed; for example 4m from Flamsteed corresponded to 5m from Celsius, 7m became 10m etc. (ibid.).

He worked hard to find funds for the construction of the Observatory and published a booklet on this issue:

> He published a booklet entitled "On the usefulness of the Swedish observatory" where he exposed the four important missions of this observatory. Firstly he argued the importance of the art of navigation for Sweden as a maritime power, an art dependent on astronomy. The second mission of the observatory would be to participate in the accurate mapping of the country, which had been decided by the Riksdag [the Swedish parliament] as early as 1734. Thirdly, he emphasized the improvement for the state that the observatory would bring to the time service as well as the reform of the calendar, since many countries had adopted the Gregorian calendar. Fourthly, he expressed his conviction that the observatory would be useful for the education of the nation, not to mention the meteorological observations that it would provide (ibid.).

Erik Benzelius, at the time librarian of the University of Uppsala, and Erik Burman, the astronomer who had held the chair of astronomy before Celsius, had desired the creation of an Observatory at the University of Uppsala, but the project remained in a draft state. Celsius had built an observatory on top of his house, but it was only an amateur work. Celsius obtained from the University Council the necessary financing to transform a professor's house that had been put up for sale into an observatory. A contract was signed in April 1739 with a building contractor (ibid.). The new Observatory was completed in 1741, and Celsius began moving in during the summer of that year. Upon his return from his European tour, Celsius had only three modest instruments: the one-foot reflector telescope made in Edinburgh purchased with his own money during his stay in London, the three-foot quadrant financed by the University of Uppsala purchased during his stay in Paris and the small portable quadrant donated by Maupertuis after the Lapland expedition (ibid.). After his return, he acquired a 12-foot zenithal sector and a Graham pendulum clock, similar to the instruments purchased in 1736 in London for the Lapland expedition. Celsius wrote to Delisle on March 13, 1742: "My observatory is now in quite good order, and I am very comfortable there. Six months ago I began observations to confirm the aberration of the fixed stars, found by Mr. Bradley." At the beginning of 1743, Daniel Ekström donated a five-foot transit telescope with a micrometer to the Observatory (Viik 2012). Celsius died from sudden phthisis in May 1744, and thus did not benefit for long from the favorable working conditions provided by the Observatory. We will see in the following that he was interested during these years in the irregular variations of the magnetic needle, in conjunction with the aurora borealis, and in the manufacture of mercury thermometers, according to a temperature scale which today bears his name.

But he also made astronomical observations. The sustained correspondence between Celsius and Delisle about their astronomical observations between 1741 and 1744 gives us information about Celsius' subjects of study. On February 3, 1741, Delisle, faithful to his approach as a collector of observations, asked Celsius to publish a collection of all the old observations made in Uppsala, and to send him a complete list of these observations, adding that he would "wish that all astronomers would publish all their observations during their lifetime. They could expose them better than one can do after their death, when one ignores many things which they did not often put in writing". On March 13, 1742, Celsius asked Delisle for his observations of the eclipses of Jupiter's satellites made in 1740 and 1741. His correspondent at the Académie des Sciences in Paris, Pierre Charles Lemonnier, could not indeed make these observations, and could not obtain those made by the "Gentlemen of the *Observatoire de Paris*". He attached a table of his own observations (dates and times of immersion/emersion, to the nearest second). He wanted to determine the differences of longitudes between the observatories. On June 5, 1742, Delisle asked Celsius if he received the observation of the comet of 1739 made by Zanotti in Bologna that he had sent him. He said he appreciated Wargentin's dissertation on Jupiter's satellites (following the thesis that he defended in December of the previous year), and wished that Wargentin published his tables. He did not receive from the bookseller of Stockholm the *Acta literaria* that Celsius had sent him, and asked Celsius to send him also other astronomy treaties published in Sweden. About his observations of the eclipses of Jupiter's satellites in 1740 and 1741, he said that he could not make any because of his trip to Siberia, but that his replacement Heinsius may have observed them. In the same letter, he was pleased that the Observatory of Uppsala was in good working order and provided with good instruments, and insisted that Celsius followed well the passages of the moon at the meridian. He joined his estimate of the trajectory of the comet of 1742 by comparison with the close stars appearing in Flamsteed's catalog. Celsius' letter to Delisle of January 24, 1743 was still devoted to the eclipses of Jupiter's satellites, with a request to send to Wargentin and to himself the observations made by Delisle of Jupiter's satellites, and the occultations of the fixed stars by the moon. In his letter of February 25, 1743, Delisle indicated to Celsius not to have had time to finish the calculation of the true times of the eclipses of Jupiter's satellites which he observed. He said he was anxious to receive from Celsius his calculations of the positions of the small stars still missing from Flamsteed's catalog, and the information he obtained on the trajectory of the comet. He said he was surprised that Celsius saw the comet until April 19, whereas he lost it on April 9, and Wagner in Berlin on April 12. It was also a question of Lemonnier's treaty on the comet, and other works. On December 12, Celsius answered that he did not determine yet the positions of the small stars, because his telescope of the transits was not yet well regulated, and advised Delisle to have the same type of instrument made by Ekström. It was still a question of Jupiter's satellites, comets and various works of Lemonnier and Zanotti. In the last letter that Delisle wrote to Celsius on February 22, 1744, he continued to ask him to send the memoirs of the Academy published in Swedish from a few years previous, as well as maps and plans of Sweden. He wanted to have Lemonnier's treatise on the comet quickly. Thus, the question of the determination of longitudes by astronomical methods, and the calculation of the trajectories of comets according to the rules of Newtonian mechanics, were the main subjects of discussion between the two astronomers in their epistolary exchanges (they never met).

4.3. Three European networks for the observation of natural phenomena

We now examine three networks of observations in which Celsius was involved. The first network concerned the observation of the aurora borealis, for which we can consider that Mairan, by his collection of observations intended to validate his system of the aurora borealis, was the center of gravity, since all the measurements converge towards him. The second network, clearly centered on Celsius, who was the initiator, was the one dedicated to the monitoring of the irregular movements of the magnetic needle, in connection with the observation of the aurora borealis. The third network concerned the meteorological observations where we focus on the measurement of the temperature, and on Delisle, a craftsman and supplier of thermometers that he distributed to his European contacts in order to realize coordinated and inter-calibrated measurements between the different countries. The light that we give hereafter to the atmospheric network science which was practiced at the time is obviously partial, concentrating on some individualities, which alone are far from covering the totality of the community involved.

4.3.1. *The observations of the aurora borealis around de Mairan*

The year following the publication of his great treatise on the aurora borealis, de Mairan published in the *Mémoires de l'Académie Royale des Sciences* a diary of observations of the aurora borealis which occurred in 1734, those which he was able to observe in Paris or in its surroundings, and also other aurorae observed in various places in the north of Europe or Russia, even Italy (de Mairan 1738). He reviewed, month by month, these observations. He quoted in particular an aurora borealis hardly seen in Paris on April 10, by an overcast weather, of which an astronomer from Bologna, Eustachio Manfredi, sent to him thereafter a detailed report. After having reviewed the aurorae observed in Paris, he testified of his gratitude to Anders Celsius, this year of passage in Paris, after having stayed in Italy the previous year. Indeed, wrote Mairan:

> Towards the beginning of this year, I learned that he had given in the previous 1733, & about the same time that my Treatise on the Aurora Borealis appeared, a collection of observations of this Phenomenon, numbering 316, made throughout Sweden, & up to under the Polar Circle, either by himself, or by his correspondents, from 1716 up to and including 1732. Among these observations, which cover 224 different Aurorae, there are 188, that is to say, about 5 of 6 [that is to say, five-sixths], which have not appeared here, or which were unknown to me. One can judge of the eagerness I had to recover a Book so interesting for me, & which I considered no less as a treasure, in what it could contain of little favorable to my ideas, & give me by that reason to rectify them, than in so far as it was able to support them. The Work was not yet widespread except in Italy, where Mr. Celsius had then been for several months, & where he had brought it first after the printing. Marquis Poleni, to whom he had presented it, gave me the first knowledge of it, & in a detail which is too important on the matter in question, & which does too much honor to my hypothesis, to be passed over in silence (de Mairan 1738, pp. 783–784).

Celsius was thus, as early as 1733, and thus even before Joseph-Nicolas Delisle, whose observations of the aurora borealis in St. Petersburg were made between 1734 and 1736, a major contributor to the large sample that Mairan gathered for his statistical study of the second edition of his treatise (de Mairan 1754). The fact that it was the Marquis Giovanni Poleni of the University of Padua, who was the first to analyze and classify Celsius' observations, before transmitting them to Mairan, is emblematic of the European dimension quickly taken by the observation of the aurora borealis after their reemergence at the beginning of the 18th century. The aurora borealis was one of the objects of observation around which developed in the first half of the 18th century the networks of astronomers of various countries and various academies, those of Paris, London, Berlin, St. Petersburg and Bologna, in particular. Because Mairan's hypothesis – namely the precipitation of solar matter in the Earth's atmosphere – could be tested by the accumulation of observations, which he thought were likely to highlight "the greatest frequency of the Phenomenon, for example, at the Perihelia of the Earth, at its times of ascendance, or when it is in the ascending northern Signs & so on for the rest" (de Mairan 1738, p. 785). While Poleni analyzed the observations made by Celsius, it was because Mairan sent him his own tables following the publication of his treatise in 1733. And it was precisely because Poleni and Celsius discussed this question in 1733 in Padua, that Celsius, during his visit to Paris the following year, met Mairan and discussed this subject with him:

> But I have still more direct obligations to Mr. Celsius; for since he has been in Paris, he has shared with me his Book, & an infinite number of observations or materials that he has collected since on the same subject, both in relation to the years contained in his Work, and for more distant times, or for the following years 1733 & 1734, & this with a generosity & a politeness that are uncommon. By means of the Phenomena supplied to me by Mr. Celsius, and of all those which I have collected elsewhere, or observed myself since the publication of my Treatise, I can apply the method and the principles which I adopted there, to more than 600 appearances of the Aurora Borealis, instead of the 229 only which I had employed there. This is also what I intend to do in a supplement, which I will be able to give much sooner than I had dared to hope. I will do justice to the other persons who have provided me, or who will provide me in the future with their observations; among whom the learned Mr. Kirch is one of those to whom I am most indebted, having received from him on this subject a Letter filled with reflections and curious research, in which he goes back to the Aurora Borealis which have appeared for two centuries (ibid., p. 785).

Thus, Christfried Kirch, of the Berlin Academy, whom Celsius met in 1733 at the beginning of his European journey, was also, from the beginning of the 1730s, one of Mairan's main contacts on this question. The latter also mentions the observations made by Pieter van Musschenbroek in Utrecht, and the first observations made by Joseph-Nicolas Delisle in St. Petersburg during the fall of

1734, which he tells us were all the more reliable since "Mr. Delisle […] had a kind of azimuthal Quadrant placed on the top of his Observatory, with the sole purpose of determining the size, the positions, & all the appearances of these Phenomena" (ibid., p. 792).

Although located in the south of Europe, and less likely to offer the spectacle of the aurora borealis than the northern countries, Italy saw a certain number of its astronomers taking an interest in this phenomenon. The great aurora of December 16, 1737, which we have seen was observed in Montpellier by François de Plantade, in particular, was the subject of numerous descriptions by Italian astronomers. Of all the aurorae observed in Bologna, recorded in the registers of the Academy of this city, Eustachio Zanotti, professor of astronomy and director of the Observatory of Bologna, tells us in the *Philosophical Transactions Abridged* (1747) that this one was the most remarkable. He gave a detailed description of it, illustrated by three drawings. He limited himself to describing the phenomenon, without seeking to provide an explanation: "I shall only relate what is entered upon the Register of Astronomical Observations, leaving to those who are fond of philosophical Hypotheses, to investigate it's Cause according to their Fancy" (Zanotti 1747, p. 533).

In the same volume of the *Transactions*, we find a report of the same aurora observed in Naples by Prince Cassano. Contrary to Zanotti, Cassano gave his opinion on the origin of the phenomenon. He reviewed the existing hypotheses. He quotes, in particular, the explanation by Mairan of which he said, not without a certain irony: "Others, in Love with Authority, and *French* names, have endeavoured to establish the Meteor as a Mixture of the Two Atmospheres of the Sun and Earth; therein tenaciously adhering to the new Opinion of Monsieur *de Mairan*, of the *Academy of Sciences* at Paris" (The Prince of Cassano 1747, p. 528). But he preferred the hypothesis of a "bituminous and sulphureous" matter raised by its weak specific gravity towards the high regions of the atmosphere, where it was crushed and ignited because of the "clashing of contrary winds". In support of this explanation, he cited the last eruption of Vesuvius, where "the Contrariety of the moving forces, the Readiness of the Matter to take Fire, the unequal Intenseness of the Light, the Streaks, and all the other Circumstances, observed in this Meteor" are similarly manifested. And he concluded, saying of the aurora borealis: "Wherefore I think it rather deserves the Name of a *Northern Light*, or *Fire*, than that of an *Aurora*: But I leave the further Consideration thereof to better Heads" (ibid., p. 529). The same dawn is still reported by Giovanni Poleni in Padua, him concluding that he observed on other occasions this kind of light, but that this aurora there "surpassed all that preceded in Magnitude, Light, Figure, Colours and Duration" (Poleni 1747, p. 532). In the same issue of *Philosophical Transactions*, we find reports of the same aurora observed in Rome, in Edinburgh, and in Rosehill in the south of England. We learn in the article "Aurore Boréale" (Aurora Borealis) of the *Dictionnaire de*

Physique (1793) that the height of this aurora was estimated by Father Roger Joseph Boscovich to be of the order of 275 leagues (i.e. 1,300 km), which explains why it was seen from many places far from each other in Europe.

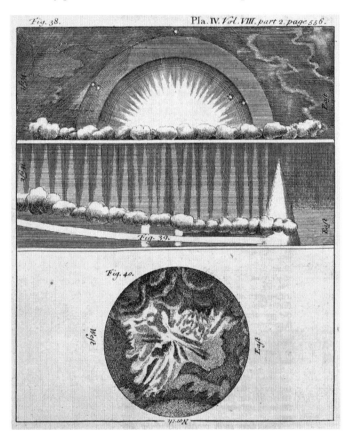

Figure 4.1. *Plates illustrating the aurora borealis of December 16, 1737 observed from Bologna by Eustachio Zanotti (Zanotti 1747, p. 535)*

About the observations of aurorae made in Italy, Mairan gives in the second edition of his treatise in 1754 some details on the way in which he obtained these observations, while passing by Francesco Maria Zanotti, then Secretary of the Academy of Sciences of the Institute of Bologna (de Mairan 1754, pp. 505–509). Zanotti provided him, on behalf of his brother Eustachio Zanotti, professor of astronomy and Director of the Observatory of Bologna, a file containing: (i) his own observations of 21 aurorae qualified to be "great", (ii) the observations of 40 aurorae made by Bartolemeo Beccari, (iii) 67 observations of aurorae communicated to

Beccari from various places of Italy, judged doubtful, and "on the reality of which Mr. Beccari suspends his judgment, until, by the confrontation of those which will have appeared in the more northern countries of Europe, at the same day & approximately at the same hour, the parts which were seen in Italy cease to be equivocal" (ibid., p. 506). Mairan said that he was able to make, "in accordance with the instruction of Mr. Beccari" this verification, and that he found 36 of these aurorae quite real, of which he sent the note with the detail of the places and the names of observers to Beccari. "These 36 appearances of the Aurora Borealis in Italy, being added to the 52 observed by Mr Zanotti & Mr Beccari [some of the 21 observed by Zanotti and the 40 observed by Beccari are identical], make in all 88, & change almost nothing in the ratio of frequency of the Perihelion to the Aphelion, which results from the 52" (ibid., p. 507).

In the same collection of *Philosophical Transactions Abridged* (1747), we find the description of several aurorae observed between 1732 and 1734 by Johann Friedrich Weidler in Wittenberg, Germany (Weidler 1747, pp. 547–548). Weidler, too, made a statement, albeit only in passing, about the physical origin of the aurora. For him, he said, "from these and other Observations, which I have made in former Years of that Light, I am more and more persuaded that it has it's Seat entirely about the magnetical Pole, or at least that it's Motion is thereby in some Measure ruled and determined, which was first of all apprehended by the great Sagacity of the illustrious *Halley*" (ibid., p. 548). Weidler went on to point out that nothing was known about the effects of the aurora borealis, and he put forward some consequences he thought were possible:

> We have not yet any certain Knowledge of the Effects of the *Aurora Borealis*. I have only observed, that some serene Days have always followed that Deflagration. The *Suedes* and *Norwegians*, to whom this *Phenomenon* more frequently appears, are said to have learnt from long Experience, that the Northern Light, when it shines frequently about the Beginning of Autumn, promises a more temperate Air, and plentiful Harvest, for which Reason they call the *Aurora Borealis*, […], or ripening of Corn. They also look upon the frequent Returns of them in Winter as a Token or Presage of a sharper Cold […]. The Experiments made in our Climate in 1731 agree with this *Hypothesis*; for the Northern Light was very frequent and bright in that Year on *Oct.* 4, 7, 8, 10, and 23, and was followed by so fruitful a Season, that we had a great Plenty both from the Fields and Gardens in 1732 (ibid.).

Celsius met Weidler in 1733 in Germany, and the two men exchanged observations of the aurora borealis. Weidler's observations were already published, and it was precisely in Germany that Celsius published his own, which were later tabulated by Poleni when Celsius met him the following year, before being transmitted to Mairan.

Thus, we can speak of a real network activity, based on the system proposed by Mairan, which called for the constitution of a catalog of aurorae, precisely dated and characterized, intended to allow the validation, or on the contrary, the invalidation of his theory. Celsius, traveling throughout Europe at that time, and having the opportunity to work with almost all the members of this network: Kirch in Berlin and Weidler in Wittenberg, Poleni in Padua and Zanotti in Bologna, Mairan in Paris, constituted a key figure in the study of the aurora borealis at that time. Joseph-Nicolas Delisle, although physically far away, since he was in St. Petersburg at the time, maintained at the same time, as we have said, a close correspondence with Celsius and all the astronomers previously mentioned who met Celsius. Thus, each in their own way, Mairan, because of his project to validate his theory, Celsius, because of the opportunity to make contacts provided by his European trip, and Delisle, because of his network of epistolary exchanges, contributed to tightening the ranks of the community of European and Russian astronomers mobilized around the observation of the aurora borealis.

Celsius did not at any time comment on the mechanisms at the origin of the aurora borealis, of which we have seen that three were favored by astronomers: (i) the circulation of magnetic matter according to Halley, favored in particular by Weidler in Wittenberg and Plantade in Montpellier, (ii) the precipitation of solar matter according to Mairan, defended by most of the scientists of the Académie Royale des Sciences in Paris at that time, and foreign scientists like Poleni, and (iii) the ascent and the ignition in altitude of the sulfurous matter emanating from the earth or from the volcanoes, a theory supported in particular by Cassano in Naples, or Mayer in Saint Petersburg. But he was at the origin of a major discovery, that of the link between the aurora borealis and magnetism at the beginning of the 1740s, thus shortly before his death, which probably explains the relative ignorance in which we find ourselves of this essential part of his scientific career.

4.3.2. *Monitoring the variations of the magnetic needle according to Anders Celsius*

It was George Graham, clockmaker and manufacturer of astronomical instruments in London, who realized towards the end of 1722 and the beginning of 1723 the first systematic observations of the regular diurnal variation of the magnetic needle (van Swinden 1780). Graham, born in 1673 near Carlisle, England, moved to London at an early age to work in the workshop of his uncle Thomas Tompion, famous for his chain of mass-produced watches. When Tompion died in 1713, he took over his uncle's business and perfected the technology of precision watches. He was appointed a Fellow of the Royal Society in 1721. As an astronomy enthusiast, he financed and realized several high-precision instruments, notably for Edmond Halley and James Bradley, the second and third directors of the Greenwich Observatory. It was, in particular, Graham who provided the zenithal sector of the Lappish expedition of Maupertuis, and it was Celsius who was in charge of obtaining the instrument during his stay in London, as we have seen. After his return to Uppsala, he ordered from Graham a zenith sector and a high-precision pendulum (whose length was invariant with temperature) for the Uppsala observatory. Celsius also arranged for a Stockholm craftsman, Daniel Ekström, to stay at Graham's in London and to be trained in the manufacture of astronomical instruments. In a letter to Delisle dated March 13, 1742, he wrote to him that the following summer he would have a transit telescope, designed to measure the time of passage of a star on the meridian, made in the manner of Graham, but in Stockholm. We learn in a letter from Wargentin to Delisle dated June 12, 1745 that Graham's clock, and Ekström's transit instrument, were indeed realized during Celsius' lifetime, but that Ekström's construction of a large Quadrant Mural for the Uppsala Observatory was interrupted when Celsius died.

The description of Graham's experiments on the magnetic needle, and its regular diurnal variation, is provided in two articles published in *Philosophical Transactions* in 1724, the first on the variations of the declination of the needle (Graham 1724a), the second on the variations of its inclination (Graham 1724b). He used three 31 cm long needles, two made of glass, one of crystal, of different weights and different widths, to make sure that there was no bias due to the inertia of the needle or to its friction on the pivot. The needle was placed in a copper case allowing an angular excursion of 20° on each side of the meridian direction. The pivot was made of hardened steel, and the angular measurement was achieved with an accuracy of 2 arc minutes. He observed important variations of the direction of the needle with the time of day, these being independent of the needle used:

> After many Trials, I found all the Needles I made use of, would not only vary in their Direction upon different Days, but frequently at different times of the same Day; and this Difference would sometimes amount to upwards of half a Degree in the same Day, sometimes in a few Hours. And this Alteration I observed, whether the Needles were drawn aside immediately before the Observation, or suffered to remain undisturbed. For I have left the Box standing for several Days together, without ever disturbing the Needle, only have taken notice what it pointed at, and the Time of the Day, and I could sometimes perceive in a few Minutes a very sensible Alteration (Graham 1724a, p. 99).

Graham made many tests by artificially disturbing the position of the needle, or by taking it out of the box and putting it back in several times during the same hour. He also tested the effect of the vibrations of the floor under his steps. In parallel, he did the same experiment in two boxes, one made of copper, the other of wood, and found that the two needles moved in exactly the same way in both cases. He noted that these variations did not seem to depend on temperature, humidity or cloud cover, nor on the wind, nor on the height of the barometer. "The only thing", he says, "that has any appearance of Regularity, is, that the Variation has been generally greatest, for the same Day, between the hours of Twelve and Four in the afternoon, and the least about six or seven in the Evening" (ibid., p. 101). After presenting the details of his observations, he concluded that between February 6 and May 10, 1722, he made more than one thousand observations and that the greatest variation towards the west was 14°45', and the smallest 13°50', the variation being rarely lower than 14° or higher than 14°35'. The second article deals with measurements of the needle's inclination made between April 1 and May 27, 1723 (Graham 1724b). The two parameters measured were the inclination of the needle and its frequency of oscillation in the vertical plane passing through the magnetic meridian, which was a measure of the intensity of magnetic force. He found a relatively small variation in the inclination, between 74°20' and 74°58', but a large variation in the frequency, with a time for 100 oscillations (measured when their amplitude became less than 10°) varying between 4 min 58 s and 6 min 12 s. In those years, Graham did not under any circumstances relate certain variations of the compass needle to the occurrence of aurora borealis. But, as we shall see, it is with his help, and thanks to the relationship that Celsius had established with him during his stay in London in 1735, that the first indisputable evidence of the link between the needle and the aurora borealis took place in April 1741.

But before coming to the discovery of Celsius, and his assistant Hiorter, it is necessary to insist on the interest given to the measurements of the magnetic needle in the 20 years that separate Graham's work from that of the Swedish scientists we are about to mention. The interest in magnetism, at that time, was essentially oriented towards the mapping of the declination and inclination of the needle on a worldwide scale. In a letter dated November 1, 1723 to Johann Gabriel Doppelmayr, professor of mathematics in Nuremberg, who undertook collecting all existing measurements of magnetic declination, Joseph-Nicolas Delisle communicated to his interlocutor information on measurements made in the 17th century, especially in Paris. Speaking to Doppelmayr about his geographer brother Guillaume Delisle, he told him that he had "the same intention as you, to collect all the observations of the variation of the magnetic needle, of all times and places, & he has a very large number of them, having needed to go through all the Journals of voyages by sea & by land for the construction of his maps". He informed him that his brother would be happy to communicate all the observations in his possession, provided that Doppelmayr sent him all his own in return. And it was still the cartographic objective which motivated the measurements of the orientation of the magnetic needle carried out during the voyage in Siberia of Louis Delisle de La Croyère between 1727 and 1730 (Klein 2001). For these measurements, he had a "compass", of which we do not have a precise description, and a "machine for the inclination", the same material being reused during the second great expedition to the north of Bering from 1733 (ibid., p. 55). La Croyère made numerous observations of the magnetic needle during his voyage. We can deduce from the excerpts of his journal of observations that the precision of these measurements was rather modest. For example, he wrote about an observation made in Oustioug that "the declination of the magnetic Needle was observed to decline from North to West by half a degree by one Needle and by 45 minutes with the other Needle" (ibid., p. 81). Or again, in Archangel: "on May 16, 1727 [...] I traced on a slate table set up horizontally a meridian by means of a plumb line and I observed that the declination of the magnetic needle was [...] of 3/4 of a degree from North to East. On the 19th, by a new meridian, I found that the compass was declining by just one degree from North to East and with a second Needle only by half a degree" (ibid.). The accuracy of the angular measurements does not appear to be better than half a degree, of the same order of magnitude as the typical amplitude of the needle's agitation during certain aurorae (which can be much greater at times). This probably explains why, despite his daily readings of magnetic declination, and the frequent aurora borealis that he observed during his trip, Louis Delisle did not note any particular agitation of the needle during the aurora.

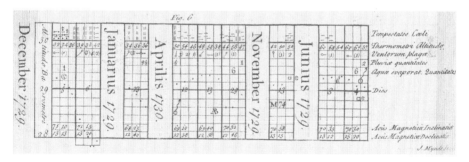

Figure 4.2. *Meteorological (sky condition, temperature, wind, precipitation, evaporation) and magnetic (magnetic inclination and declination) measurements made by van Musschenbroek in 1729 (van Musschenbroek 1731, p. 386)*

Pieter van Musschenbroek published in 1731, in Latin, in the *Philosophical Transactions* an article reporting his "Meteorological, barometric, epidemic, and magnetic observations made in Utrecht" during the years 1730 and 1731 (van Musschenbroek 1731), translated into French a few years later (van Musschenbroek 1741). He mentions the numerous aurorae that he observed, 13 in total, among which five took place in October 1731, most of these aurorae being also the object of reports by other observers, in St. Petersburg, Copenhagen, Uppsala, etc. He gave, for each month of these two years, the maximum and minimum declinations and inclinations of the needle. He noted some cases of atypical variations, like the considerable variation of nearly 10° of the inclination in one night, between March 25 and 26, 1730, making him write: "What of Discoveries on the Magnet are left for our successors to make? Of all the Months there have been few where I have seen so many varieties on the Needle of Inclination, & of all the Months at the same time, there has not been one where the Needle of Declination has been more constant" (ibid., 243), but in no place does he mention any link found between magnetism and the aurora borealis. The great Bering Northern Expedition, a few years later, certainly also resulted in many observations, both of the magnetic needle and of the aurora borealis, but apparently, again, without the link between declination and aurora being discovered. In his memoirs of 1738, Joseph-Nicolas Delisle delivers all the observations of the declination and the inclination of the needle made, "both in Petersburg and in various other places in Russia, which compared with those of the rest of Europe, & with those observed on the sea, & which are marked on the map of Mr. Halley will serve to mark the direction & the figure of the magnetic lines over the whole extent of Northern Asia" (Delisle 1738, p. 18).

Celsius was probably the first scientist to have noticed, with his colleague at the Observatory of Uppsala, Olof Peter Hiorter (Biot 1820), and this as early as 1740, the agitation of the compass needle during the aurora borealis. With Hiorter, they proceeded, using the magnetic needle developed by Graham, to measure the temporal variation of the orientation of the needle. Following the example of Graham's experiments in London 20 years earlier, they took care that there were no metallic objects in the room at the time of the measurements, banishing even keys in their pockets or metallic buckles from their shoes (Viik 2012). Graham's needle provided them with an accuracy on the order of minutes of arc. Between January 1741 and January 1742, Hiorter, at Celsius' urging, made 6,638 observations of the magnetic needle (Brekke 2018), made in a chilly room by candlelight, and with the aid of a magnifying glass (Viik 2012). When, in 1747, four years after Celsius' death, Hiorter reported on his work in the Proceedings of the Royal Society of Uppsala, he had made 10,000 observations, and expressed himself thus: "But who could have imagined that the aurora borealis had something in common and a relationship with the magnetic needle, and that the aurora borealis, when it passes the zenith to the south or accumulates near the west of the eastern horizon, would cause a considerable disturbance of the magnetic needle of several degrees in a few minutes" (Brekke 2018). The first evidence of the relationship between the aurora borealis and the position of the magnetic needle by Hiorter dates back to March 1, 1741. At the very beginning of the same year, Celsius had written to Graham in London asking him to follow the needle regularly, hoping that a measurement of the coincidental variation of the needle between Uppsala and London during an aurora borealis would establish an unequivocal link between the two phenomena. The hoped-for event occurred on April 5 and, according to Hiorter (1747), Graham wrote in his observational diary: "The alterations that day were greater, than I had ever met with before. Tho' no alteration of any thing in the Room could occasion it [...] the only thing in which I am certain, is, that there was no change of position of any thing in the Room, that could cause it, being alone the whole day" (Hiorter 1747, cited in Brekke 2018). In his 1747 paper, Hiorter credits Celsius with the discovery: "Thus I am glad to ascribe this discovery completely to the late Professor Celsius alone, [...] and who made it possible for me to continue these researches and to publish these discoveries which otherwise would have been buried with him." (ibid.)

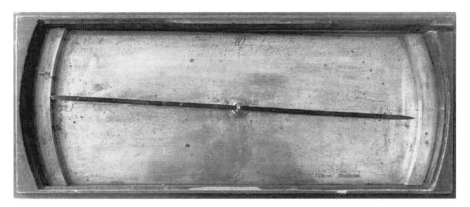

Figure 4.3. *Contemporary copy of the magnetic needle (compass) commissioned by Celsius from Graham in London in 1736 and used to study magnetic field changes (Ekman 2016, p. 113)*

Here is what Hansteen (1827) wrote about this observation made simultaneously by Celsius and Hiorter in Uppsala and Graham in London:

> The experiment was made as early as the 5th of April, 1741, by Professor Celsius, in Upsal, and the instrument maker, Graham, in London. Celsius found on that day the needle becoming restless at 2 o'clock P.M., so that at 5 o'clock it was 1° 40' more west than it had been at 10 o'clock A.M. At 5 o'clock 18 minutes it had receded to the E. by 20', and 6 minutes later it went again 18' westward. From that time till half-past 8 o'clock next morning it went back to its usual position. In the evening an aurora borealis was seen. A few weeks previously Celsius had requested Mr. Graham in London, also to observe his needle on the same day, in order to ascertain whether the

same irregular motions were observable in two places so far apart at the same time: on the same day, viz. Sunday, 5th of April, Mr. Graham, observed in London such extraordinary and frequent irregularities in the needle as he had never seen before, sometimes even at intervals of from 2 to 3 minutes. But Mr. G. makes no mention of any aurora borealis (Hansteen 1827, p. 334).

The same simultaneous measurements, made in conjunction in Uppsala and London with other aurorae, showed that the movements of the needle were identical in both places: "the stronger the aurora borealis seemed to be and the more widespread it was in the sky" (Biot 1820, p. 345). The needle was all the more calm, and its movements of weak amplitude, that the aurora consisted of a simple gleam, "low and quiet, towards the horizon of the north". And "it was still so, when the meteor, although high, had its principal focus in the extension of the vertical plane where the needle is directed, and which one calls the magnetic meridian" (ibid.). In the *Memoirs of Stockholm* (1772), we find interesting details on the irregular variation of the needle between 1741 and 1747, which could reach 30° in extreme cases, according to Hiorter, and on the announcing effect of a variation of the needle prior to the triggering of the corresponding aurora borealis, according to Wargentin:

> Since the month of March 1741, until January 1747, various aurorae observed at the same time as the compass have deviated the needle from one degree twenty-four seconds, to twenty-nine degrees thirty-two seconds. All those who use the compass, either for observations or in the practice of the arts, such as navigation, surveying, research & mining, must carefully observe the effect of the aurora borealis on this instrument. Olav. Pet. Hiorter.
>
> The compass can announce an aurora borealis a few hours in advance. On February 28, 1749, the magnetic needle experienced an extraordinary movement around four o'clock in the evening. As soon as it became dark, a very bright aurora borealis was seen. On April 2 at the same hour the needle was in a movement which lasted until the fourth around six o'clock in the evening; these two nights were illuminated by a very bright light. A great number of other observations confirmed the effect of this meteor on the compass. Pierre Vargentin (Collection académique 1772, p. 191).

In 1752, Wargentin, then secretary of the Royal Swedish Academy of Sciences, published in *Philosophical Transactions* a letter written in Latin dated May 1, 1750 to Cromwell Mortimer, then Secretary of the Royal Society, on the variations of the magnetic needle (Wargentin 1752). Mairan, in clarification XVIII of the second edition of his treatise on the aurora borealis, discusses the contents of this letter:

> Mr. Wargentin, without touching the systematics of the question, and focusing only on the facts, first notes that "Mr. Halley had suspected some correspondence between the Boreal Light and the magnetic needle. He adds that Mr Celsius & Mr Hiorter had noticed that this Needle was sometimes disturbed, & as if worried, when the Boreal Light went up to the zenith, or passed beyond it towards the southern part of the Sky, in such a way that its declination seemed to follow this Light & vary sometimes of three or four degrees in a few minutes of time". On what having wanted to try the observations with a Needle of one Swedish foot of length, he had found them in conformity with what these learned Astronomers had said about it (de Mairan 1754, p. 450).

And Mairan continued (ibid., pp. 450–456), indicating that Wargentin obtained the magnetic needle that was to be used for his experiments at the beginning of February 1750, and then, having begun the measurements, noted its declination every day at different times. Like Graham, Celsius and Hiorter before him, he noticed that between 7 a.m. and 2 p.m., the needle deviated by a quarter or a third of a degree from east to west, then returned to the east to regain its initial position around 8 p.m., before remaining more or less stationary during the night, from February 6–15. On February 15, in the evening, a faint aurora occurred, and Wargentin observed the following day rapid variations of the needle reaching 3°, which were superimposed on the regular variations. The same phenomenon of irregular agitation of the needle in the presence of an aurora took place on February 28, then nothing more occurred during all of March, even on March 6 when, in spite of an aurora, the needle did not move. On April 2 and 3, rapid variations of the needle around 5° were observed in conjunction, again, with an aurora borealis. Mairan tried to reconcile these observations with his hypothesis of a precipitation of solar matter, by defending the idea that the magnetic disturbance was a consequence of the fall of "zodiacal" (solar) matter. "Also", he said, "we see by all that is said here, both by Mr. Celsius & Mr. Hiorter, and by Mr. Wargentin himself, that these variations occur mainly when the greater part of the sky appears or has appeared to be covered with the matter of the Phenomenon, from the Pole to the zenith, and beyond towards the South". He noted that the aurora borealis cited by Wargentin were also seen in Paris, which shows an extension of the zodiacal matter "far beyond the zenith of Stockholm towards the South, since a part of it manifested itself between the zenith of Paris & that of the Hague, in this luminous Band whose height we have calculated". He did not fail to point out that, in certain cases, as on

March 6, the needle did not suffer any variation in spite of the appearance of an aurora, and that in other cases, the needle varied without concomitant manifestation of an aurora, the aurora having manifested itself several hours before, or after, the variations of the needle. He noted that it would be "appropriate to observe, if in countries much less northern than Sweden, such as France, Italy & Spain, the variations of the Magnetic Needle, in the presence or at the approaches of the Aurora Borealis, also take place, if they are not enclosed within narrower limits, or if they do not cease altogether". He noted, finally, and in charge against a possible link of causality between magnetism and aurora borealis, that "the Aurora Borealis visibly has some action on the magnetic needle, but that this action is very little compared to the one exerted by the Earth where the origin of magnetism seems to be". In support of his argument, he cited the considerable differences, slowly varying (on the scale of decades), between the western declinations of the needle in 1750 between Stockholm (7°) and Paris (17°), or between Europe and America or Asia, with deviations that could reach 20° or 25°, "while the focus of the Aurora Borealis goes by jumps & without rule from the West to the East, & sometimes stops directly under the Pole, although commonly it declines towards the West, & all that in the same year, in only one month". And Mairan concluded that the aurora borealis "has never seemed to be affected by Magnetism, which does not resemble it in any way, neither by the visibility of the parts that compose it, nor by the variety of its colors, nor by the diversity of its Phenomena, nor by the vicissitudes of its recurrences and appearances, nor by the region that it occupies, nor by the place from which it comes", and, consequently, "will it be independent of Magnetism".

Let us note that, while striking a serious blow to Mairan's theory, the discovery of Celsius and Hiorter did not give credit to Halley's magnetic theory, because this discovery came at a time when the hypothesis of an electrical origin of the aurorae was rapidly gaining in popularity, as we have already mentioned. However, there is a link between electricity and magnetism as underlined, for example, by Father Cotte in his treatise on meteorology (1774), when he explains that the agitation of the magnetic needle in stormy weather indicates that the same cause acts "& on the electric conductor, & on the magnetic needle" (Cotte 1774, p. 25), noting that the variations of the needle were greater and more frequent in summer, and at the approach of a storm. In his *Éléments de Physique Terrestre et de Météorologie* (1847), Becquerel expresses precisely this state of affairs:

> Halley supposed that the aurora was due to magnetic vortices crossing the earth with excessive speed from south to north, and being able to become luminous by themselves or by their contact with the terrestrial substances which they meet. Mairan, who collected all the observations made up to him on this subject, started from the fact that there exists around the sun a kind of luminous matter of extreme tenuousness, and admitted that the aurora borealis was only a portion

of this vapor, or rather a portion of the solar atmosphere which the earth met on its way and carried with it into space.

This theory, presented with talent, was adopted by scientists until 1740, when Celsius and Hiorter discovered that the magnetic needles experienced an extraordinary agitation during the appearance of the aurora. By comparing this fact with the luminous effects of the aurora borealis, which are similar to those produced by electricity in a vacuum, it was assumed that electricity must play some role in the production of the phenomenon. But it was not enough to find an identity between the electric light and that of the aurora, it was still necessary to demonstrate the existence of a sufficient quantity of electricity in the atmosphere; this was done later by Franklin and other physicists (Becquerel 1847, pp. 603–604).

In his treatise on meteorology, Cotte mentions that he himself observed the link between the agitation of the needle and the aurora: "this is what happened to me on September 17, 1770, a day when I noticed a continuous agitation in my magnetic needle; from one moment to the other it varied from 15 & 20 minutes; I believed myself authorized consequently to predict an Aurora Borealis for the evening, there was one indeed" (Cotte 1774, p. 26). From the observed link between the aurora borealis, whose origin was without question for Cotte of an electrical nature, and the variations of the needle, he concluded precisely a link between magnetism and electricity. And he said, about the second edition of Mairan's treatise of 1754: "The discoveries that one made on electricity since the publication of this Work, made me abandon this system, to look at the phenomenon of the aurora borealis, as an effect of the electric emanations" (Cotte 1774, p. 93), showing clearly that Mairan's hypothesis did not resist to the electric interpretation of the aurorae, which imposed itself quickly from the middle of the 18th century.

4.3.3. *Thermometry and meteorological records around Joseph-Nicolas Delisle*

From the beginning of the 1720s, a dynamic was created in Europe to observe the temperature and other meteorological variables in different places and at different times. In 1723, James Jurin published in *Philosophical Transactions* a letter in Latin inviting the scientists to carry out meteorological observations over a long period of time (Jurin 1723). However, meteorological observation began well before this date. Thus, the Parisian physician Louis Morin, from February 1665 to July 1713, kept a daily bulletin, with readings taken three or four times a day, of temperature, barometric pressure, hygrometry, the strength of the wind and its orientation, the direction of the movement of the clouds and their quantity, the rain, snow and fog (Legrand and Le Goff 1987). The Académie Royale des Sciences decided in 1688 to make the meteorological observation a service task of its Observatory, with an astronomer permanently in charge of the daily summaries: Sédileau until 1696, then Philippe de La Hire until 1718, briefly succeeded by his son, then Jacques Philippe Maraldi from 1721 to 1725, replaced by his nephew Jean-Dominique Maraldi until 1743, then Jean-Paul Grandjean de Fouchy until 1753, as attested by the meteorological reports published annually in the *Mémoire de l'Académie Royale des Sciences* until this date (Cotte 1774, pp. xviii–xix). In England, from the beginning of the 1660s, Robert Boyle and Robert Hooke developed meteorological measuring instruments and pointed out the interest of continuous observations of pressure and temperature, Boyle insisting, in particular, on the connection between meteorology and public health (Cornes 2010, pp. 13–14). Hooke took regular readings for more than a year in 1672–1673, as did Boyle during 1685. John Locke took intermittent surveys from 1666, which became more complete from 1692. These surveys included measurements of pressure, temperature, humidity, and wind direction and strength. On certain days, during strong meteorological disturbances, Locke measured these parameters every hour. The meteorological follow-up in this period was thus in England the result of uncoordinated individual actions, even though the Royal Society played a catalytic role through the initial investments of Boyle and Hooke, contrary to what happened in France where a real service was instituted by the Académie Royale des Sciences as early as 1688. It was not until 1774 that the Royal Society instituted a daily record of meteorological observations (Cornes 2008, p. 230).

Giovanni Poleni, in Padua, was one of the first to put Jurin's recommendations into practice. He published in 1731 in the same journal and in Latin the account of six years of observations (Poleni 1731), translated a few years later into French (Poleni 1741), while being careful not to establish hypotheses on the physical cause of the variability of meteorological quantities, in terms very similar to those used by Fontenelle in his foreword to *Histoire de l'Académie Royale des Sciences* of 1699 (see Chapter 2):

> In Physics it seems to me very dangerous to make systems, when one has not collected all the observations that can serve as a basis for establishing the truth. There is certainly no subject so sterile (it must be admitted) as Meteorology; it is therefore necessary to collect Observations, & there will come a time when they will be used: certainly it will be very useful afterwards to know the history of the Meteors & to be able to rise from the certain effects to the causes, without running the risk of falling into error (Poleni 1741, p. 260).

Anni	Men-fes.	Dies S. V.	Hora h.	Maxima Barometri Altit. Dig. Dec.	Minima Barometri Altit. Dig. Dec.	Thermometri Altit. Dig. Dec.		Venti	Tempeſtas.
1725	JAN.	19	15	30 28		48	45	W.	Coelum ſudum.
	DEC.	8	4 15		28 56	47	98	SW.4.	Coelum nubibus fere obductum.
1726	NOV.	28	15	30 18		48	70	N.	Coelum ſudum.
	FEB.	13	15		28 92	48	62	S W.	Coelum nubibus fere obductum.
1727	NOV.	20	15	30 24		48	88	NW.	Coelum ſudum.
	OCT.	29	3		28 80	49	6	S. 2.	Coelum nubibus obductum.
1728	DEC.	2	15	30 20		48	98	N.	Nubes rarae.
	DEC.	12	15		29	48	60	N.W.	Pluvia tenuis.
1729	DEC.	20	15	30 30		48	88	W.	Coelum nubibus fere obductum.
	NOV.	10	15		28 90	49	30	N.	Pluvia.
1730	DEC.	20	15	30 40		48	20	N.	Coelum ſudum.
	FEB.	27	15		28 98	48	78	S E.	Sol et nubes alternatim.

				Barometri Altitudo. Dig. Dec.	Maxima Ther. Altit. Dig. Dec.	Minima Ther. Altit. Dig. Dec.		Venti	Tempeſtas.
1725	JUL.	9	15	29 04	52 50			S. 2.	Coelum ſudum.
	DEC.	23	15	29 25			47	82 N E.	Sol et nubes alternatim.
1726	JUL.	15	15	29 74	52 40			S.	Coelum ſudum.
	JAN.	14	15	29 68			47	68 W.	Coelum nubibus fere obductum.
1727	JUL.	13	15	29 60	52 18			E.	Sol paucaeque nubes.
	JAN.	2	15	29 68			48	15 S E.	Aer nebuloſus.
1728	JUN.	22	15	29 68	52 54			S.	Sol paucaeque nubes.
	DEC.	26	15	29 30			48	8 N. 2.	Coelum nubibus fere obductum.
1729	JUN.	25	15	29 70	52 28			N E.	Coelum ſudum.
	JAN.	14	15	29 50			47	82 SW.	Coelum ſudum.
1730	AUG.	4	15	29 76	52 28			N.	Sol et nubes alternatim.
	DEC.	23	15	30 30			47	58 W.	Coelum ſudum.

Figure 4.4. *Meteorological records published by Giovanni Poleni showing at different dates (Julian calendar) pressure (in inches of mercury), temperatures (Amontons thermometer), wind directions and conditions of humidity and cloudiness (Poleni 1731, p. 212)*

He underlined that many scientists did not see the interest of meteorological measurements, but it was necessary, according to him, "to work for the glory and the utility of those which will follow us". He described the instruments he used, in particular: the mercury barometer, with a vessel wide enough in relation to the pipe so that the level of mercury in the vessel was invariable, the thermometer built according to the principles of Guillaume Amontons, the rain gauge, the anemometer that provides him with the direction of the wind, and its force quantified according to three levels of intensity, while noting that the force of the wind depended on the height at which the measurement was made above the ground. He provided in his article the tables of measurements taken between 1725 and 1730, by year, by month and by season, and drew general conclusions on the years, and the months, the driest and those the wettest. He attempted to quantitatively highlight the link between the height of the barometer and the occurrence of rain. He was also interested in snow days, and the amount of water provided by the melting of this snow.

At the same time, Joseph-Nicolas Delisle built thermometers in St. Petersburg. He sent them with their method of manufacture to many collaborators in Europe (Delisle 1738, pp. 267–282). He first used thermometers made with spirits of wine (ethanol), thermometers with which his brother was equipped during his Siberian voyage of 1727–1730, and which he himself used in St. Petersburg for his measurements. These thermometers were used at the time for more than 70 years, when Philippe de La Hire had a prototype made by the English-born enameller Louis Hubin (Delisle 1749). They enabled the observation of very low temperatures, notably in 1695, 1709, the year when the Adriatic Sea froze near Venice, and 1716. In 1724, Delisle, still in Paris, made several large thermometers with the spirits of wine for the measurement of the temperature during the total eclipse of the sun of May 22, 1724. But he could not take his thermometers to Russia, and made new ones in 1732 by replacing the spirits of wine by mercury, which allowed him to fix on the boiling point of water (ethanol, unlike mercury, boils below the boiling point of water). He divided the volume of mercury expanded by the heat of boiling water into 100,000 parts for his large thermometers (10,000 parts for his small thermometers). Thus, for example, on January 27, 1733, at 7:00 a.m., he found that his thermometer fell to its lowest point at 2004 parts: "thus by the cold of that day, & at that hour when the cold was the greatest of the whole day, the mercury was condensed a little more than 1/50 part of the volume that it has in boiling water" (Delisle 1738, p. 274). This thermometer made it possible to measure all temperatures below the freezing point of water. Delisle established a table of correspondence between the graduations of his thermometer and those of Réaumur's thermometer. The freezing point of water was the zero point, and the boiling point was at 80 degrees. The temperature of 0°, in the Delisle scale, corresponds to the boiling point of water. When we "lower", according to the Réaumur scale, or Celsius, the temperature below 100°C, the temperature on the Delisle scale increases: 94° for 37.5°C (30° Réaumur), 150° for 0°C (0° Réaumur) and 206° for -37.5°C (-30° Réaumur). Delisle published his work on the creation of his thermometers in the Proceedings of the Royal Society of Berlin, and in *Philosophical Transactions* (Delisle 1735), and wrote a detailed memoir, which was reproduced in the *Memoirs* of 1738, at the Academy of Saint Petersburg in February 1733. He specified in the *Memoirs*:

> And finally, in order to be able to compare my thermometers with those which have been made up to now, such as those of Desaguliers,

Fahrenheit, & lastly those of de Réaumur, one will find here in abbreviated form the description of all these thermometers, & the principles on which they are composed with the manner of reducing them to each other, as far as it will be possible to do so; by which one will be able to compare together all the observations which have been made up to now, & which are found spread in the Memoirs of the Académie Royale des Sciences of Paris. This will serve to show all that has been learned so far on this subject, and what still remains to be known.

The observations of the barometer have too much connection with those of the thermometer, to be omitted here, & especially having the advantage of being able to correct by my mercury thermometers, the variations that heat & cold make on the mercury of the barometer (Delisle 1738, pp. 15–16).

In St. Petersburg, Delisle observed in 1732 a temperature of 27° below zero in the Réaumur scale, that is, -33°C, a temperature much lower than the -19°C recorded in Paris in 1709. He measured a comparable cold period in 1740 and 1747, and his small thermometers that remained in St. Petersburg went down to -37°C after his departure. Delisle, during his stay in Russia, made every day at noon an observation of the temperature. He equipped his brother with small mercury thermometers for the second northern expedition, which led him to Khamchatka in the middle of the 1730s. The opportunity offered by this expedition to make meteorological observations on a large scale throughout Siberia constituted, wrote Delisle, his main motivation to realize his thermometers. During this expedition, which started in 1733, his brother measured on January 16, 1735, at 6 o'clock in the morning, the lowest temperature ever recorded at the time, in the village of Yeniseisk, namely, 280° on the Delisle scale, that is, around -70°C, far below the lowest temperatures recorded on the European continent during the Lapland expedition of 1736–1737, that is, -46°C. During his Siberian voyage of 1740, whose objective was the observation of a transit of Mercury, Delisle had "three large thermometers delivered by boat, with a small one, which he had not wanted to take care of during his voyage by land, preferring to have them brought by water, because they ran less risk of being damaged" (Kœnigsfeld 1768, p. 125). These thermometers also enabled the observations of the barometer of the effect of the variations of volume of mercury related to the temperature to be corrected:

> Because as the mercury thermometers that I propose, show at all times the quantity of the condensation of mercury by the present temperature of the air, it is only necessary to mark what is the height of these thermometers each time that one makes the observation of the barometer: & if one wants to reduce the observed height of the

barometer to that which it would have in boiling water, it is only necessary to add to the observed height of the barometer the proportional part suitable for the condensation of mercury below the boiling water shown by the thermometer. For example, in the very cold of Jan. 27, 1733, my simple barometer was at 28 inches 8 lines 2/5 of a foot of France. The 1/50 part of this height is about 6 lines 9/10: thus the height of the barometer would have been in that time about 29 inches 3 lines 3/10, if the mercury of this barometer had been exposed to the heat of boiling water (Delisle 1738, p. 276).

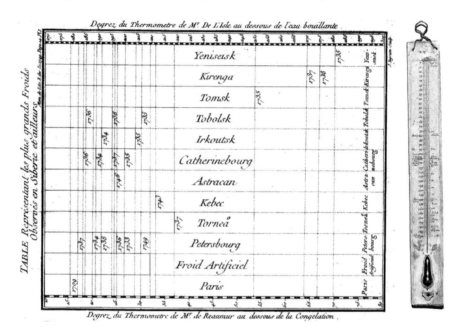

Figure 4.5. Left panel: temperatures corresponding to the greatest cold observed in Russia according to the thermometric scales of Delisle and Réaumur (Delisle 1749, p. 15). Right panel: thermometer made by Delisle sent to Celsius, with Delisle's scale on the left side, and Celsius' scale, added in 1741, on the right side (Ekman 2016, p. 104)

Delisle noted that Amontons made the same type of correction, but in a smaller range of temperature variation. He suggested that "the curious & delicate observation that was made of the height of the barometer at the bottom & top of the mountains, where I see nowhere that attention was paid to the effect of the temperature of the air, although it is certainly known that this temperature is very different in high & low places" (ibid., p. 277), which we have seen gave rise to a considerable revision of the estimation of the height of the atmosphere, should have been analyzed taking into account this effect.

Delisle frequently mentioned his thermometers in his correspondence. In particular, he asked his interlocutors to compare the temperatures given by his "universal" thermometer with the temperatures they obtained with their own thermometers, in order to make comparisons between countries. Thus, on June 16, 1735, he wrote to Christfried Kirch about two small thermometers that he sent him:

> I do not doubt that you have received them in good condition and that you will not make the use of them that I asked you in my previous letters; when Mr. Weidler will have received the one of these two thermometers that is intended for him, I hope that he will send me the comparison that I asked him for of the degrees of the thermometer with the degrees of the thermometers that he has used up to now in the meteorological observations that he made in Wittenberg with regard to the 2 thermometers that I had intended to you previously among all those that Mr. Bernoulli took away from here; I learned that they remained in the hands of Mr. Klein, secretary of the town of Dantzik who uses them to make the meteorological observations of which he has a mathematician from Dantzik note in a journal that he sends to us here.

Sometimes thermometers broke during transport, so he suggested to some of his interlocutors to build their own thermometers on the basis of the manufacturing method he published in 1738, as he developed it in the following letter to Giovanni Giacomo Marinoni of December 25, 1742:

> I am sorry to learn that all the thermometers that I sent you to spread throughout Italy have been broken, without excepting one [...] I wish, now that the construction of these thermometers is published [...] that all those who wish to have them, would take the trouble to build them themselves. The method is easy, and can be practiced in several ways. They were easily made in England and in Denmark, and they were found to be perfectly in agreement with those that were then received from me. I beg you to propose the same thing to Mr. Martino, and also to Mr. Zanotti, as I have little reason to believe that I can bring thermometers from so far away without the danger of being broken on the way. If you take the trouble to build one yourself, you may be able to make Mr. Poleni hold one without accident, and you will learn whether it will agree with the one he received from me. But in the absence of a comparison with thermometers that I have built myself, you can always be sure if you have succeeded, by comparison with

any other thermometers that you may have; provided that they have been regulated according to the principles of Mr. Fahrenheit, or those of Mr. Réaumur &c.; because I relate the degrees of these thermometers to my own.

Indeed, in his report of the aurora borealis of December 16, 1738, Poleni mentions his measurement of the temperature, realized, he says, with the Amontons thermometer, and also with the one Delisle sent from St. Petersburg (Poleni 1747, p. 529). Some of Delisle's interlocutors, like Peder Horrebow in Copenhagen, built their thermometers according to Delisle's method, and compared their measurements with those realized with a thermometer directly realized by Delisle, with success if one judges the postscript of a letter written by Delisle to Horrebow on March 7, 1744, in which he directly compares temperatures measured at the same time in Copenhagen and in St. Petersburg:

> P.S.: I was charmed to learn from your letters that you received one of my thermometers, and that you found it so well in agreement with the one you took the trouble to build yourself following the same method. I will receive with pleasure the observations you will communicate to me in the future on this instrument. In the greatest cold you experienced on February 5, 1740, which you believe to have been one of the greatest that has happened in Copenhagen, your thermometer only went down to 183¼ at 8:½ a.m.; mine was also that day at 7:½ a.m. at one of the lowest lows where it arrives in Petersburg being very near 200. The cold is still so great here that on the 2nd of this month […] at 7:¼ a.m. the therm. was at 185.

But the comparison also needed to involve pressure measurements and, on this subject, Delisle asked Horrebow for clarification in order to be able to compare truly comparable quantities:

> You have given me the pleasure of communicating your observations of the greatest and least height of the barometer. But you forgot to tell me according to which feet and inches these barometers are divided. If you could also know by an exact leveling how high your barometer is above the level of the sea, it would make your observ. useful & more suitable to be compared with those I have made here without interruption for 14 years, & in which I know the precise height of my barometers above the level of our river which is not far from the sea.

It is possible that Celsius' interest in thermometers was aroused by his correspondence with Delisle and Lemonnier, whom he had met on Maupertuis' Lapland expedition (Viik 2012, p. 16). Celsius experimented in 1741 with various thermometers, and studied the dependence of the boiling point of water on atmospheric pressure. In July 1742, he presented to the Swedish Academy of Sciences a report, which was published the same year, and in which he assigned a temperature of 0° to the boiling point of water, and a temperature of 100° to the solidification point of water. This scale resembled that used by Delisle, in that the zero was set to the boiling point of water, but instead of assigning a temperature of 150° to the freezing point of water, he chose 100°. It is said in Hiorter's journal of observations that in 1747 the Stockholm craftsman, Daniel Ekström, presented a centigrade thermometer, in which the temperature of 0° was attributed to the freezing point, and a temperature of 100° to the boiling point, thus inverting the Celsius scale. It seems that this scale was introduced in 1737 by Carl Linnaeus evidenced on the first page of his *Genera Plantarum* published in 1737 by the image of a cherub holding a 100 degree thermometer where the zero point apparently corresponds to the freezing point of water (Viik 2012, p. 17). It was at Linnaeus' request that Ekström made the first thermometer graduated according to the so-called Celsius scale. This was presented by Linnaeus to the Swedish Academy of Sciences in December 1745, one year after Celsius' death. This thermometer was called "Ekström", then "Swedish", and, according to Viik, it was to Jöns Jakob Berzelius, who wrote in his treatise on chemistry published in 1818 that "Celsius, a professor at Uppsala, was the first to draw the attention of physicists, about 1741, to the necessity of graduating the thermometer according to two definite temperatures, without regard to the dilatation of the liquid", that we owe the paternity of the present graduations, used in Sweden, France and all over Europe, to Celsius.

In the foreword to his *Traite de Meteorologie* published in 1774, Louis Cotte says he owes to Delisle's portfolios of observations, kept at the Navy depot, and which were entrusted to him on the recommendation of Lalande, a large number of the meteorological observations that he used, some from distant lands. In his description of these portfolios, Cotte first mentions the experiments carried out by Delisle to construct his thermometer, and the corresponding handwritten memoir which was read by Louis Godin in 1734 at the Académie des Sciences in Paris. One of the portfolios was dedicated to aurora borealis sightings. Concerning the actual meteorological observations, here is the general description given by Cotte:

> I also found a series of observations of the barometer & of the thermometer, made by Mr de l'Isle in Petersburg, from June 22, 1727 to March 31, 1747; & in Paris, from October 4, 1747 to December 1760.
>
> These portfolios still contain a rather considerable collection of meteorological Observations, made in various places, & particularly in

Siberia, & in the other countries of the North, during the Kamtchatka expedition, undertaken by the orders of the Czarina, in the years 1736, 1737, 1738 & 1739. I have not been able to make much use of these observations, because they were made during the course of a voyage which did not allow the Observers to stay long enough in the different countries they were travelling through, to make sustained observations, from which one could obtain general results. I have used these observations in part to draw up the Table which ends this article.

One also finds in these portfolios, the comparison of more than fifteen different thermometers, with that of Mr de l'Isle. This work of Mr de l'Isle was very useful to me in Book II where I treated this subject (Cotte 1774, pp. 383–384).

He mentions several cities for which he had complete sets of observations including, in addition to Paris and St. Petersburg, the city of Stockholm. His contact in Stockholm was Peer Wilhelm Wargentin, successor of Celsius and then secretary of the Royal Swedish Academy of Sciences. Wargentin made three daily temperature observations between 1756 and 1772, and established tables providing the evolution of the temperature averaged over 5 days, with the calculation of monthly, seasonal and annual averages. From the tables provided by Wargentin, Cotte made comparisons between cities in different regions of the world, for example:

The average degree of heat throughout the year is only 4 or 5 degrees above the freezing point in Stockholm. The observations of Réaumur & Duhamel give it for Paris of 8 or 9 degrees; those of the Correspondents of Mr de Réaumur, give it for Algiers of about 15 degrees, & for Pondicherry of 20 degrees; thus the difference between the climates of Sweden & France, is not as great as that which is found of France, Algiers & Pondicherry.

In the Memoir that Mr Wargentin was kind enough to send me, he remarks that the thermometer in Stockholm is usually only 2, 3 or 5 degrees lower in the morning than at noon in winter, that often the cold is even or stronger at noon; but that in the other seasons, the daily variation is much greater, and that it often goes up to 10 or 12 degrees; that, usually, the height observed at eleven o'clock in the evening, is about the average of the whole day (in Paris, this average height takes place at nine o'clock in the evening); that in the month of July, for example, the average heat of this month being 13 to 14 degrees, that of noon is 17 or 19; often it goes to 20 or 21, very rarely up to 22, & at the most up to 24 (ibid., pp. 370–371).

Cotte also analyzed barometric data, in particular the series of pressure measurements made in Paris, covering 67 years (1670–1709 by Morin, 1747–1760 by Delisle, and later years by Messier), and pluviometric data, in particular those of Delisle collected in St. Petersburg between 1738 and 1746, which he compared with the Parisian data obtained at the same time:

1) The rainiest month in Petersburg, as well as in Paris, is June.

2) The least rainy month in Petersburg is January, and in Paris February.

3) Much more water falls in winter in Paris than in Petersburg, because in the latter city, all the water that falls in winter is provided by the snows which give very little water.

4) The amount of rain is almost equal in spring in both cities.

5) It is much greater in summer in Petersburg than in Paris, because storm rains are more frequent in Petersburg. In 1742, for example, the month of June alone provided 63 1/2 lines of water in Petersburg, while it provided only 14 lines in Paris.

6) Fall is rainier in Paris than in Petersburg; it is for the same reason as the one I mentioned above when talking about winter, because the snows come back to Petersburg from the month of November.

Etc. … (ibid., p. 385).

A long chapter of the book is devoted to instruments, and in particular to thermometers developed by many physicists, including Guillaume Amontons, Daniel Gabriel Fahrenheit, Joseph-Nicolas Delisle, René-Antoine Ferchault de Réaumur, Francis Hauksbee, Giovanni Poleni, Isaac Newton, Stephen Hales, Anders Celsius, Luigi Ferdinando Marsigli, etc., to name only the best known, and provides a table of correspondence between the graduations of 16 different thermometers. Concerning Celsius' thermometer, he notes that it was the one used by Wargentin for the measurements reported in his meteorological tables. He emphasized that Sweden is a country "where the taste for observation is making daily progress useful to Physics" and that as such he "thought it necessary to give a place to the scale of the thermometer of Mr. Celsius" (ibid., p. 136). At the time, in fact, the Celsius scale was not yet the standard, as it would become in the following century in continental Europe.

In the foreword of his work, Cotte defines the spirit which presided over its composition. Like Poleni several decades before him, and maintaining the course set by Fontenelle in 1699, he said that he preferred not to say anything about the physical causes of temperature variations than to hazard necessarily uncertain explanations. His objective with this treatise was not to put forward the set of observations used for its own sake, but rather to "make one feel how much these small Observations, these apparently minute remarks, are nevertheless necessary, since it is only by bringing them together & comparing them together, that one can hope to draw useful results on Agriculture" (ibid., pp. ix–x). Cotte considered his work, which he wanted to be useful to the optimization of the productions of the earth, and also to medicine, as the provisional link of a vaster historical enterprise, "a term of comparison for the years which have passed since one deals with these kinds of Observations, until now". For him, "it will only be a question of bringing the later Observations closer to those which are contained in my Treatise, of examining well the circumstances which accompanied the ones & the others; & if one glimpses some resemblance, one will be bolder then to pronounce on the consequences & the results which they will present" (ibid., p. x). In the preliminary discourse of his book, he quoted Mairan, who in the *Histoire* of 1743, wrote that "the most brilliant works, & which require the most penetration & finesse do not always become the most useful to men, & especially to posterity" and for whom "all these Observations [meteorological] made with care during several years, during several centuries & in each country, will probably produce some day a more perfect & more sure Agriculture & Medicine" (ibid., p. xvii). He mentioned the first meteorological observations of Picard in 1666, and recalled that from 1688, there was permanently a member of the Académie Royale des Sciences in charge of meteorological measurements: La Hire, Maraldi, Cassini, etc. He underlined that Mairan supplied for a few years his observations of the aurora borealis, and that physicists such as van Musschenbroek in the Netherlands, or Gautier in Quebec, ensured meteorological follow-ups. He quoted major observers, including Delisle, Celsius and Wargentin, and brought back an extract from Fontenelle's foreword to the *Histoire* of 1699, already mentioned, supporting the importance of measurements made in different places and times, in other words, of the constitution of meteorological observation networks over the long term:

> One would think, said Mr de Fontenelle, "that it is quite useless to keep an exact record of the wind which blows each day, of its force and duration, of the quantity of rain. However, the changes that occur in all this great mass of air may seem even more bizarre than they are in fact, for lack of observers who have applied themselves to it long enough and carefully enough to discover regularity; and if it is possible that there is any, it will only be discovered by a long series of observations made in different places" (ibid., p. xxv).

4.4. The Royal Society of Uppsala and Celsius' legacy

During Celsius' European journey, the activities of the Royal Society of Uppsala, of which he had been Secretary before his European trip, almost completely ceased. On his return, Celsius took over and gave a new impulse to the Society. On the subject of Celsius's high conception of the role of a scientific society, and before going into further detail on the history and functioning of the Royal Society of Uppsala, let us quote Olof Glas, a physician who became the secretary of the Society in the middle of the 19th century and who wrote a historical essay about it (Glas 1877):

> We can see from the following how seriously he considered the importance of a scientific society. Before the founders of the Royal Academy of Sciences in Stockholm had finally constituted themselves as a society, they had thought of choosing the members who should be part of it; their votes had gone to the secretary of the Society of Sciences of Uppsala, André Celsius, among others. The latter answered from Uppsala on May 29, 1739 that "no member should be admitted who is not a lover of useful sciences, even if he possesses profound knowledge in one of them, because some people might wish to be part of the Society for the sole purpose of adding to their titles".
>
> [...] CELSIUS' activity as a member of the Royal Society of Sciences is attested by the memoirs he inserted in the Acta Societatis, and his untiring zeal as secretary is proven by the minutes of the sessions (ibid., pp. 18–19).

4.4.1. *Historical context of the Enlightenment in Sweden*

Uppsala University was founded on February 27, 1477, only three years after the introduction of printing in Sweden (Stempels 2011). It is the oldest university in northern Scandinavia. Astronomy was taught there from its inception, as evidenced by reading notes dating from the 1480s. It was first a Catholic university, and teaching there was interrupted after the Lutheran reform, the two churches separating in 1527. A first effort to re-establish the university was made in 1566, and astronomy was taught there for part of the 1570s. Then, at the very end of the 16th century, astronomy figured again in the constitution of the university. Its first teacher was Laurentius Paulinus Gothus, who studied at German universities and was familiar with the work of Copernicus. He discussed in his course of 1600 the three systems of Ptolemy, Copernicus and Tycho Brahe. Although subsequent teachers incorporated the works of Kepler and Galileo, Uppsala still remained a hotbed of geocentrism, mainly because it was the religious center of Sweden. The education given in astronomy at the beginning of the 17th century was above all centered on the preparation of calendars and ephemeris, rather than on actual observation. The development of this science in 17th century Sweden was hampered by the lack of observation instruments, as evidenced by an inventory of 1678, mentioning only terrestrial and celestial globes, a quadrant, two astrolabes and a geometric compass (ibid.). Mention was made of the invalidation by the university authorities, at the end of the 17th century, of the thesis prepared by Niels Celsius, the father of Anders, in which he affirmed the pre-eminence of empirical observation over the scholastic doctrine, a thesis which undoubtedly led to him not being appointed professor of astronomy following Spole's death in 1699.

Lutheran orthodoxy, nevertheless, slowly liberalized at the turn of the 18th century, and a demand for a new worldview to replace Aristotelianism, challenged by the new Cartesian physics, emerged (Sten 2014, p. 5). In Sweden, the quarrels between Aristotelians and Cartesians subsided in 1689, when King Charles XI issued a letter establishing the freedom to philosophize. Many Swedish university professors stayed at the universities of Leiden or Utrecht in Holland, precisely where Descartes had lived and taught, and in addition Descartes spent his last days in Stockholm, invited by Queen Christine. Thus, Cartesianism was very present in Sweden at the beginning of the 18th century, when the Royal Society of Sciences of Uppsala was founded in 1710, then the Royal Swedish Academy of Sciences in 1739. The latter was based in Stockholm, a city that did not have a university. There were only three universities in Sweden at the time: the historic University of Uppsala and the universities of Lund and Abo, founded in the mid-17th century. The main mission of the university was then to provide state servants and priests for Lutheran parishes, research not being encouraged as such, because it carried heresies and revolution. Aristotelianism was still only slowly giving way to Cartesianism, and the mathematical physics of Newton and Leibniz, as well as Newton's theory of gravitation, of which Celsius and Hiorter were already followers, only entered the curricula of Swedish universities in the middle of the 18th century, a little later than in other European countries.

Another important determinant of the creation of these scientific societies was the catastrophic economic state in which Sweden found itself at the beginning of the 18th century. In the previous century, Sweden had established itself as a great European power, but this position of strength collapsed abruptly at the beginning of the 18th century as a result of the wars waged by Russia in search of access to the Baltic Sea. The Battle of Poltava (July 8, 1709), where Peter the Great defeated Charles XII, put an end to Sweden's status as a great power and marked the beginning of Russian hegemony in Eastern Europe. The territory of Sweden was reduced to its natural borders, where "one speaks the same language, obeys the same law and professes the same religion" (Glas 1877, p. 4). The country, ruined by war, saw famines and epidemics which decimated the populations. The situation in

Uppsala, and in its university, at the time of the birth of the Royal Society, was not better:

> Uppsala was not in a less deplorable and desperate situation than the rest of the country. The aftermath of the fire which, on May 16, 1702, reduced two-thirds of the city to ashes and caused considerable damage to the cathedral and the castle – which is not even repaired today – was still being deeply felt when the plague of 1710 broke out: the University was deserted by the studious youth. "In order to be able to forget at least for a few moments," says Prosperin, "in this idleness, the sad objects which presented themselves to the eyes and mind on all sides, Dr. Eric Benzelius the younger, then librarian of the University of Upsal, invited some of the most famous men who were at the University, to assemble a few times a week in the Library of the Academy, in order to converse there about literature and science, as well as to correspond with Christopher Polhammar and Emmanuel Svedberg" (Glas 1877, p. 5).

The parliamentary era that began in Sweden in 1718 coincided with the Enlightenment. It ushered in "the century of freedom" and marked the beginning of a period of gradual recovery in living conditions, which improved "through the encouragement of entrepreneurship, agriculture, mining, and small industry" (Sten 2014, p. 9):

> In the same period, the idea that science should serve the economy and be useful to society became dominant. In particular, the success of Linnaeus's taxonomy (*Systema Naturae* 1735) and the numerous scientific expeditions of his "apostles", as his disciples were called, suggested an extensive search for uses for the natural sciences – botany, zoology and mineralogy to the benefit of man, the more so as this pursuit was not seen to be in conflict with the divine origin and purpose of the world. Characteristically, in every university in Sweden, a Chair of Economy was established in the mid-eighteenth century, focusing on natural history and agriculture. Astronomy was also considered a prime subject for the Swedish universities owing to its importance in cartography, chronology and navigation. The French-Swedish expedition led by Pierre Louis Moreau de Maupertuis to Swedish Lapland to discover the general shape of the Earth was not only a triumph for "Newtonian science", but an important step for Swedish astronomy as well (ibid., pp. 9–10).

4.4.2. Birth and development of the Royal Society of Sciences in Uppsala

It is in this context that the Regia Societas Scientiarum Upsaliensis was born under the initial impulse of Erik Benzelius (Glas 1877). Benzelius was an important figure in the history of Sweden, both for the breadth of his scientific interests, and for the active role he played in the dissemination of knowledge, both nationally and in Europe. Born in 1675, he became university librarian in 1702, professor of theology in 1723, and was archbishop and pro-chancellor of the University of Uppsala from 1742. He was interested not only in theology, but also in archeology, literature, political history, linguistics, natural sciences and mathematics. For 50 years, he maintained an extensive literary correspondence with scholars from all over Europe, and was considered by his contemporaries as one of the most eminent men in Sweden. He was at the origin of projects of public interest, such as the collection and classification of statistical data on the Swedish population. Glas reported the following excerpt from Emmanuel Svedberg's foreword dated October 23, 1715, to the *Dædalus hyperboreus*, the Society's quarterly bulletin, printed in Uppsala in 1716:

> Some scholars of Uppsala had exchanged for five years their thoughts with Mr. POLHAMMAR and received answers containing profound thoughts and mentioning new inventions and machines, to enlighten mechanics in general as well as general and special physics, astronomy and even political economy (ibid., p. 7).

Christopher Polhammar was a Swedish scientist, inventor and industrialist, who played a large part in the dynamics of the constitution of the Society, first called Collegium Curiosorum, officially founded in 1710. In astronomy, it was Pehr Elvius who held the chair of astronomy at the university, joining the society at its creation. The scholars of the new society, most of them devoted to the study of mathematics or natural sciences, set themselves the primary objective of opposing the superstitious beliefs that were omnipresent in Swedish society at the time, in the spirit of the Enlightenment. We do not have the minutes of the works of the society during the first years of its existence. Svedberg published the *Dædalus hyperboreus*, or "New mathematical and physical essays and remarks", a bulletin that can be considered as the first attempt of Proceedings of the Society of Sciences, and which appeared quarterly during the years 1716, 1717 and 1718 in six small issues, written mainly in Swedish. Mathematics, physics, political economy and astronomy were the main subjects treated in these bulletins. To give a new impulse to the Society, which felt the state of misery the country was experiencing, Benzelius organized on November 26, 1719 a meeting of the members, including the astronomer Burman, during which it was decided to form a Literary Society (in Swedish *Bokvettsgillet*), and to publish an academic journal in Latin on the model of the German *Acta*

Eruditorum Lipsientia and the French *Journal des Sçavans*, newsletters on current scientific activity and discoveries:

> The Society was to take note of everything printed in Sweden concerning literature and to give a brief account of it; everything new and useful was to be quoted, such as the search for truth in science and history; and finally, information on the life and writings of scholars, news of their death, etc. The sessions were to take place every Friday at 5 o'clock in the evening; the member who would have missed three times in a row without major reason was to be considered as excluded by himself from the Society. Every three months a part of the Acta was to be published in Latin under the title of Quarter, and on February 15, 1720 the first volume was delivered to the public, of which E. BENZELIUS wrote the foreword (ibid., pp. 10–11).

The *Acta literaria sveciæ* contained (i) reviews of works of interest (gradually disappearing to make room for scientific memoirs), (ii) news relating literary innovations and announcing the publication of scientific works and (iii) notices mentioning recent publications. These proceedings appeared in two volumes: one for the years 1720–1724, and the other for the years 1725–1729. Anders Celsius joined the new Society as soon as it was founded in 1719. In 1725, he became its Secretary, succeeding Erik Burman. The success of the Society abroad in the following years encouraged its members to enlarge its circle of activities to become a mathematical-literary Society. To this end, Benzelius wrote a proposal to the king in which he praised the merits of the Society:

> The works published during five years by the Society had conquered the esteem of the foreigner, to the point that they had been appreciated there especially by the Société Royale des Sciences of France and that of England, because these societies hoped for our assistance for the astronomical and physical observations, which are necessary to them as coming from our regions more close to the pole ... The astronomers can expect to be able to measure in our country the degrees of latitude of the earth to compare them with the measurements already carried out in Italy, in France and in England, in order to be able to evaluate the exact dimension and the shape of our globe ... The members of the Royal Society of London had welcomed with particular satisfaction the meteorological observations which had been sent to them by the Society and desired their continuation (ibid., p. 13).

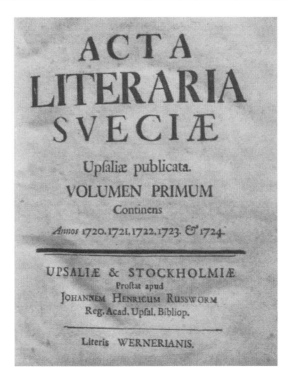

Figure 4.6. *Cover page of the first volume of* Acta literaria sveciæ *for the years 1720–1724 (Acta literaria sveciæ 1720–1724)*

On October 1, 1725, the project was presented to the king, who approved it in its academic dimension, but did not grant the requested means, the finances of the Kingdom being at their lowest. The scholars obtained only postal exemption for their national and international exchanges. Understanding that the lack of political support for the Society was a handicap for the accomplishment of its projects, Benzelius proposed on August 18, 1727, during the last Assembly of the Society which he attended after taking office as bishop in Gothenburg, to elect a senator as Honorary President of the Society (*Præses illustris*). It was Count Arvid Horn, then Chancellor of the University of Uppsala, who was chosen for this role. The Society was named Societas Regia literaria et scientiarum, and its draft statutes were given royal assent on November 11, 1728. The royal decree indicated, in particular, "the objects on which the activity of the Society must be focused, namely not only letters in general, but mathematics, physics, natural history, chemistry, the culture and the economy of the country, the improvement of the foundries, the progress of the mechanical arts, the astronomical and meteorological observations, the geography and antiquities of Sweden, the Swedish language, as well as botany, mineralogy and geology" (ibid., p. 16).

Anders Celsius became a member of the Royal Society and retained his position as Secretary. In 1730, the *Acta literaria* became annual (instead of quarterly), and focused on the publication of scientific papers. During Celsius' European trip, the position of Secretary was entrusted successively to two professors of the university. Less dynamic than Celsius, they let the activity wither away. Only one meeting of the members was organized in 1732, five took place in 1733, but none in the following years. The Proceedings ceased to appear. Upon his return, Celsius revived the Society, with the help of Hiorter who officially became his assistant in this task. At the Society's Assembly of 1738, Celsius requested that his Honorary President call a singular meeting to discuss the future activities and development of the Society. The latter, Count Gustaf Bonde, invited the members of the Society to a two-day seminar in his house to identify ways to revive the Society. The main conclusions of this meeting, which led to the creation of the Royal Swedish Academy of Sciences in Stockholm the following year, were:

> Among the many other problems to be dealt with, Captain Triewald suggested that the *Acta* should henceforth be published in Swedish (until then the journal had been published in Latin), arguing that the Royal Society in England published its reports in English. Bishop Benzelius objected, pointing out that the Swedish language was less widely known than English. Triewald put forward another proposal: to leave the *Acta* in Latin but to found a Swedish journal in parallel. Although the Society did not reject this proposal, the Swedish-language journal did not see the light of day and soon the Swedish Academy of Sciences was founded which began to publish its own bulletins.
>
> Triewald's proposal was still of interest, however, because Celsius took it up again in 1741, suggesting to follow the example of the Danes and to publish the journal in Swedish. This proposal was accepted and Celsius' uncle, Olof Celsius, became the editor. Unfortunately, the journal was not published until 1742.
>
> At the meeting at Count Bonde's house, Triewald had also proposed that the members living in Stockholm found a branch of the Society there. This proposal was accepted, and Triewald, Linnaeus, von Höpken, Bielke, Alströmer, and Cederhielm met on June 2, 1739 to found the Swedish Academy of Sciences (Viik 2012, p. 13).

On January 15, 1742, the Royal Society of Sciences of Uppsala presented to the king a project signed by eight of its scholars, among them Celsius and Hiorter, aiming to allow the election, in addition to its *Præses illustris* and its 24 members, of "great lords of the Kingdom as honorary members and twelve foreign scholars as corresponding members" (Glas 1877, p. 20). Among the foreign scholars chosen, several Frenchmen were to be counted: Maupertuis and Clairaut, companions of Celsius during the expedition to Lapland (the first taking a few years later the presidency of the Academy of Berlin), Mairan and Delisle, and two botanists, Bernard de Jussieu and François de Sauvages, member of the Royal Society of Montpellier. The Society had three officials: the Secretary, the Treasurer and the Librarian. The Secretary was permanent, as in the other great European academies. His role was defined as follows:

> The roles of the secretary consisted in writing the minutes of the meetings and in addition to maintain, on the order of the Society and in its name, internal and foreign correspondence, to preserve and file all the proceedings and documents belonging to the Society, by providing them with the registers and journals which concern them. Should the Society receive from the Province communications written in Swedish which were deemed worthy of being delivered to the printer, he had to arrange for their translation into Latin. Moreover, he had to give an account of the works of mathematics and physics appearing in Sweden. Finally, it is his duty, as well as that of another member of the Society, to receive and keep the funds of the Society, to use them for current needs according to the decision of the assembled members, but not without the authorization of the *Præses illustris* when it was a question of large expenses; every year he had to give an account of the use of these funds. In order that the secretary may more easily fulfil the duties incumbent upon him, the Society appointed an assistant (ibid., p. 31).

The work of the Society was published in the annual journal *Acta Societatis Regiæ Scientiarum Upsaliensis*, which was published until 1751, and then ceased for the next 20 years under the secretariat of Carl von Linnaeus, who took over from Celsius after his death, due to a lack of both scientific papers in sufficient number and financial means. The lack of means constituted a recurring situation that Benzelius, then Celsius, had to face. In 1716, Benzelius had obtained from the Queen the authorization to dig up and sell cast iron water pipes that had become useless, which had brought in 9000 thalers to the society. In 1725, Benzelius tried to convince the king to grant the Society the exclusive right to manufacture and sell almanacs, a monopoly that would result in "greater accuracy in the indications of these almanacs and the elimination of all the baseless conjectures leading to superstition that one usually encounters in them, errors that cannot be erased to the honor of the nation without harming the authors or publishers, as long as these almanacs were published by several people" (ibid., p. 39). But this request, taken up again in 1738, never succeeded, and the financial situation of the Society remained extremely precarious, the Proceedings being published at the expense of its members.

4.4.3. *Relations between the Royal Society and the University*

Olof Glas pointed out the dual role of the University in Sweden: "to instruct the youth in the knowledge which is necessary to enter the service of the State" and to aim also "at the interests of the scientific investigation, to the progress and to the maintenance of which it must devote itself" (ibid., p. 49). For that, the university was organized in faculties. These faculties were in charge, not only of "the distribution and the teaching of the various branches of instruction, so that it satisfies what the State requires of the higher instruction of the young people", but also of the scientific research in the taught disciplines, and the development of the "reciprocal relations" between the various fields of research, in a perspective that we can qualify as interdisciplinary. Thus, "a learned Academy and a scientific association have only one goal, the progress of science, and it is for this reason that the members who are part of it can be considered as collaborators of the teaching body of the University in scientific investigation" (ibid., p. 50). The proximity of the University and the Royal Society in the same city facilitated the exchanges between the two institutions, and explains the strong influence they exerted on each other. The influence of the University on the Royal Society was directly reflected in the fact that the members of the Society were professors at the University. In the other direction, the first example of the University's contribution to the Society was the Astronomical Observatory. The first attempt to set up an observatory in Uppsala was made by Erik Benzelius in 1716. He wrote on April 2 of that year in a letter to Svedberg:

> As for the observatory, things have reached the point where the Governor has promised to recommend to His Majesty that the best keep of the Castle be repaired for this purpose: there are enough bricks to be taken from the neighbourhood, and the communal wood will provide us with the beams and other construction wood that may be necessary. As for the means to cover the costs of repair, I found them in the ground, I mean the big underground cast iron pipes which were used to lead the water of the Fyris from the mill to the Castle: they are only rusting now. There are also some excellent metal pipes that are of considerable value and that can also be used. The first pipes will be used at the Wattholma factory, the last ones at the Stockholm foundry. We will receive from the Library all that it possesses in the way of

instruments, while waiting for better. As for the rest and the annual purchases, I thought to provide for them thanks to the monopoly of the almanacs, that is to say that only one author will publish them (ibid., p. 54).

It is known that some money was finally obtained from the sale of cast iron pipes, but the request for a monopoly on almanacs did not succeed. In 1723, the astronomer Burman tried again to obtain means for the construction of an Observatory. He obtained from Jacques Cassini a description of the Observatory of Paris that he added to the report presented to the State. There again, the approach was unsuccessful, but it had the merit of attracting the attention of the government and the Academic Council of the University on the project. The project finally saw the light of day at the end of the 1730s, under the impulse of Celsius, as we saw, and the Observatory of Uppsala, finally realized, was operational in 1742.

A second contribution of the Society was a project for a chair of Physics at the University, a discipline previously taught only in the framework of the chairs of medicine and mathematics. The Society proposed that the chair of poetry, vacant in 1729, be transformed into "a chair of history and natural philosophy (or of physics)". The main reasons in favor of the creation of this chair, as inscribed in the project presented to the Academic Council of the University on May 17, 1729, were:

> 1) Considering that the economy and culture of the country are based not only on mathematics and mechanics but also on physics, it seems that students should also be educated at the Academy in such a useful science;
>
> 2) Considering that the prosperity of the Kingdom does not depend essentially on poetry, this one can be cultivated at the University with other disciplines less necessary and be combined with another chair that has some affinity with it;
>
> 3) Considering that there is no great university in Europe, except Uppsala, where there is no chair of history and natural philosophy;
>
> 4) Considering that it is not appropriate for physicians to treat physics, as is customary in Uppsala, since they already have enough to do with the teaching of the branches which are especially their responsibility (ibid., p. 56).

The university blamed the Society for this project, believing that it arrogated too many rights to itself, and the project was not retained. We have seen the difficulties encountered by Celsius and Klingenstierna in affirming their adherence to Wolff's natural philosophy in this period, and the censorship instituted by the Chancellor of the University at that time, and this failure was therefore perfectly foreseeable. The chair was re-proposed in 1736 by Anders Celsius, the University being then still under the control of the same chancellor hostile to natural philosophy, who would end his term of office the following year. Celsius demanded for the project "to be carried out three years after the chair of poetry became vacant, in order to use the salary saved during this time for the purchase of indispensable instruments". He proposed that the chair of poetry be abolished, or merged with the chair of eloquence, but the approach was not successful either. It was not until about 20 years later that the teaching of physics was finally recognized at the University of Uppsala by the creation in 1755, 11 years after Celsius' death, of a specific chair. This chair was entrusted to Klingenstierna, precisely the introducer in Sweden of Wolff's thought, marking the victory of the new philosophy. As Glas noted:

> Remarkably, the time between the Society's first projects and their realization was about the same in both cases. Indeed, the Society's first proposal for the establishment of an astronomical observatory (in one of the towers of the Château) was submitted to the States in 1716, and the observatory was not completed until 1741; on the other hand, the first project for the creation of a special chair for the teaching of physics was presented in 1729, and this chair was not created and occupied until 1755 (ibid., p. 57).

The Royal Society also contributed to strengthening meteorological observation, and to give it a scientific rigor, following the efforts of the Royal Society of London to encourage a coordination on a European scale at the beginning of the 1720s, as we have seen. Numerous scientists had learned about thermometers, and considerable efforts were made so that the temperatures provided by the various thermometers, having different systems of graduation, could be compared on a single scale. Burman, then Celsius, and Wargentin, produced long series covering the years 1722–1731, then 1739–1769 and then all years beyond 1774. The Royal Society also worked from 1726 on a project to improve the teaching of mathematical and physical sciences in schools. Samuel Klingenstierna wrote a report on the method to be followed for treating physics and related subjects in the classroom. The departure of Benzelius, a supporter of the project, who left in 1727 the University of Uppsala to join the diocese of Gothenburg, delayed the realization of it. Glas still quoted to the credit of the Society the realization of an astronomical calendar published by Hiorter in 1739, the coordinated campaign in all Sweden of the observation of the eclipse of the sun of May 1733, the works on the geography of Sweden, as well as linguistic studies of northern dialects. The Society also financed trips abroad, such as those of Linnaeus, in 1729 and 1732, or of Hiorter in the central provinces of Sweden.

4.4.4. *Celsius' legacy*

From June 1744, Delisle resumed with Hiorter the correspondence interrupted by the death of Celsius. The main subjects of discussion remained the same, concerning the estimation of longitudes by the observation of eclipses of Jupiter's satellites, or occultations of stars by the moon, and the calculation of the trajectory of comets according to the rules of the Newtonian mechanics. In his letter of August 15, 1744, in which he agreed to take over from Celsius as Delisle's correspondent, Hiorter said that he had identified a parabolic trajectory for the comet that did not deviate by more than one minute of arc from the observed position, both in longitude and latitude. In his answer to Hiorter, Delisle asked him to communicate all the elements relative to the orbit of this comet: time of passage to the perihelion, inclination of the orbit on the ecliptic, place of the node, distance of the perihelion, his goal being to supplement the table of comets established by Halley.

In November 1744, Pehr Wilhelm Wargentin, who had completed his thesis on Jupiter's satellites under the direction of Celsius, and was then 27 years old, sent his first letter to Delisle, marking the beginning of a sustained correspondence that would last nearly 15 years. Wargentin, in this letter of November 24, confided to him that Celsius had the intention of sending him to the Observatory of St. Petersburg to work with him, a step that the rumor according to which Delisle was departing to return to Paris dissuaded him from undertaking. He said that if Celsius had lived longer, he would have made the Observatory of Uppsala one of the best in Europe, and noted that the successor of Celsius at the direction of the Observatory, his associate and brother-in-law Hiorter, was rather a mathematician than an astronomer. A first exchange of observations on Jupiter's satellites and the eclipses of the sun began. In a letter of March 27, 1745, Wargentin confided his fears of a decline of astronomy in Uppsala, following the death of Celsius:

> Since the beginning of this year, I have not been in Uppsala, if only for a few days. But I am quite sure that nothing was observed there, because Mr. Celsius is no longer there. I am afraid that astronomy will be reduced to the same state in Uppsala, after the death of Mr. Celsius, as it was in Copenhagen after the death of Roemer, and in Berlin after the death of Mr. Kirch.

He obviously did not have access to the instruments used by Celsius (which had to pass, as well as his library and his observation books, into the hands of Hiorter, his official successor), and said he did not have the means to acquire his own instruments. These exchanges show the extreme fragility of the European astronomical observatories, for some of them built on the only impulse of a renowned astronomer who, after his death, did not leave an heir able to take succession. In his letter of June 12, 1745, Wargentin pointed out to Delisle that

Celsius' observations of the comet of 1742 could not be printed, because of the positions of small stars not included in Flamsteed's catalog necessary for an accurate reconstruction of the trajectory. However, the manufacturing of the large wall quadrant ordered by Celsius to Ekström was interrupted by the death of Celsius, and even "supposing that Mr. Hiorter had enough instruments, I believe that it will not be easy for him to recover these small stars". He said about the latter:

> For the last great comet, Mr. Hiorter assured me that he would write to you physically after my departure from Uppsala, and that he would communicate to you the principal points of the observations of Mr. Celsius on this comet, with the true situation of its orbit, determined on the theory of Mr. Halley. You can judge from this the industry of the one in observing, and the strength of the other in astronomical calculations. It takes a Celsius and a Hiorter to a well-stocked observatory. That is why Celsius needed Hiorter: but Hiorter also needs a Celsius. It is not that Mr. Celsius could not calculate, or that Mr. Hiorter could not observe.

In the same letter, Wargentin confirmed to Delisle that it was Hiorter who officially took the direction of the Observatory, the position of professor of astronomy at the University left vacant by Celsius being occupied by Martin Strömer, who was not an observer. The same month, Hiorter wrote to Delisle that he would send him the literary proceedings of the Society of Uppsala that the latter had asked Celsius for before his death. Hiorter also told him that he would establish the true orbits of seven comets: 1698, 1702, 1706, 1707, 1718, 1729 and 1739, and try to determine the periodic times of these comets, as Halley did. Wargentin wrote to Delisle, on January 19, 1746, that Hiorter continued his research on the last comet, whose parabola he converted into an ellipse and estimated its period at 345 years. He also told him that he obtained, supported by Strömer, a small sum of money to visit the main observatories of Europe, and that he wished to start with the one of St. Petersburg, so he wanted to know when, exactly, Delisle would leave Russia. The activity of the Observatory, under the direction of Hiorter, degraded quickly, as Wargentin explained it, about the observations made in Uppsala, in his letter to Delisle of February 27, 1748:

> But since you are pressing me, we must confess to our shame that we are doing almost nothing. Mr. Hiorter has taken on a commission, which almost entirely prevents him from making observations. Mr. Strömer's health is too delicate to try the practice of observations. As for me, I have enough leisure and strength to observe, so that I would do nothing better than to occupy myself with such useful work. But it is not permitted to be at the observatory alone. I can only enter when Mr. Hiorter goes there himself. Thus, the instruments, which Mr.

Celsius collected with such care, remain without use. But it seems that Mr. Strömer's health will recover, and that Mr. Hiorter will have more leisure to attend to his duties. While waiting for this blessed time, a large wall quadrant is being made, to be in a better condition to start something of importance.

After the troubled period which preceded his departure from Russia, and the few months of his return to Paris, Delisle resumed in June 1748 his correspondence with Hiorter, concerning Jupiter's satellites, the orbits of comets, and also the occultation of May 1737 of Mercury by Venus, as well as the last eclipses of sun and of the moon. In August 1748, Delisle complained to Wargentin of the inconstancy of Hiorter's responses, and indicated to his interlocutor to have asked and obtained that Wargentin become the correspondent of the Academy of the Sciences of Sweden (created in 1739 in Stockholm) to the Académie Royale des Sciences of Paris. In his letter of November 6 of the same year, Delisle asked about the relations between Wargentin and Hiorter:

> I have not been less satisfied with the few that you sent to me in Uppsala, but I believe that their scarcity comes only from the fact that Mr. Hiorter is not very communicative. But since you are back in Uppsala and have begun to practice astronomy, can you not make your observations in Mr. Hiorter's observatory, which would put you in a better position to make more exact observations of all kinds and at the same time more useful, and which would also provide me with the means of obtaining through your correspondence those which I would need most; but in whatever way you trade or reside with Mr. Hiorter, if he is not in a position to do so, I would like to ask you to do so. Hiorter, if he is so reserved with regard to you that he communicates to you only with difficulty what he does, I promise you to reserve to myself alone without making any public use of what you may get from him.

In a letter written at the end of 1749, Delisle asked Wargentin if, as a French member of the Royal Society of Uppsala as well as of the Swedish Academy of Sciences, he could not intervene officially with the Society of Uppsala to ask that Wargentin have access to the Observatory when Hiorter was not there, and could make observations there, taking full advantage of the good instruments that Celsius had installed there. But an important event puts everything in question. On November 28, 1749, Wargentin announced to Delisle his nomination as permanent secretary of the Swedish Academy of Sciences. He told him that he had settled in Stockholm, where he was more comfortable and felt he was better able to serve him. He pointed out to him the construction in progress of an astronomical observatory in Stockholm:

> The Academy is building an observatory in Stockholm, whose situation is infinitely more advantageous than that of the observatory in Uppsala. The academy will not spare any expenses either, to provide me with the instruments necessary to observe the stars well. Thus I dare to flatter myself, that I will be able one day to contribute something to the advancement of astronomy. It is true that my observatory will not be habitable in two years, but I do not pretend to remain idle until then. With the few instruments that I already have, I will not fail to do all that I can.

In the same period, around the middle of November 1749, Delisle received the official diploma of his association with the Academy of Sweden, for which he thanked Wargentin in a letter dated December 18 of the same year. Each man was thus an associate member of the Academy to which the other belonged, this link between scholars sealing an institutional link between their Academies:

> Our academy of sciences in Paris is very satisfied with the association it has just contracted with yours and it has asked me to assure you that it will send you a copy of what it produces from now on. It eagerly awaits the collection you have promised of all the memoirs of your academy and those of Uppsala with the other books you judge suitable for it, to which I beg you to enclose a copy of all the maps engraved in Sweden up to now. All that you will send it will be carefully kept in its archives for the use of all its members.

In the same letter, visibly tired of Hiorter's unproductivity and of the communication difficulties that he experienced with him, Delisle suggested to Wargentin that he take in hand the collection and publication of all the observations made in Sweden:

> I was pleased to see from the observations you kindly communicated to me in your last letter that they are being continued in Uppsala at the same time as in Stockholm and that Mr. Strömer is in charge of those in Uppsala. Are we thinking of replacing Mr. Hiorter? Or will Mr. Strömer be able to do the two jobs that were shared with Mr. Hiorter? As the latter was not very communicative, as you yourself have admitted to me, and as I have recognized by his letters, can we not hope that his observations will be collected and published with those of Mr. Celsius? The latter had once written to me that he intended to collect all the old observations made in Sweden in order to publish them; this is a work that would suit you and your academy in Stockholm perfectly.

This letter also speaks about the voyage of Nicolas-Louis de La Caille, or Lacaille, to the Cape of Good Hope and the campaign of coordinated observations which had to be organized in Europe, a subject that occupied a significant part of the correspondence between the two men over nearly three years. The purpose of these concerted observations was to specify several important quantities, such as the parallaxes of the Sun, Moon and some planets, the obliquity of the Ecliptic, etc. Wargentin, in his letter of October 30, 1750 to Delisle, welcomed the association between the Royal Academy of Sweden and that of Paris, "the mother of the academies". He said about this association that the Swedish academy "will try to make itself worthy of it, and although it is still very weak, it will endeavor to contribute more and more to the growth of science". But the fields of research in the northern countries were specific, oriented towards the economy, and it took some time for the natural sciences to acquire a more significant place:

> The main goal of our academy has been since its foundation to encourage the economy in our country. That is why our memoirs are mostly on economic subjects, which are of little interest to you, being almost exclusively applicable to our climates. It has, however, tried to intermingle something of the other sciences in its memoirs, and when the taste for science is more universally introduced in the North than it has been up to now, our memoirs may become more interesting. However I will make my efforts, that the *Académie Royale* of France will not have cause to repent of its generosity: for I will always join to the memoirs of our academy, those of the society of Uppsala and other books which I will judge worthy of your curiosity.

Wargentin ended his letter by mentioning the death of Hiorter in April. In a letter of January 1751, Wargentin points out that the Observatory of Stockholm would not be in working order for two years, not allowing him to envisage many observations of quality in support of the Lacaille expedition, and that in addition, Strömer in Uppsala was not well equipped. He asked Delisle if the Academy in Paris would be ready to support the request of exemption of teaching made by Strömer to observe. He mentioned other observers in other places in Sweden. Two months later, on March 16, 1751, Wargentin wrote to Delisle that the steps taken by himself and Strömer with the Swedish Academy and the government to obtain support for the campaign of observations coordinated with those of Lacaille were crowned with success. "The king has just granted the Academy more than it had dared to ask. He promised to provide all the necessary expenses for the instruments and for the trips that the academy would judge necessary to make on this subject, and he even promised to reward graciously all those among the astronomers who will distinguish themselves in making good observations". But this political success was not accompanied by a fleet of instruments ready to operate at the highest of stakes, continued Wargentin. Ekström worked alone, and his ability to deliver high-performance instruments on time was limited.

A dramatic event came to thwart the preparation of observations. A fire ravaged a part of Stockholm on June 8, 1751, and reduced to ashes the house where Wargentin had installed his small personal observatory, while waiting for the official observatory to be ready. Wargentin wrote to Delisle that the library of the Academy and the instruments had been saved, but that he had no place to observe. Delisle, impatient to receive the Swedish observations of Wargentin, and unaware of the fire in Stockholm, was sharp in his letters. Wargentin, who also had to face bad weather conditions, answered him on October 8, 1751, in a letter which said a lot, as well on his abnegation as his perseverance to honor his engagements:

> Indeed, far from scolding me, as you do in your letter of August 15, you are right to pity me when you know that I have been obliged to change my place of residence three times during this summer, and that I will change it again on the fourth. Judge for yourself, Sir, the disorder in which the disastrous fire has put my affairs and if it has not been possible to be very exact in maintaining correspondence. In spite of all this, I have not missed any important observation, which clouds or other obstacles of the same nature have made possible. It is true that there are many phenomena between those that Mr. Delacaille has marked, which I have not observed, but the last two months, which are usually the most beautiful months of the whole year in Sweden, have been so cloudy and rainy, that most of the phenomena that we had to observe, have entirely escaped us, to our great regret. For me in particular, not having had the convenience to make the observations with the shelter of the insults of the wind, I was obliged to observe in a garden, with open sky, where the winds have contributed only too much to reducing the number of my observations and to make doubtful a part of those that I have managed to make.

The bad meteorological conditions, the lack of quality instruments and the Stockholm fire are all reasons that led to the observation campaign carried out in Sweden in support of the Lacaille observations not being a clear success, even though the support obtained from the Swedish state was already a success in itself. Several Swedish observers operated. The observations made were "curious and useful", noted Delisle in his letter of March 2, 1753.

It was then on the next transit of Mercury that the correspondence rolled. On May 8, 1753, Wargentin wrote to Delisle to be finally installed in the new observatory. He wrote that he had done everything possible to encourage all his correspondents to observe the transit of Mercury. He instructed them on how to do it with the few instruments they had: Gadolin in Abo, Hellant in Tornea, Gissler in Hernogard, Strömer in Stockholm, Schenmark in Scania. He did not know what the observation gave for his other correspondents. As far as he was concerned, the weather was unfavorable. He saw the sun only around 11 a.m., when Mercury was about to leave the disk. He did not have time to process his observations, nor to reduce them to true time. But, he said, as a convinced Newtonian, "I was at least for my satisfaction, convinced of the preference of the tables of Mr. Halley, on those of Mr. Cassini at least for what concerns Mercury". And he concluded, to the address of Delisle, by these few words, which embody well alone the dream of these pioneers of astronomy that were the astronomers who built the first observatories at the beginning of the 18th century:

> Nothing more useful for the advent of astronomy, than the design you have of publishing all the astronomical observations of all species, each species separately, according to the order of time, with remarks and results on each. This will be the true history of astronomy. I wish with all my heart that heaven will keep you strong enough to accomplish it. It is such a work that I have always desired, and I have long been convinced that it is only from you that it can be expected.

4.5. Conclusion

Thus, it was only in the middle of the 18th century that the Observatories of Uppsala and Stockholm finally came into existence, after 30 years of initially fruitless efforts. It took about as long to achieve the creation of the chair of physics at the University of Uppsala, showing the difficulty that experimental sciences had in imposing themselves in a context strongly marked by religious imprint, despite the Cartesian reaction which appeared at the beginning of the century, and the influence exerted by the Newtonian theory on Celsius and his entourage. The first half of the century was marked by the constant efforts made by Celsius to bring the Royal Society of Uppsala, created by Erik Benzelius at the turn of the 18th century, into existence and to give it the international dimension necessary for its development. The study of astronomy and more generally the observation of nature, during the first part of the century, were very brief; the scientists, and the astronomers in particular, operated only with minimal, even rudimentary means. Thanks to his European connections, and in particular to the expedition of Lapland, Celsius succeeded in building up, in addition to a solid reputation which helped him in his enterprise, a stock of some astronomy instruments, some financed from money from his travel grant, or directly by Uppsala University, others by its foreign partners. He also sent a craftsman from Stockholm to train in London with the instrument maker Graham, thus preparing for the future of astronomical observation in Sweden. In this context of chronic shortage of resources, Swedish astronomers such as Celsius or Wargentin deployed considerable energy to give the small community of Swedish astronomers a role within the concert of the great European nations, while facing the particularly difficult conditions of exercising their profession in the severe climatic conditions of the northern countries, and they succeeded. At the same time, through the academic societies in which they participated, and in which they occupied eminent roles, they tirelessly appealed to politicians to obtain from them the means which they lacked, and their efforts were in certain cases crowned with success. The short life of Anders Celsius did not allow him to develop the Uppsala Observatory as much as he would have liked, but he paved the way, and the work was continued by his successors. Swedish astronomy in the middle of the 18th century was doing better than a few decades earlier.

5

Genesis of the Academies of Bologna and Berlin, the Involvement of Women in Astronomy and the Gender Issue

5.1. Introduction

Anders Celsius, during his European trip, visited several large observatories, including the one in Berlin, where he spent the winter of 1732–1733 in the company of Christfried Kirch and his family. About this visit, Alphonse des Vignoles, member of the Royal Prussian Academy of Sciences in Berlin, wrote in his Eulogy of Christfried Kirch:

> At the end of this year 1732, Mr. André Celsius, Royal Astronomer in Uppsala, & another Swedish scholar, named Mr. Meldercreutz, arrived in Berlin, & were lodged in a House, which almost touched that of Mr. Kirch: which gave us occasion to see them often, to converse familiarly, & to eat frequently together, over several Months, that they remained in this City (des Vignoles 1741, p. 335).

Thus, Celsius, as an immediate neighbor, got to know the entire Kirch family, including the Kirch sisters Christine and Margaretha, who lived in the same house and also practiced astronomy. After Berlin, he moved to Italy in stages and spent seven months in Bologna, where he worked and observed with Eustachio Manfredi, the founder of the Accademia degli Inquieti (literally the "Academy of the Restless") in 1690, whose merger with the Institute of Bologna created by Luigi Ferdinando Marsigli led to the creation of the Academy of the Institute of Bologna, that we will sometimes call for simplicity Academy of Bologna, in 1714. Like Kirch, Manfredi was helped in his work as an astronomer by his two sisters

Maddalena and Teresa, and, as in the case of Kirch, the siblings lived under the same roof.

After Bologna, Celsius went to Paris where he met the mother and sister of Joseph-Nicolas Delisle, who, as we have seen, was stationed in St. Petersburg. In a letter he wrote to Kirch in 1734, he said:

> By chance or by astronomical instinct I find myself lodged in the same house where the mother and the sister of Mr. De L'Isle live, this young lady is very learned in astronomy and she is worthy sister of such a famous astronomer. She even attends the Assemblies of the Academy of Sciences. I spend the evenings very pleasantly in her company […] I begin to believe that it is a plan, that all the astronomers that I have the honor to know in my journey, have their sisters, and those learned ones (quoted in Lémonon Waxin 2019, p. 274).

Celsius was struck by the presence of women around the astronomers with whom he had met in Berlin or Bologna, or in Paris in the Delisle family. The baron of Zach, in his astronomical correspondence (1819), evoked the passage of Anders Celsius in Bologna, by insisting on the interest that he carried in the presence of many women whom he judged learned in the entourage of his astronomical colleagues:

> *Manfredi* was then 60 years old, *Celsius* 33, he respected in *Manfredi* his master, to whom he was tenderly attached. *Celsius* had made his first steps in practical astronomy in Berlin, under the famous *Christfried Kirch*. He wrote from Bologna to his friends, that he had found there a second Berlin, and in *Manfredi* a second *Kirch*. In Paris he had been very well received by M. *De l'Isle*. The latter had a sister devoted to Astronomy, *Kirch* had some very learned ones, who helped him to calculate ephemerides. *Manfredi* had two sisters who followed celestial movements, which made *Celsius* say in one of his letters to *Kirch*: "I am beginning to believe that it is destiny that all the Astronomers I have the honor to know in my journey have their learned sisters; I also have a sister, but she is not very learned, it is therefore necessary to make her an astronomer too, to preserve harmony" (von Zach 1819, p. 463)

We note, through these lines, the image of a big family that Celsius gave of this community of astronomers, women and men, gathered around his three colleagues, Kirch, Manfredi and Delisle, whom he met during his trip. Delisle had in fact passed through Berlin during his trip to St. Petersburg, which Vignoles noted in connection with Kirch, the visits of Delisle and Celsius to Berlin being the only two he mentions in his long eulogy of the German astronomer:

> At the end of the Winter & of the year, Mrs. les Frères de Lisle & de la Croyère, going to Petersburg, stopped, a few weeks, in Berlin, where I had the pleasure to talk & eat with them, more than once. On the 18th of January 1726, the sky being very serene in the evening, they went up, with Mr. Kirch, on the Terrace of the Observatory. But the cold being excessive, they went down into one of the rooms, from where they observed, that at 9 o'clock, the Planet of Mars came out of the Moon, which had hidden it. This Observation was also reported in the Memoirs of the Académie des Sciences de Paris (des Vignoles 1741, pp. 325–326).

At least in the cases of the Kirch and Manfredi families, the women were closely associated with skywatching, and as such also observed the aurora borealis. This fact is proven in the case of the Kirchs, and more precisely in the case of Christfried Kirch's mother, Maria Margaretha Winkelmann, wife of the astronomer Gottfried Kirch, and herself an astronomer. On November 4, 1707, she wrote a letter to Gottfried Wilhelm Leibniz, president of the Berlin Academy founded in 1700 by Frederick I of Prussia, to describe her observation of an aurora borealis. In his letter, Winkelmann drew Leibniz's attention to the aurora borealis "as my husband has never seen them" (Schiebinger 1987, p. 182). Vignoles, in his eulogy of Mrs. Kirch, specified the conditions of the observation of the aurora on March 6, 1707. On that night, as on all others, Gottfried and Maria had divided the sky, she watching the northern side, he the southern. And it was Maria who discovered the aurora. Vignoles explains that in the first published account of the aurora borealis, Kirch did not dare to name his wife, "as he did afterwards, when it was made clear to him that he should have no qualms about naming her" (des Vignoles 1721, p. 178), probably for fear of reproach from his employer, the Royal Prussian Academy of Sciences. And indeed, at the international level, the discovery is attributed to Gottfried Kirch, as attested by the report of Jacques Philippe Maraldi:

> There is still in these Memoirs [those of the Royal Society of Sciences of Berlin] another Observation of the same phenomenon, made the same day 6 March 1707 in Berlin by Mr. Kirchius. This astronomer observed it at 8 o'clock in the evening in the form of a Rainbow, but wider, whose length occupied at the horizon about 100 degrees. The superior part of this Arc was elevated from 8 to 10 degrees on the horizon, from where came out luminous rays that were directed towards the Zenith. Then on the first arc he saw a second one appear at the height of 30 degrees, but it was not well finished nor well continuous (Maraldi 1716, p. 105).

Maraldi, from observations of the same aurora made in different places – Berlin, by Winkelmann, Copenhagen, by Ole Christensen Rœmer, Paris, by himself – deduced a height of the top of the auroral arc included between 15 leagues (60 km) and a value three times larger. The famous aurora of March 1716, which was observed by Edmond Halley in London, as we have seen, and inspired his aurora system, was seen in Danzig, from the Observatory of Johannes Hevelius, where Maria Winkelmann-Kirch and her son Christfried, as well as her daughters Christine and Margaretha were then living, the father, Gottfried, having died in 1710. The Norwegian naturalist Joachim Frederick Ramus in his dissertation provided a series of drawings based on observations made precisely in Danzig (Brekke and Egeland 1983, p. 59).

Figure 5.1. *Some examples of Ramus' drawings of the aurora borealis of March 17, 1716 as seen from Danzig (Brekke and Egeland 1983, p. 59)*

Vignoles specifies in his eulogy of Christfried Kirch, about this aurora borealis:

> The Great Czar of Muscovy was a spectator in person, with all his Court. This Prince did not disdain to attend other Observations of Mr. Kirch; & was so pleased with the work of this young Astronomer, that he made him offer an honorable Vocation for Moscow. Mr. Kirch would have gladly accepted it: but his Mother could not bring herself to go to Muscovy, nor to separate herself from her Son (des Vignoles 1741, p. 311).

Vignoles mentions the tsar's proposal also in his eulogy of Mrs. Kirch (Vignoles 1721), but presents it in substantially different terms, indicating that it was to Maria Winkelman that the invitation was first addressed:

> But she received there [in Danzig] a thousand honesties from the Daughter, & other Heirs of the famous Astronomer Mr. Hevelius. Her stay in Danzig was only eighteen months, or thereabouts, during which time the Czar of Muscovy tried to attract her to his States, and made her an offer. But Mr. Jean Henri Hoffman, Astronomer of the Academy of Sciences of Berlin, having died in the meantime, & Mr. *Christfried Kirch*, son of his predecessor, having been called to fill his place; Mrs. Kirch returned to Berlin, with all her Family, not having been able to resolve to separate from her Son (des Vignoles 1721, p. 183).

We can conjecture that the observations made in Danzig in the presence of the tsar were conducted jointly by Maria and her son Christfried, who was then only 22 years old and much less experienced than his mother, and perhaps even his older sister Christine, then 18 years old, and that the tsar's invitation to join Moscow was addressed to both mother and son; Christfried's recall to Berlin to fill the vacancy left by Hoffmann in fact led the family to return to the Berlin Observatory rather than accept the Russian proposal.

There is little doubt that the Manfredi sisters, initiated from their youth by their brother Eustachio to the practice of astronomy, were also involved in the observation of the aurora borealis seen in Bologna by their brother, as well as by his pupil and successor at the direction of the Observatory of Bologna, Eustachio Zanotti. It was Zanotti who provided to Jean-Jacques Dortous de Mairan the collection of the aurorae seen in Bologna, but Manfredi observed and described himself many auroras. In his *Journal d'Observations des Aurores Boréales*, Mairan mentions the very detailed description that Manfredi sent him of the aurora of April 10, 1734:

> But I was well compensated for these obstacles by the exact description that Mr. Manfredi sent me of what he had seen in Bologna on the same day and at the same time. He observed, undoubtedly, by a very favorable time, since from 8:34 am that he began to see the Phenomenon, until midnight, he noticed all the appearances of it almost every minute. I keep this detail preciously to use it on occasion, & I content myself with transcribing here what it contains of more essential in relation to the plan that I have made for myself in this Journal (de Mairan 1738, p. 775).

There are other mentions of observations of aurorae made by Manfredi, for example, in 1726 (*Dictionnaire de Physique* 1793, p. 352), or in the spring of 1728 (Cavazza 2002, p. 13), an aurora of which Manfredi informed the Royal Society. Manfredi, like Zanotti, Kirch, Delisle or Celsius, was, because of the long periods spent observing the night sky, an observer of the aurora borealis, and the family character of the practice of astronomy implies in his case, as in that of Kirch, a female participation of fact in the observation of the aurora borealis. As we said, this activity was regarded for all these astronomers as a "side activity", as was meteorology. But the point that we raise here, for the observation of the aurora as for those of the stars, is the family character of the process of producing observations, proven in the case of Kirch, implicit in that of Manfredi. It is different in the case of the Delisles, since the observation of the aurora borealis by Joseph-Nicolas and Louis was done in Russia, far from their sister Angélique, who remained in Paris with their mother.

In section 5.2, we describe three "astronomical households" (see, for example, Mommertz 2005, p. 161): the Kirchs, the Manfredis and the Delisles. We devote section 5.3 to the description of the astronomical academies and observatories of Bologna and Berlin, linked, respectively, to the Manfredi and Kirch households, and to their genesis. Section 5.4 examines, under the prism of the question of gender, the astronomical households and institutions analyzed in the two previous parts.

5.2. Three examples of "astronomical households"

The "astronomical household", as illustrated by the Kirchs or the Manfredis, represents an old form of scientific work inherited from the 17th century, which persisted in some cases during a large part of the 18th century, coexisting with the new forms such as the academies and the universities on which they relied (Mommertz 2005). This took different forms depending on the economic, social and cultural context, and national traditions. It was within households that women scientists were found at this time, as they were mostly excluded from the university system, and therefore from the institutions that were linked to it. Women of the literate urban middle class, like the Manfredi sisters, were educated by men of their family, a father, brother or husband, following the humanist ideal of the *puella docta*. Others, such as Maria Winkelmann, educated by her father during childhood, was then trained in the field by self-taught amateurs practicing astronomy as craftsmen, and she perpetuated her activity by marrying a professional astronomer, her husband trained by master astronomers, whom they helped to carry out his scientific work, thus developing their own scientific competence. Women such as Angélique Delisle, educated in families of scientists from father to son, acquired from their male environment general knowledge that allowed them to assist them by ensuring the stewardship of their life as scientists, replacing them when necessary when they were absent. Thanks to the scientific household, some women had access to education, and even to a more or less recognized research activity. The household was the basic cell, at the same time of production and reproduction, of the society of the time. It was a social unit, to which the community as a statutory corporation, like guilds or universities, can relate. It constituted the primary labor force of the members, men and women, of the family:

> Through kinship, trade, and loan relationships, as well as neighborhood ties, the household formed a network with other households in mutual dependency. Depending on the type of the household, different forms of sociability were typical. An organization form among household members that was based on mutual dependency was only possible as long as they worked within a hierarchical structure based on gender, age, and class. A king's court, a businessman's shop, the workshop of a craftsman's, a convent, etc. could be referred to as a "household" and function along similar principles (Mommertz 2005, p. 163).

The notion of a network of neighbors applies, for example, to the Manfredi and Zanotti families, whose male representatives, Eustachio Manfredi and Eustachio Zanotti, succeeded each other as head of the Observatory of Bologna in the first half of the 18th century. The Zanottis were a family of artists and scientists close to the Manfredis, whose father was a poet, painter and art historian, and whose brother was secretary of the Academy of Bologna. The sisters of the two Eustachio brothers were very close, companions in the art of embroidery and the composition of poems in the dialect of Bologna. The two households were in a cooperative situation, both from their entrance into the good society of Bologna, on the side of the women, and of the assembly of the astronomical science in the Academy of Bologna, on the side of the men. On a larger scale, the astronomical community formed at the European level by the network of households such as those of Kirch, Manfredi, Zanotti and Delisle, appeared strongly interconnected, as shown, for example, by the observation activity of the aurora borealis and the establishment of catalogs of observations, then collected and classified by Jean-Jacques Dortous de Mairan in Paris, or the meteorological follow-ups assured by these families, collected later by Louis Cotte to write his treatise on meteorology (Cotte 1774). Within this European network of households involved in the observation of the sky, and of the aurora borealis, as well as of meteorological variables, the labor force represented by the contribution of women was important, at least equal, if not superior to that provided by men.

We will now examine the forms of organization within the three astronomical households visited by Celsius, namely a household of artisan astronomers in the guild tradition represented by the Kirchs, an urban household, both artisanal and literary, represented by the Manfredis, and finally the Delisle family, where Joseph-Nicolas' stay in St. Petersburg and the premature death of his brother led the women to manage the scientific patrimony thus left behind.

5.2.1. *The Kirchs: an artisanal-type household inspired by the guild tradition*

This particular form, in Germany, was inherited from the tradition of the guilds, still alive at the time, in which women were officially admitted, and had some rights, such as the right to take over, under certain conditions, their husband's trade after his death (Schiebinger 1987). But, unlike men, women were not entitled to the years of companionship that allowed men to travel and train with different masters, and they had to train at home, often under the authority of their father. Astronomy, at that time, was more related to arts and crafts than to academic disciplines. Astronomers were both theoreticians and technicians of observation. In addition to theories of the cosmos or mathematics, they had to master the techniques of glass grinding, copper engraving and instrument making. They were more like guild masters and apprentices than scholars, and women, who were effectively excluded from universities, could thereby gain access to the techniques of astronomical observation, and thus to the science associated with it; in this case, neither Gottfried Kirch nor his second wife, Maria Winkelmann, learned astronomy and mathematics at university (Mommertz 2005, pp. 163–166). Kirch was the son of a shoemaker, was a schoolmaster, and trained in astronomy with masters such as Hevelius in Danzig. Winklemann was the daughter of a Lutheran pastor, orphaned at 13 and initiated herself to astronomy by entering the service of a self-taught farmer-astronomer, Christophe Arnold, who had a certain reputation in Leipzig. In addition to their taste for astronomy, they shared pietistic convictions, which animated them with a strong religious feeling, and constituted an important element of the agreement that was to unite them for nearly 20 years. Married in 1692, they had six children, a son, Chrisfried, and five daughters, of whom only Christine, Johanna and Magaretha reached adulthood, and whom they introduced to astronomy at the age of 10, in a tradition directly inherited from the craft culture. The women of the Kirch family were already working on astronomical observation before the foundation of the Royal Prussian Academy of Sciences in 1700 and continued to do so for more than half a century. Christine Kirch continued her observations and calendar work for the Academy until the end of her long life on behalf of the Academy. Mommertz emphasized that "the household as a way of life, economy and production retained its importance as a place of scientific work" (Mommertz 2005, p. 161), operating in parallel with the Berlin Academy. The work of the women, in symbiosis with that of the official "male" representative of the household within the institution, in this case Gottfried and then Christfried Kirch, contributed to the development of academic science in a situation described as the "invisible economy of science". We shall see that Maria Winkelmann, like her daughters, was excluded from any official representation of astronomy within the Academy, but that the Academy did not exclude them from its sphere of activities, the income from the sale of calendars being absolutely necessary for its survival.

The highlight of Maria Winkelmann's story is her meeting with Gottfried Kirch, who gave her access to the skills and practice of astronomy. Gottfried Kirch was born in Guben, Saxony, in 1640 and became an astronomer in Leipzig. He trained with Eberhard Weigel in Jena and Johannes Hevelius in Danzig. Hevelius was a lawyer and brewer in addition to his profession as an astronomer. During his apprenticeship with Hevelius, Kirch witnessed the collaboration of his mentor with his wife Elisabetha Koopmann, who was 36 years his junior and who acted as her husband's chief assistant both in astronomy and in the beer business. Since Hevelius' observatory was private and located outside his house, Elisabetha's role also included domestic work. She worked with him for 27 years and, after his death, published a catalog of almost 2,000 stars with their positions (Schiebinger 1987, p. 195). It is very likely that this astronomical household was a model for the young Kirch, which he later reproduced by marrying Maria-Winkelmann. Kirch's first wife, Maria Lange, gave him eight children, seven boys, all of whom died young, and one girl. He had to support this large family and earned a living, as best he could, by making calendars, an activity that he started at the age of 27. Alphonse de Vignoles mentioned that in 1681 Gottfried "began to publish ephemerides, which contained not only what is seen in the ordinary calendars but also the movements of all the planets, their rising and setting, their situation in the sky, their aspects, etc., to which he added various observations made in previous years" (des Vignoles 1721, pp. 174–175). Kirch introduced his daughter from his first wife, Theodora, to the practice of astronomical observation. There is no evidence of Kirch's first wife's collaboration in the observations, but his daughter's participation is attested by the observation books she kept. In the 1680s, a self-taught astronomer, "a peasant from the Village of Sommerfeld, who was not ignorant of Astronomy, & whose mind was above birth" (de Chaufepié 1753, p. 36), made a name for himself. It was Christophe Arnold. In July 1683, Arnold observed a comet eight days before Hevelius in Danzig, and informed the academic circles of Leipzig, who referred to the discoverer of the comet simply as "someone from the country". In July 1687, a new comet was observed by Arnold, and these observations were published in the Leipzig newspaper, along with Kirch's observations of the same comet. In October 1690, Arnold was among those who observed the transit of Mercury in front of the sun, a particularly difficult observation to realize, with Kirch in Leipzig, and Johann Philipp von Wurzelbau in Nuremberg. It is probable that Maria Winkelmann entered farmer Arnold's household as a maid (Mommertz 2005). Interested in astronomy, she learned the techniques of astronomical and meteorological observation. She recorded her observations and began what would later become the intensive meteorological research of the Winkelmann-Kirch family. "Thus, education and work, socialization and first contacts with scientific research were combined in this household" (Mommertz 2005, p. 164).

Maria Lange died in 1690, and two years later, Kirch, who had met Maria Winkelmann at Arnold's house, married her. She was then 22 years old, 30 years younger than him, but already had a solid experience as an observer. Born in 1670, she lost her father, a Lutheran pastor, in 1682, and her mother the following year. She was educated by a fellow pastor of her father, Justin Toellner, showing early a taste for literature, then for astronomy, under the influence of Christophe Arnold. Through this marriage, which secured the practice of astronomy for Maria over the long term and strengthened astronomical activities and the production of calendars for Gottfried, the two spouses sealed the birth of a new astronomical household, strongly resembling that of Hevelius a few decades earlier and falling within the direct heritage of Arnold's home, where the two spouses met, as shown in the following passage from the *Godefroi Kirch* entry in the *Nouveau Dictionnaire* (de Chaufepié 1753, p. 36):

> This rare man was called Christophe Arnold [...]; he was a subject of the Senate of this City [Leipzig], which, in recognition of his Observation on Mercury, gave him a present in cash, and discharged him for the rest of his life from the fee he had to pay every year. After his death, which occurred on April 15, 1695, his Portrait was placed in the Public Library of the Senate. Besides his printed Observations, he had made several others, Astronomical & Meteorological, since the year 1688 until April 6, 1695. He bequeathed by his Will the first six years to Mr. Kirch, to whom they were given in Original, with a Copy of the others, whose Original must be in the Library of the Senate of Leipzig.

Following the arrest of Pietists in Leipzig, in a period of repression of the movement by the Lutheran authorities, the couple retreated to Guben, Kirch's birthplace. Christfried was born there in 1694, then his sister Christine in 1697, and the other two sisters, Johanna and Margaretha, were born in Berlin at the very beginning of the following century. Maria supported Gottfried in the creation of calendars, and made considerable progress in astronomy. Within a short time she became independent in her observations and was able to calculate the movements of the planets. In July 1700, Frederick, King of Prussia, founded an academy in Berlin. Gottfried Kirch was called to be a member of this academy and an astronomer in charge of the future observatory, with an honorable pension, allowing him to stabilize the family financially, and the Kirchs moved to Berlin as a family. About the division of tasks between Maria and Gottfried, here is what is said in the article "Marie Marguerite Kirch" of the *Nouveau Dictionnaire*: "Her Husband & she relieved each other, in the Observations which should not be interrupted, & which one person alone could do. In other [cases] they observed separately in various places & shared the work, in those which one person alone could not do exactly" (de Chaufepié 1753, p. 37). Maria systematically observed the sky every evening from 9 p.m. On the night of April 20–21, 1702, she discovered a comet. This is what Gottfried Kirch wrote in his observation journal:

> Early in the morning (around 2 a.m.), the sky was clear and starry. A few nights before, I had observed a variable star, and my wife (while I was sleeping) wanted to find it and see it for herself. In doing so, she found a comet in the sky. She then woke me up, and I discovered that it was indeed a comet.... I was surprised that I had not seen it the night before (Schiebinger 1987, p. 180).

The two spouses followed the comet until May 5. Thus, Maria, like her husband 22 years earlier, when he discovered the famous comet of 1680 which bears his name, also spotted a comet. As much as the discovery of 1680 brought Kirch great fame, which was probably the reason for his appointment in Berlin, the discovery of 1702 did not bring Maria the recognition she deserved. The observation report that was sent to the king was signed with the name of Gottfried Kirch and many historians attributed to Kirch Winkelmann's discovery. A publication in the proceedings of the Berlin Academy had to be in Latin, which may explain why Maria, who did not know Latin, did not communicate her discovery herself. Moreover, since the couple had followed the comet together for two weeks, and generally worked in close symbiosis, Gottfried could have felt as involved as his wife in the discovery of this comet. According to Alphonse des Vignoles, however, Kirch was at first reluctant to publicly attribute the discovery to his wife, probably thinking that the discovery, if attributed to him, would contribute to securing his position by the notoriety it would bring him, or even that the attribution of the discovery to his wife would directly threaten the retention of his position. But later, an authority in the Academy told him that "he could feel free to acknowledge [his wife's] contributions", and in the first volume of the Journal of the Berlin Academy, *Miscellanea Berolinensia*, published in 1710, he opened his report with the words: "My wife [...] beheld an unexpected comet" (Schiebinger 1987, p. 182).

The only works that Maria Winkelmann published under her name, in German, were two pieces dealing with astrology. The first one, from 1709, is about the conjunction of the sun, Saturn and Venus on July 2, 1709, the day of the meeting of the kings of Prussia, Poland and Denmark in Potsdam, and its influence on meteorology. We read in the *Nouveau Dictionnaire* that "she had applied herself very much to this part of Astrology, which is called Meteorological, without however adding much faith to Judicial Astrology". In 1711, she published another piece, in preparation for the conjunction of Saturn and Jupiter, suggesting that these conjunctions were often followed by the appearance of a comet. This treatise was very well received by the academic community, and no objections were made to the astrological connotation of the treatise. It is not known to what extent Winkelmann really believed in astrology. Vignoles said of her: "She made, nevertheless, some Horoscopes, at the request of her friends: but always against her will, & in order not to disoblige the people who protected her" (des Vignoles 1721, p. 185). The practice of astrology by astronomers was common at the time, allowing them to live materially when they had no patrons. We have seen that Joseph-Nicolas Delisle had to practice astrology in the 1710s to earn his living, a discipline he said in his letters he despised to the highest degree. The calendars made by the Kirch men and women mixed astronomy and astrology, a practice that ensured the popularity of the almanacs and their sale, as astrology was used to predict the weather: "The conjunction of planets was thought to influence winds and temperatures. Thus, the conjunction of 'cold and dry' Saturn with 'cold and moist' Venus forebodes much snow and hail in winter, while 'warm and moist' Jupiter with 'hot and dry' Mars brings warm and thundery weather" (Schiebinger 1987, p. 185). In addition to meteorological astrology, which was perfectly accepted by the Academy because its income from the sale of calendars depended on it, the Kirchs devoted part of their scientific life to daily surveys of meteorological variables. Between 1697 and 1774, several members of the Kirch family carried out meteorological monitoring: "Kirch and Winkelmann traded off observing the weather from 1697 to 1702. Winkelmann then observed daily from 1702 to 1714 and 1716 to 1720. Her daughter Christine carried on after her mother's death, observing the weather daily from 1720 to 1751, 1755 to 1759 and 1760 to 1774. The offices of astronomer and meteorologer were split after Winkelmann's death in 1720" (ibid.).

At that time, the Berlin Observatory was not yet fully operational. Although it was partially operational in 1706 and ready for occupancy in 1709, it was not officially opened until 1711. During this period, the Kirch family continued to observe mainly "at home", as did many astronomers of the time, even after the opening of the observatory. The observation books of Gottfried and Maria attest to this (Mommertz 2005, p. 167). The family observed from the window, the yard, the garden, or the attic. Sometimes the hanging of laundry in the attic, or the loss of a key, interrupted the observations. The time was read on the clock in the living room. One of the family members would run to the living room to read the time during an astronomical event, or they would exchange signals by knocking on the partitions. All the family members took turns watching during full nights. Outsiders from all walks of life sometimes helped with the observations for hours or even days. Because of Gottfried's declining health, which led to frequent downtime, the couple's children were often called upon to help. Maria was in charge of providing the scripts for the calendars on time, Gottfried was the only mediator recognized by the Academy, which made him the head of the family.

The death of Gottfried Kirch in 1710 disrupted the life of the household (ibid., pp. 168–170). Johann Heinrich Hoffmann succeeded him as head of the Berlin Observatory. Maria Winkelmann asked the Academy several times if she could keep responsibility of the calendars, but this request was refused, whereas Hoffmann was known to lack competence in the matter. After two years spent in the apartment allocated by the Academy, she was welcomed in 1712 in the Observatory of Baron de Krosick, private adviser of the king, which he had built in 1707. She could make observations there. At the same time, Christfried left to study astronomy with a rich private individual in Nuremberg (Histoire de l'Académie Royale de Berlin 1752, pp. 61–65). In 1713, he was in Leipzig and the following year, at the age of 20, published ephemerides of all planets for 1714, as well as some observations made by his mother in Baron Krosick's observatory. Kirch joined his mother in Berlin, and made observations with her. Krosick's death in 1714 caused the Kirch family to move to Danzig in March 1715. There they received the support of the son-in-law and family of Johannes Hevelius, the astronomer who had trained Gottfried Kirch. Chrisfried reorganized Hevelius' observatory (Hevelius had died in 1687) and published new ephemerides, some for the year 1715, to which he added observations of his mother, and others for the year 1716, which were supplemented by observations made by himself and Johann Wilhelm Wagner, mathematician and astronomer from Berlin. When Hoffmann died in 1716, Christfried was called back to Berlin to succeed him, but because of his youth, he was only named "Observer" (and not Astronomer), and Wagner was added to him, the two Observers being in charge of running the Observatory. From 1717, Maria Winkelmann, disappointed by the lack of recognition of the Academy towards her, closed in on herself. She observed at home, doors closed. In 1720, when Wagner left to work in another city for a while, Christfried was given the official position of Astronomer and Director of the Berlin Observatory, of which his father had been the first director two decades earlier. This was also the year of his mother's death. The Winkelmann-Kirch household was replaced by Christfried's household, which included his sisters.

The new household reproduced in all points the practices of the old one, with a well-established distribution of roles, and this in spite of the ongoing institutionalization of astronomy within the Academy:

> Of the three sisters of Mr. Kirch, the second named Jeanne, recently married, had the principal direction of domestic affairs. The others were involved in their turn, but they were often busy helping their brother in the exercise of his profession, both having some knowledge of Astronomy. They nevertheless hid themselves from it, and never spoke of it, except with their most familiar Friends. By the same modesty, they avoided going to the Observatory, when there was some Eclipse, or when some Observation was to be made, which had been announced, & attracted foreign Spectators. But, on other occasions,

they often went up there with their Brother. Marguerite, the youngest, as having the finest sight, usually held some Telescope. Christine, the eldest, did the same, when necessary, or used other instruments. But most often she stood by the Clock, pen in hand, to mark the exact moment of each particular Observation. Besides, being in the Logis, she often made calculations, for the relief of her Brother: or even redid some of those he had made, for greater certainty (des Vignoles 1741, pp. 349–350).

Chrisfried Kirch, in his notebooks, rarely mentioned family life, and there is little information about the period from 1720 until his death in 1740 (Mommertz 2005, p. 169). His rare allusions show, however, that he continued to observe at home and that his sisters were involved in his work. The Academy provided him with assistants, but he judged them to be useless, unable to replace his family effectively. His sisters not being allowed to work with him at the Observatory, he naturally observed at home, following smooth and proven cooperation protocols. However, the Academy only recognized the Observatory as a place of observation and research, and the astronomer and his assistants, necessarily men, as its official representatives. The household work system was perceived as competing with the institution, and the Academy criticized the fact that Christfried worked at home. Thus, observational practice had not fundamentally changed, but a gendered division of labor was evidenced by this de facto conflict between "private sphere" and "public sphere". While at home, men and women could easily exchange tasks, it was quite different at the Observatory, where only men were allowed to honor the state with their scientific work, which introduced a gendered division and effectively placed men in the position of head of the household, the household in which the research was carried out, from which the Academy also directly derived scientific fruits.

The death of Christfried in 1740 did not interrupt the relationship between the Academy and the household. The Kirch sisters immediately offered to take over the production of calendars, which had previously been officially assigned to their brother, and since the sale of calendars was essential for the support of the Academy, they were granted a salary for the production of them (ibid., p. 170), which had been denied to their mother. Between 1740 and 1742, Frederick II conquered the populated provinces of Silesia, which increased the income of the institution through the sale of calendars in this new part of the kingdom. Christine became responsible for the production of calendars for Silesia. From 1741, Joseph-Nicolas Delisle, from St. Petersburg, resumed with Christine the correspondence of over 20 years that Christfried's death had interrupted (it can be found in the *Bibliothèque Numérique – Observatoire de Paris* from which, unless otherwise mentioned, the analyzed letters are borrowed). This correspondence, which extended over a period of five years, concerned essentially the observations of Christine's father, mother and brother, of which Delisle wanted to recover the entirety. Christfried had already communicated to Delisle some of his mother and father's observations, but some were still missing, and some of them already translated into Latin. The problem was that, like his mother, Christine did not know Latin. He suggested to Christine to call upon Wagner, his brother's successor as director of the Observatory, whom he had known for a long time, who could collect, classify and, if necessary, translate into Latin the remaining observations of the Kirchs, because Delisle did not know German (the letters from Christine Kirch to Delisle, written in German, were transcribed into French by a translator). The following years' exchanges of letters between Delisle and Christine Kirch were largely devoted to Wagner's hoped-for help, and to the difficulties Christine had in obtaining anything from the new director of the Observatory, whom "calendars, his age and all sorts of fatalities [...] sometimes make him ill-tempered", she confided to Delisle in a letter of July 24, 1744. Euler, who had left St. Petersburg for Berlin in the summer of 1741, served as an intermediary between Delisle and Christine Kirch. Thanks to the help of Wagner and Euler, the latter was able to gather and classify her brother's observations, which she attached to her letter. In the same letter, she explained that she had observed with her younger sister, from the window of their house, with the help of a 2 ft telescope equipped with a micrometer, the comets of 1742 and 1743, and she attached a drawing of their trajectories, for which Delisle thanked her, without elaborating, in a letter of January 16, 1745. In the same letter, he told her that he expected from Euler a small treatise in German of his father on the comet of 1680, to enrich his collection of observations of comets that he planned to deposit in a public place in Paris "to provide posterity with the means to work for the advancement of astronomy and at the same time that it will contribute to preserve the memory of the astronomer authors of these collections".

At that time, the Kirch sisters recorded important events, such as comet transits, in their observation notebooks; these notes seemed to be intended for publication, although there is no material evidence that they were made and it is not known for which audience they were intended (Mommertz 2005, p. 171). They organized joint observation sessions with other members of the Academy. It is known from reliable sources that the sisters made visits to the Observatory and that they maintained correspondences with astronomers. Thus, the astronomical household continued to function at the margin of the institution. From 1759, Christine alone continued to receive a salary. Sources show that she then became the official head of the household, and it was in this capacity ("Me and my family", she wrote about her household in her letters) that she asked the Academy for her salary increases. She played a large part in the production of the calendars and as such received money to pay a clerk in charge of the administrative work under her supervision. From 1769, the household received a total salary that compared favorably with that of the academicians, higher than that which Christfried received when he was director of the Observatory. In 1772, the Academy officially thanked Christine for her services and gave her a pension until her death. In the same letter, the Academy asked her to train Johann Elert Bode, a future director of the Observatory, married to one of her nieces, in the realization of the calendars. The knowledge developed in the household for nearly a century was thus finally exported to a male astronomer, who used this knowledge in the new institutional environment of the Academy's working system.

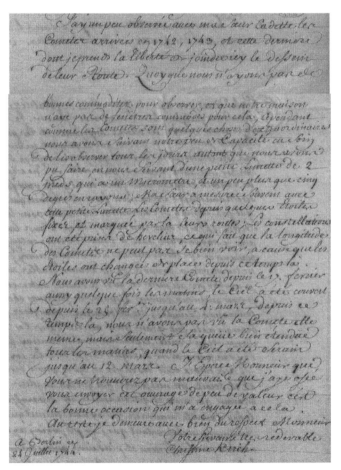

Figure 5.2. *Extract from the letter of July 24, 1744, from Christine Kirch to Delisle (Bibliothèque Numérique – Observatoire de Paris)*

5.2.2. The Manfredis: a household with a humanistic coloration inherited from the Renaissance

The context in which the Manfredi family developed is very different. Eustachio Manfredi came from the Bolognese petty bourgeoisie, the son of a notary, and he attended, like two of his three brothers, the University of Bologna, where he obtained a doctorate in canon law in 1692 (Robertson and O'Connor 2012). The family did not have the means to provide for all of the children in the long run, and the sons were expected to acquire, via their university career, situations that would allow them to support themselves. His two sisters Maddalena and Teresa, on the other hand, received only a very traditional elementary education in a convent of nuns. In 1690, at the age of 16, Eustachio Manfredi founded a club of friends to which he gave the name of Accademia degli Inquieti. Manfredi invited his friends, as well as his brothers and sisters, to meet regularly at his home to discuss science and literature, or to conduct experiments in physics. At the beginning, Manfredi oriented his activities mainly towards astronomy, which he was passionate about. He and his close friends, like Vittorio Francesco Stancari, studied astronomical publications and built an observatory in the Manfredi family's house. Lacking money to buy instruments, they built themselves their own quadrants and rudimentary telescopes. The group, little by little, grew and extended to a wider public than the one of the city of Bologna, and to other disciplines than astronomy, like anatomy or mathematics. This academy soon developed considerably, to give birth in 1714 to the Academy of Sciences of the Institute of Bologna, by merging with the Institute created by Luigi Ferdinando Marsigli in 1712, Manfredi then becoming the first director of the Observatory of Bologna. It was within the Accademia degli Inquieti, of which they were members, that Manfredi's two sisters, like his brothers, were trained in astronomy. The astronomical household was not only of a craft type, as in the Kirch family, but also had an explicit literary dimension. Manfredi, like his sisters, was interested not only in astronomy but also in poetry and literature. He was a member of a literary academy, the Accademia della Crusca. As mentioned by Fontenelle in his eulogy of Manfredi (Fontenelle 1766b), he rejected the pompous poetry of his time and approached the ancient models, in a tradition inherited from the Renaissance. The presence of women in a circle of scholars was not unacceptable in Italy; on the contrary, it was part of the humanist tradition of those women that the Italian city-states of the 14th and 15th centuries encouraged to study, especially languages, philosophy and literature, "exceptional women" who contributed to the prestige and vitality of these city-states (Frize 2013, pp. 11–17). There were in the 17th and 18th centuries some women graduates of the University, or members of learned academies, in Italy, and the Manfredi sisters, members of the Accademia degli Inquieti, were among them. The academic positions held by three of the Manfredi brothers, the fourth becoming a Jesuit, were poorly paid, and only Gabriele, the mathematician brother, got married. Eustachio stayed with his sisters, becoming the head of the "household" in charge of the relationship with the institutions (the Academy and the University), the family moving into the palace of Count Marsigli in 1701. Maddalena and Teresa devoted all their energy to the management of the "family business" (Scienza a Due Voci 2004–2010), collaborating in the scientific works of their brothers and producing literary works for the market of the educated bourgeoisie of Bologna (Bernardi 2016, pp. 103–106). Their notable contributions to observations and ephemeris calculations, to the service of Eustachio, and beyond that to the development of the astronomy at the Academy of the Sciences of the Institute of Bologna, are, as in the case of the women of the Kirch family, although in a less marked way, part of an "invisible economy of science" (Mommertz 2005). Although they were members of this academy, they only appeared in the published works of their brother, who was at the forefront.

The Manfredis were educated bourgeois. Eustachio was born in September 1674, one year after his sister Maddalena. In 1679, Teresa and Emilio were born, then Gabriele and Eraclito in 1682. Eustachio studied first with a priest, then with the Jesuits between the ages of 11 and 16 (Baldini 2007). He followed a course of study on the arts at the University of Bologna, which resulted in a thesis in 1691, which has not been preserved. He then turned to law, obtaining his degree in canon law in April 1692, but did not practice. He was interested in history and geography and, after obtaining his law degree, took Domenico Guglielmini's course, who had been appointed professor of mathematics, especially astronomy, at the University of Bologna two years earlier, and was destined to occupy the new chair of hydrometry created especially for him in 1694. What characterized Eustachio Manfredi was his wide range of interests and activities. In his youth, he was the author of an anatomical collection, a collector of observation instruments and natural curiosities, a lover of the French language, which he studied, but above all, he was interested in poetry, which he practiced in contact with personalities such as Ludovico Antonio Muratori, historian, as well as writer, linguist and grammarian, considered the founder of Italian historiography, or Pier Jacopo Martello, poet and playwright, and Giovanni Giuseppe Orsi, the main animator, together with Eustachio Manfredi, of the *Renia* cell in Bologna, which they set up in 1698 within the Accademia d'Arcadia, founded in Rome in 1690. With his friends of the *Renia* cell, Manfredi published sonnets, stories and songs. His last song, in 1700, was dedicated to Giulia Catarina Vandi, the only known love of his life, who became a nun. He integrated in this song elements of Descartes' vortices, recalling his taste for the sciences. In his poetry, Manfredi took up again the forms of the ancient lyric poetry, in particular that of Petrarch. At the same time, Manfredi cultivated the sciences in the field of the astronomical and hydrological knowledge given by Guglielmini to the University. Others than Manfredi followed Guglielmini's course, among whom his brother Gabriele and his friends Vittorio Francesco Stancari, who also intended to study astronomy, and Giuseppe Sentenziola Verzaglia, the group of young people being introduced to Cartesian geometry and differential calculus. In 1695, Jean-

Dominique Cassini and Guglielmini restored the meridian of the Basilica of San Petrone in Bologna, "marking the degrees of the distance to the zenith and their tangents, the signs of the zodiac, the hours of the night, the seconds and thirds of the circumference of the earth, and the width of the sun's image in summer" (Lalande 1790, p. 56). This event definitively determined Stancari's astronomical vocation, being four years younger than Manfredi, and with whom he was very close. From 1696 on, the members of Manfredi's group of astronomers began to make observations from their homes. Manfredi was made a public lecturer in mathematics at the University of Bologna in late 1698. In 1699, Guglielmini was called to a professorship in Padua, and the University of Bologna opened a chair of mathematics specific to the subjects taught by Guglielmini, a chair that was given to Eustachio Manfredi. It was a difficult period for Manfredi, due to his father's poor financial situation; his business was not doing well, and he had to leave Bologna, leaving Eustachio in charge of the family. His salary as a teacher and the money he earned from his poetry collections were not enough to support the family, and he had to turn for a while to the Marquis Orsi, whose financial support helped him to get out of this bad situation.

The key event that founded the astronomical household was not, as in the case of the Kirch family, the marriage of two astronomers pooling their skills and work force, but the foundation of a circle of friends, the Accademia degli Inquieti, which met in the house and included women, such as Maddalena and Teresa, Eustachio's sisters, who were to learn astronomy, mathematics and Latin in the family circle, which was regularly extended to the Academy. It was in 1690 that Eustachio Manfredi founded this cultural circle dedicated to literature and experimental sciences (Bernardi 2016). We will return to this academy at greater length in the section on Institutions. In addition to the poet Martello, this circle had some notable members, including Giovanni Battista Morgagni, physician and anatomist, Ferdinand Antonio Ghedini, poet and naturalist, the philosopher Francesco-Maria Zanotti and the physicist Francesco Algarotti, author of the famous "Newton for ladies" (Baldini 2007). Apart from Emilio, Eustachio's youngest brother, who became a Jesuit, the other two brothers had a brilliant academic career, both studying medicine, and then becoming for one, Gabriele, a professor of mathematics and author of renowned works on differential calculus, and Heraclitus, a professor of hydrometry. A particular community link united the Manfredis to the family of Zanotti. Giampietro Zanotti, poet, painter and art historian, was of the same generation as Eustachio Manfredi. It was him who, in 1748, nine years after Manfredi's death, published the complete poetic work of his friend. His younger brother Francesco-Maria Zanotti, writer and philosopher, member of the Inquieti, was secretary of the Academy of Sciences of Bologna in 1723, then its president from 1766. Giampetro had son Eustachio Zanotti, future astronomer and director of the Observatory of Bologna, where he succeeded Eustachio Manfredi, and daughters Maria Teresa and Angiola Anna Maria, born, respectively, in 1693 and 1703, who were linked by close friendships to the Manfredi sisters, their eldest by about 20 years. We have seen that Eustachio Zanotti was the main Italian collector of observations of the aurora borealis, which he collected and communicated to Mairan in Paris. There is no evidence to suggest that the Zanotti sisters contributed to the scientific work of their astronomer brother, as was the case in the Manfredi family. Belonging to the generation following that of Eustachio Manfredi, they could not participate in the Inquieti circle, within which their Manfredi elders were trained in astronomy, which may explain their probable lack of involvement in the sciences. On the other hand, the Zanotti sisters, with the Manfredi sisters, contributed, in close connection, to the translation into Bolognese dialect of Neapolitan stories, to which they incorporated original elements of their composition: songs, proverbs, rhymes and allegories. These women did not sign their works, whether scientific or literary, probably because of the humility that women had to show at the time to maintain their respectability. But everyone in good Bolognese society knew that they were the authors of the translations of Bertoldo and Chiaqlira that they published, if only through the many social receptions in which they took part, which was worth signing (Scienza a Due Voci 2004-2010). It is possible that the Manfredi family's collaboration in the production of high-level scientific and literary works served as a model for members of the Zanotti family, the two households being linked by close relations both scientific, around astronomy, and literary, around poetry. The reunion of the two families can certainly be described as an extended household, encompassing the astronomical household of the Manfredi, within which a shared economy of production of intellectual works provided the group with the income necessary for its survival, both in the form of salaries from the university or academy, and royalties from literary productions.

In 1711, when Eustachio Manfredi was appointed Astronomer of the Academy, his sisters followed him to his new apartments in the Observatory. The contributions of the two sisters to the scientific side of the family business were mainly due to the help given to Eustachio in the calculation of ephemerides. In 1715, he published two volumes of ephemerides dedicated to Pope Clement XI. The first volume was an introduction to ephemerides in general, and to all astronomy. The second volume contains the ephemerides of 10 years, 1715–1725, calculated using Cassini's unprinted tables. Here is what Fontenelle said in his eulogy of Manfredi:

> His Ephemeris embraces much more than an Ephemeris used to embrace. One finds in it the passage of the Planets through the Meridian, the eclipses of Jupiter's satellites, the conjunctions of the Moon with the most remarkable Stars, the Maps of the Countries which must be covered by the shadow of the Moon in the solar eclipses (Fontenelle 1766, p. 550).

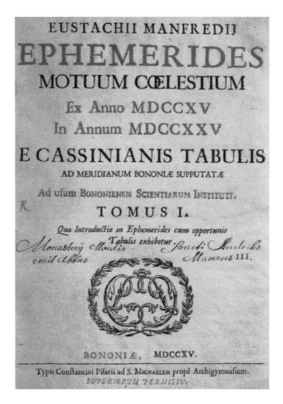

Figure 5.3. *Cover page of the Ephemerides published by Eustachio Manfredi in 1715 at the Academy of Bologna*

At the end of the manuscript of his Introduction to the *Ephemerides*, we can read the following note, absent from the printed edition of the collection: "I began the ephemerides in December 1712 in Bologna. With many interruptions, they continued in the following years with the help of my two sisters Maddalena and Teresa, and of Mr. Giuseppe Nadi, and again of Mr. Cesare Parisij… The table of longitudes and latitudes was calculated by my sister Maddalena in 1702 or 1703" (Bernardi 2016, p. 92). Shortly thereafter, he published new ephemerides for the periods 1726–1737 and 1738–1750. Here is what Fontenelle says of it:

> Mr. Manfredi did not fail to inform the public of the names of those who had helped him in the tiresome composition of his Ephemerides. However, he certainly received help that he concealed; and he would be justly reproached for it, if the reason he had for concealing it did not present itself as soon as one knew from whom it came. It was from his two sisters who did most of the calculations in his first two

volumes. If there is anything quite directly opposed to the character of women, especially those who have spirit, it is the relentless attention, & the invincible patience which calculations very unpleasant by themselves, & as long as unpleasant, require; & to add to the marvel, these two Calculatrices (for a word must be made for them) sometimes shone in Italian Poetry (Fontenelle 1766, p. 551).

Thus, the merit that Fontenelle attributed to these women was to be, according to him, different from the common women of spirit, in other words, not to be women, whom he judged unfit for patient and disinterested effort, but equivalents of men, hard-working and hard at work. For Fontenelle, a woman could only be exceptional by distancing herself from her very condition as a woman, a judgment that was specific to certain men of the time who accepted the idea that women could do science, as we will detail later.

The Manfredi sisters also helped their brother to arrange and sort the astronomical and geographical observations made by Francesco Bianchini after the astronomer's death in 1729, a difficult task from which Manfredi drew a selection of observations that he published. In 1704, Manfredi was appointed Superintendent of Water, a position he held until his death. The rivers of the region were indeed subject to frequent overflows, flooding the land. As these rivers crossed different states (Bologna, Ferrara, Mantua, Modena, Venice), a coordinated policy of water management and major works was necessary, and this is what Manfredi occupied himself with for part of his life, to the detriment of his astronomical activities. A last work related to his function as superintendent occupied him until 1735, concerning a conflict between Ferrara and Venice which he was charged to arbitrate. His sisters helped him to go through the voluminous file of deeds and maps related to this case, resulting in an academic work entitled *Compendiosa informazione di fatto sopra i confini della Communità ferrarese d'Ariano, con lo Stato Veneto* ("Concise information on the borders of the community of Ariano, near Ferrara, with the Venetian state"). The work, anonymous and without typographical notes, was printed in Rome in 1735, but the annexed maps of the Po Delta region were drawn and printed in Ferrara in 1736 (Baldini 2007). It was intended to demonstrate that since the Middle Ages the Ariano region was the domain of the Este family. Fontenelle described the role of the Manfredi sisters in helping with these two operations:

> In this affair of the Ferrarais, as well as in the unraveling of Mr. Bianchini's papers, we still find his two sisters, who were infinitely useful to him, especially for all the unpleasant maneuvering of these kinds of works. With a lot of spirit, they were suitable for what would require almost a complete deprivation of spirit.

Without this domestic help, he would never have succeeded in all that he did in the last five or six years of his life, during which he was tormented by kidney stones [from which he died] (Fontenelle 1766, p. 556).

The Manfredi sisters, members of the Accademia degli Inquieti, were well-known figures in the scientific and literary world of Bologna at the beginning of the 18th century. Prospero Lorenzo Lambertini, the archbishop of Bologna, his native city, from 1730, and who would become pope under the name of Benedict XIV in 1740, played an essential role in supporting the Academy of Bologna, and in particular women scientists. Fascinated by science and a man of the Enlightenment, in 1741, he had the Vatican authorize the first complete edition of Galileo's works, and in 1757, all works that supported heliocentrism. Lambertini was in fact the true patron of the Academy of Bologna, to which he allocated considerable resources during his pontificate. He was an admirer of women who expressed their knowledge in books, especially the Manfredi sisters whom he knew and appreciated. In 1732, he supported the physicist Laura Bassi for a professorship at the university and a place at the Academy's institute. As Findlen notes, "the scientific climate of Bologna in 1732 was particularly conducive to public and institutional recognition of a learned woman, especially one versed in the latest mathematical and experimental philosophies" (Findlen 1993, p. 447). The decline of the international reputation of the University of Bologna at the end of the previous century, which the reforms of the university's curriculum had failed to halt, had been effectively halted at the beginning of the century by the creation of the Institute by Marsigli, as we shall see in detail. In 1728, Algarotti redid Newton's optical experiments in front of the members of the Institute, thus stimulating the diffusion of Newtonianism in Italy. But the death of Marsigli in 1730, and the difficulties encountered by the Institute to develop, due to the lack of vision of the Senate of the city, did not allow Bologna to fully restore its scientific image. "Thus the university, the academy, and the city needed Bassi as much as she needed them. Publicizing Bassi's accomplishments, and enhancing them beyond anything achieved by earlier learned women, would add luster to the reputation of Bologna" (ibid., p. 448). At the urging of Lambertini, Bassi agreed to participate in the public debates. In this way, she left her status as a scientific curiosity to become "the symbol of the scientific and cultural regeneration of the city". Her thesis defense, which did not take place at the University but in the Public Palace, in the presence not only of the professors but also of the whole good Bolognese society, made her emerge as a public figure. Lambertini was her most lavish admirer, and intervened when some academicians tried to exclude Bassi from her activities at the Institute, or to limit her participation. Lambertini saw Bassi as the embodiment of the intellectual renaissance of Bologna, restoring the city to its position among Italian cities and beyond, and as such, as a lover of his city, he supported remarkable women such as Bassi, and later Maria Gaetana Agnesi, as we shall return to.

In a letter attributed to Lambertini, who in the meantime had become Benedict XIV, we read, "I assure you that, in browsing through libraries, I would take great delight in finding there, next to our learned doctors, worthy women who enshrined their knowledge in modesty" (Findlen 2016, p. 45). This term "modesty" is quite characteristic of the men's view of the role of women in science at the time, who should not question their obligations within the family unit. In the same vein, the writer Ilaria Magnani Campanacci characterizes the life of the Manfredi sisters as a balance constantly maintained between awareness of their own talent and submission to the public male figure of their brother, and expresses herself as follows:

> It is clear from many indications, which can be extracted either from the data of their eighteenth-century biographers as well as from their rhymes [...] that in the recurring male controversy against the knowledgeable woman they speak openly and lively in favor a woman's right to make use of her mind in spite of those who would like her stupid [...]. But they firmly reject as well the possible temptation of attributing them a sententious character [...] thus in agreement, even if with an unknown degree of self-irony, with the dominant negative opinion about a woman who makes a undue and harassing show of her knowledge. [...] In short, the Manfredi were the practical demonstration of the female genius [...] applied to the study, but rather as a complement instead of alternative to the commitments of a woman. Not with a competitive attitude, but in close and supportive collaboration with the more recognized intellectual figures of their academic brothers (quoted in Bernardi 2016, p. 95).

As can be seen, in spite of the difference in nature between the Kirch and Manfredi households, the same major trends emerged. In spite of the de facto equality that seemed to prevail between men and women in the Winkelmann-Kirch pair, where the spouses made all their observations together, or among the Manfredi siblings, where the literary activity of the women was equal to the scientific activity of the men, the separation between the household and the institution, to which the woman has little or no access, resulted in the installation of a gender hierarchy in the household. Women's activity was essentially limited to the private sphere, and this socially imposed norm was internalized by some women and resisted by others, but within the limits imposed by the maintenance of their respectability and credit. In both cases, these women did not consider that the norm called into question their right to intellectual emancipation, the man ensuring, for the good of the household and its financial stability, the essential nature of the representation outside. For the Manfredis, as for the Kirchs, women did not sign the scientific or literary works they published, with the exception of Maria Winkelmann's astrological treatises written in German. The reality was therefore complex. In Berlin, the Kirch sisters finally received a salary from the Academy for their work on calendars, which allowed them to continue to practice astronomy. In Bologna, the dominant humanist culture, shared by influential members of the Catholic clergy, allowed some women to gain public recognition and access to representative functions, even if the dominant codes did not allow everything. Thus, Laura Bassi was not allowed to teach within the walls of the university, and had to practice her teaching in private places. The Manfredi sisters were known and appreciated at the Academy of Bologna. The situation for women scientists in Bologna seemed to be less closed than in Berlin.

5.2.3. *The Delisle family: an artisanal household where women took care of the family scientific heritage*

The case of the Delisle family was somewhat different, though similar in social background to that of Manfredis. Joseph-Nicolas Delisle was the youngest son of Claude Delisle, who was a cartographer and "royal censor", in charge of reading works awaiting printing and deciding on the authorization of publication and the granting of royal privileges (Garcelon 2017). Of Claude's many children, only four sons and one daughter reached adulthood. The children lost their mother at an early age, and grew up under the education of Claude's second wife, a lawyer's daughter. The three brothers of Joseph-Nicolas were scientists: one, Guillaume, the first cartographer of Louis XV, the second, Simon-Claude, historian, and the third, Louis, a geographer, who accompanied him to Russia, as we have seen. Guillaume and Simon-Claude, older than Joseph-Nicolas by about 15 years, both died in 1726, the year in which Joseph-Nicolas moved to Russia with his brother Louis. Thus, like Manfredi, Joseph-Nicolas Delisle came from the urban petty bourgeoisie, but he developed very early in the scientific environment, in contact with his father and two brothers who were much older than him, all three of them confirmed and recognized scientists. We traced Delisle's life path in Chapter 2. He studied at the Collège Mazarin, a college of the University of Paris located on the Quai Conti, on the present site of the Institut de France, from which he graduated with a degree in rhetoric, and continued his studies in mathematics. It was the solar eclipse of 1706, which he witnessed at the age of 18, which, as we have said, determined him to embrace a career as an astronomer. He completed his apprenticeship under the direction of Jacques Lieutaud, assistant to the Académie Royale des Sciences and main editor of the *Connaissance des temps*, and set up his first Observatory above the entrance to the Palais du Luxembourg in 1710 (Garcelon 2017). We have seen that his material situation in the following years was precarious, obliging him to practice astrology to earn his living. In 1718, Delisle solicited abbé Bignon because he was, he said, lacking in resources, with a sick father (Claude Delisle died indeed shortly after, in 1720), and finally obtained a chair at the Collège Royal. The following year, he was appointed Associate Astronomer at the Académie Royale des Sciences. His double role as professor and member of the Academy ensured him an income, which he negotiated to be paid to his sister, Angélique, and to her mother, during his absence in Russia, thus starting in 1726, a way to allow the two women to survive, and also to remunerate his sister for the services that he asked her to provide (Lémonon Waxin 2019, pp. 273–292). Isabelle Lémonon Waxin uncovered the actual role of Angélique in the family scientific enterprise. It has not been demonstrated that she, as did the Kirch or Manfredi women, practiced astronomical observation, as the previously reported words of Celsius might suggest. The correspondence between Delisle and his sister while he was in Russia suggests that she mastered some technical terms related to astronomical instruments and that she also knew the Greek notations of the planets that her brother used in his letters to name them, but not more. The role played by Angélique Delisle during the 22 years spent by her brother in St. Petersburg was that of a scientific assistant, managing for him, in Paris, all his interactions with his colleagues of the Academy and the Collège Royal, or of the government when it was a question of the maps of Russia that Delisle communicated to Jean-Frédéric Phélypeaux de Maurepas, secretary of state of Louis XV. This role of "steward", as Isabelle Lémonon describes it, made necessary by the distance and the slowness of the mail exchanges, and which led her to meet frequently with the scholars of the Academy, gave her a real power in the management of her brother's scientific affairs.

Just after his arrival in St. Petersburg, in spring 1726, Delisle sent to his sister in Paris a power of attorney which made her manager of his financial and scholarly affairs, "affairs in Paris which [you] can take care of. You are the absolute mistress by the power of attorney that I send you" (quoted in ibid., p. 277). Angélique thus became her brother's "trusted woman". In January 1727, Delisle specified to his sister that he "took the party to have no other familiar trade than with" her (quoted in ibid., p. 278). She became his official intermediary in all his exchanges with his fellow scholars in Paris. She managed her brothers' library, which was enriched by the Academy's regular publications. She updated the catalog of the books of the library and lent some of her books to the academicians or to the members of the Collège Royal, a role which fell to her naturally insofar as she received from these two institutions the pension which would normally belong to her brother if he were present. In this way, she took his place, and, failing to carry out her research and teaching duties, served as a representative between him and the members of these institutions. Delisle gave instructions by mail, telling her to whom she could lend books with confidence, and to whom she could not, or could only do so by obtaining guarantees. She also took care of buying or having copied scholarly journals, or books on astronomy or of geography, which she sent to her brother. She mailed him the books that he asked for, and sometimes, on her own initiative, books that she thought might interest him. Moreover, Angélique Delisle was one of the only people in the confidence of his sendings of maps of Russia. About the first box sent to his sister in 1730, here is what he said to her in the accompanying letter: "you will therefore deposit it in Luxembourg with my other effects when you receive it and you will keep it secret so as not to give rise to conjectures or rumors that could harm me in this country. Only the Count of Maurepas knows what I have sent in this box" (quoted in ibid., p. 286). Other boxes reached Angélique, shortly before her brother's return from Russia. The management of the Observatory at the Palais du Luxembourg, housing allotted in 1710 to her brother who established his library and his observatory there, was part of Angélique Delisle's tasks. On her own initiative, she refused access to the dome to certain applicants, for fear that they would leaf through and borrow books without saying so. She managed instrument loans, such as the quadrant instrument, which the Academy took back for a time.

Angélique Delisle also took care of the "family's intangible heritage" (ibid., p. 289), by checking with the publishers of her father's works to make sure that the rules were respected: verification of the contents, publication of accounts in journals, requests for compensation, etc. She made sure that her brother's discoveries were well circulated in journals. For example, she asked that the party organized in St. Petersburg by Delisle on the occasion of the birth of the Dauphin in 1729, a party which he was reproached for by Bignon, who refused that the French government reimburse the astronomer for the expenses incurred, be reported in the *Mercure de France*. She provided biographical elements of her father to Antoine Lancelot, inspector at the Collège Royal, on the occasion of the publication of his *Abrégé d'Histoire Universelle*. In the foreword, Lancelot did not mention Angélique Delisle as a contributor. After the publication of the work, she pointed out the errors which were there so that they could be corrected, thus expressing herself in a letter of September 17, 1731, addressed to her brother in St. Petersburg:

> Mr. Desplaces found in it, as did other people, several errors which I am writing down as I go along. Mr. Lancelot, who was the censor as you know, did not want to give himself the time to examine it properly or was not in a position to make up for it; what is true is that he was more inclined to have the booksellers sell a lot of copies than to be attentive to what could harm the memory of my father [...] *I made him remove* this place so that one would not impute to my father the very gross errors which are there, there have already appeared on this work several criticisms I have not yet been able to have them but I will buy the *Journal des Savants* of August that I have just been warned to speak advantageously of this history (ibid., p. 290).

In these lines, she asserts herself as the guardian of her father's scientific reputation, a guarantee of the credibility of the family scientific enterprise, which in the absence of her brothers, she carried entirely on her shoulders. Marie Darbisse Delisle, the wife of Guillaume who died in January 1726, continued her husband's activities in the same way, in the tradition of craftsmen's workshops. She continued the geographical "map business", helped by Philippe Buache, a former pupil of her husband, who in 1729 obtained the position of first geographer of the king. The same year, he married Charlotte Delisle, daughter of Guillaume Delisle and Marie Darbisse, which allowed him to take over Guillaume Delisle's map business, after the transitional period of management of the business by Marie Darbisse over three years. This marriage was nothing less than prepared. When her husband died, Marie Darbisse and her daughter found themselves with no income. The perpetuation of Guillaume Delisle's workshop, a condition for the continuation of a quality cartographic production, was the object of all of Abbé Bignon's attention. The latter pointed out the need for the widow "to attach herself to her husband's pupil by marrying his daughter" (Dawson 2000, p. 80). But both Marie Darbisse and Philippe Buache were without financial resources. "This marriage", the widow and the abbot insisted, "could only be made with the grace and help of his Majesty". They asked the king "to grant this student the title of ordinary geographer of His Majesty and to grant him a pension of 1,200 pounds in favor of marriage" (ibid., pp. 80–81) with the orphan Delisle. Not only did the king accept and appoint Buache first geographer but also he was present at the signing of the marriage contract in early 1729. We find again the pattern of the artisanal household that we have encountered with the Winkelmann-Kirchs, with a transmission of the workshop from the master to the pupil, while the wives, most probably, helping their husbands at home to honor their scientific commitments.

We can wonder about Angélique Delisle's effective manoeuvring in the way she managed the family business "in the absence of Joseph-Nicolas, especially considering the slowness of communications with her brother. It is known that in the absence of her brother, Angélique went regularly to the Louvre, the heart of the institution, or to such and such an academician, to whom she communicated in return the contents of the last letter received from St. Petersburg" (Chabin 1983, p. 119). We can evoke the memory regarding the shape of the Earth that Delisle asked his sister, in 1734, 15 years after its writing and its transmission to Bignon, to communicate to the editor of the *Journal de Trévoux* to counter what he considered as plagiarism by Jacques Cassini of his idea to measure a degree of the parallel of Paris. This event illustrates the autonomy of Angélique Delisle, and her sense of responsibility, since she did not do what her brother asked of her. Mairan and Fontenelle, whom Delisle informed directly by mail from St. Petersburg, followed the affair at the Academy, and it is very likely that Angélique Delisle met with them and let herself be convinced not to risk aggravating the situation by going into open conflict with Cassini. It was not his sister who informs Delisle that she did not communicate the manuscript to the editor, probably because she feared that her brother had insisted, but Souciet, one of the editors of the journal, whom Delisle's sister "urged not to publish [the] dissertation" (Delisle 1734–1738), having even wished that he would hand it over to her (which he did a little later), evidently to avoid any risk of leakage. This event highlights the strong involvement of Angélique

Delisle in the preservation of the image and interests of her brother, and more generally of the Delisle family.

The Delisle household was thus particular with the role of steward assumed by Angélique Delisle due to the long absence of her brother, whom she represented in relation to the Parisian academic community. This absence lasted 22 years, which gave Angélique plenty of time to get to know the Parisian actors of the Académie and the Collège Royal, and to train herself fully to her task of managing her brother's scientific interests. Receiving the pension due to her brother if he had stayed, she was in a way his Parisian double, and had to fully play this role to justify her salary. In addition to her role as representative and manager of both material and immaterial goods, including those bequeathed by her father, Angélique defended the image and the place of her family in the Parisian academic system, by taking initiatives and by imposing herself as a woman of responsibility. Here, no more than elsewhere, her role as woman of an astronomical household was not publicly recognized at its true value, and she remained a shadowy figure. Like Christine Kirch in Berlin, paid by the Academy for her calendars, or the Manfredi sisters in Bologna, who contributed to the family's subsistence by selling their literary works, Angélique Delisle was paid for her work, which gave her activity an economic value. As in Berlin or Bologna, an "invisible economy of science" was at work, from which the institution benefitted fully. Concerning Guillaume Delisle, the geographer brother of Delisle who died prematurely, the defense of his interests was taken in hand by his wife Marie Darbisse. The perpetuation of cartographic heritage was a matter of great importance to the family. The perpetuation of Guillaume's cartographic heritage was achieved through his daughter's marriage, organized by his widow and the political power of the time, to Philippe Buache, a geographer to whom Joseph-Nicolas had given lessons in astronomy and whom he had tried, without success, to take with him to Russia. Here again a complex family game appears, with the Delisle scientific household relying on women, sisters or wives, in this case more as managers than scientists, to function and reproduce.

5.3. Two examples of astronomical institutions: the academies of Bologna and Berlin and their observatories

In the second half of the 17th century, scientific institutions were created everywhere in Europe, the most famous being the Académie Royale des Sciences in Paris and the Royal Society in London. In the north of Italy, three academies were created: the Accademia del Cimento in Florence in 1657, in Bologna, the Accademia delle Traccia in 1666, and finally the Accademia degli Inquieti created by Eustachio Manfredi in 1690. The members of the Accademia del Cimento, established under the patronage of Prince Leopold de Medici, set themselves the task of perfecting the Galilean school's experimental method (Boschiero 2004). The only publication of this academy, made in 1667 (the year of its dissolution), narrates the experiments carried out, and mentions that the intentions of the academicians were not to stray into speculative arguments, but simply "to experiment and report". Some historians consider this Academy, created three years before the Royal Society, and nine years before the Académie Royale des Sciences in Paris, as the first modern scientific institution. Leopold de Medici wanted the knowledge developed by the Academy to be factual, not speculative, and free from all considerations of a theoretical nature. To reinforce this neutrality, he wanted the names of the scientists not to appear in the presentation of the knowledge gathered. The observation books of the members of the Academy, and the letters exchanged between them, show, however, behind the neutrality of facade, a background of philosophical presuppositions which directed the debates within the group. This Academy was dissolved in 1667, when several of its members left Tuscany. But philosophers of other regions of Italy were inspired by it by creating their own Academy. Geminiano Montanari created the Accademia delle Traccia in Bologna in 1666, probably to continue the experimental program and the physical studies carried out previously in Florence. The choice of Bologna was largely related to its astronomical tradition, since Cassini was originally from there. Moreover, the Senate of Bologna was beginning to offer lucrative academic positions at this time. Montanari arrived in Bologna in 1664 with his mentor Cornelio Malvasia, an aristocrat and astronomy enthusiast. Between 1662 and 1664, Montanari became familiar with Malvasia's astronomical instruments. After Malvasia's death in 1664, Montanari was appointed to the chair of mathematics at the University of Bologna. A supporter of the experimental approach, he himself built optical instruments for observing the sky, and carried out physics experiments on fluids. Montanari, who enjoyed debates among his students and colleagues, therefore created the Accademia della Traccia (also known as the Accademia dei Filosofi). The academy met in Montanari's house. Its program of study revolved around experiments on vacuum, air pressure, weight and viscosity of fluids, light, sight and sound. Montanari regularly gave physico-mathematical speeches on subjects of interest. The University of Bologna suffered a financial crisis at the end of the 1670s, and Montanari accepted a better-paid professorship in Padua. Experimental activities slowed down in Bologna and did not resume until 20 years later with the creation of the Accademia degli Inquieti, built in the spirit of the precursor academies of Florence and Bologna.

In Germany, the main precursor academy of the Berlin Academy was the Accademia Naturae Curiosorum called in French the "Curieux", or the "Scrutateurs de la Nature" (Bartholmess 1850, p. xxii). It was founded in the imperial city of Schweinfurt in 1652 on the model of the Italian academies. Its founders were physicians, and its purpose, its statutes said, was "to discover what nature holds hidden in its bosom, and to exhibit it on the stage of the world" (ibid.). Contemporary of the Florentine Accademia del Cimento, it was part of the same approach to the observation of nature. In 1670, the Society began publishing a scientific journal exposing the work of its members, under the name of *Miscellanea Curiosa*. This journal was mainly concerned with medicine and its related aspects such as botany and physiology. The itinerant nature of the Academy's headquarters, which migrated according to the residence of its presidents, first in 1686 to Nuremberg, then to Augsburg in 1693, then to Altdorf, Erfurt, Halle, etc., gave it a broad geographical base, its management being done essentially by correspondence. At the time of the creation of the Berlin Academy, at the turn of the 18th century, the number of the Curious, exceeding 400, included most of the Prussian scholars. In the years following its foundation, the Accademia Naturae Curiosorum sought political protection. In 1677, Emperor Leopold 1 of Habsburg officially recognized it and took it under his protection. Ten years later, Leopold granted the Academy special privileges, guaranteeing its independence from the various reigning dynasties of the region and freeing it from any censorship concerning its publications. The Academy then took the name of Sacri Romani Imperii Academia Caesareo-Leopoldina Naturae Curiosorum, or Leopoldina in short (History of the Leopoldina 2021). Other important institutions were created in the last decade of the 17th century, under the impulse of the Elector of Brandenburg, the future Frederick 1, King of Prussia, such as the Academy of Painting, Sculpture and Architecture of Berlin in 1691, and the University of Halle in 1694 (Histoire de l'Académie Royale de Berlin 1752, p. 3). In December 1699, the Diet of Regensburg, the assembly of the sovereigns of the Holy Roman Empire, decided to adopt the Gregorian calendar to correct the gap in the old Julian calendar (11 days had to be caught up). The transition to the Gregorian calendar had been going on for more than a century in Catholic countries, but, as Kepler said, "Protestants like to disagree with the Sun more than with the Pope", and the reform had not yet been carried out in the Protestant world (see the site Calendriers Saga (2022)). However, its execution could only be done with the help of astronomers able to measure solar time precisely. Frederick decided to locate the production of the new calendars in Berlin. And this idea led to another one. Since the sale of the reformed calendars generated profits, he decided to use these profits to finance the pensions of scholars grouped in an Academy, and working in all sciences, not only astronomy. The desire to set up an astronomical observatory dated back to 1697, when Sophie-Charlotte of Hanover, Frederick's wife, had expressed her wish to her husband (Bartholmess 1850, p. 15). This idea had been taken up and amplified by Leibniz, the philosopher and mathematician, friend and correspondent of Sophie-Charlotte. The reform of the calendar, with the income it would generate, was the event that led the double project of an astronomical observatory and a Berlin Academy emerging. The project was officially launched on March 18, 1700 at Orangeburg Castle (ibid., p. 17).

5.3.1. *The Academy and the Bologna Observatory*

Eustachio Manfredi founded the Accademia degli Inquieti. The term *inquieti*, or restless, characterizes those who, in a restless quest, are busy looking for the Truth (de Limiers 1723, p. 16). From that year on, Manfredi gathered his brothers, sisters and friends at his home to discuss scientific questions and to carry out experiments in the natural sciences. It has been said that the scientific activities of Manfredi and his friends, especially Stancari, were oriented at the beginning especially to astronomy. The young men built their own instruments of observation. In 1694, however, the group expanded to include other disciplines and other scholars from nearby cities. On this occasion, the meeting place moved from Manfredi's house to the larger one of Jacopo Sandri, professor of anatomy at the University of Bologna. The following year, Manfredi and Stancari made their observations from their respective houses (Baldini 2007). From 1697, they used an observatory set up in the palace of the future cardinal Gianantonio Davia. In 1700, Manfredi became close to Count Luigi Ferdinando Marsigli, who was to play a major role in the creation of the Academy of Sciences of the Institute of Bologna. He had served as a hydraulic expert in the army of Emperor Leopold I, and had been interested in natural philosophy since his youth. During his travels, he collected books, scientific instruments, minerals and natural fossils. At the end of his service in the Austrian army and upon his return to Bologna in 1702, Marsigli made his collections, books and instruments available to the Inquieti. In 1703, he had an observatory built at his palace, which the two friends began to use for their observations. In 1704, the Accademia degli Inquieti became official. Giambattista Morgagni, a medical student, was appointed President, and Stancari took on the role of Secretary, in charge of annotating the discussions and experiments conducted at the meetings (Boschiero 2004, p. 14). The following year, the meeting place of the Inquieti was moved to the Marsigli Palace. Stancari died prematurely in 1709. That same year, Marsigli applied to the Senate of Bologna for the establishment of an Institute, to which he promised to donate his collection of "pieces that could be used for natural history, instruments necessary for astronomical observations or chemical experiments, plans for fortifications, models of machines, antiquities, foreign weapons, etc." (Fontenelle 1766, p. 412). The senate of Bologna gave its agreement in principle, but lacked the necessary means to realize the project, and asked Marsigli to address the Vatican. To convince the pope, Marsigli asked the painter Donato Creti, under the supervision of Manfredi, to make paintings representing academics observing planets with their instruments. Marsigli thus sought, through the arts, to interest the religious power in the experimental method advocated by the Inquieti (Boschiero 2004). A geographical map project of northern Italy was also presented to the pope to obtain his support. On January 21, 1712, the pope having given a favorable answer, the Senate of Bologna officially received Marsigli's donation. The Palazzo Poggi, where Marsigli's collection was to be assembled, was prepared for the Accademia degli Inquieti, renamed on this occasion Accademia delle Scienze dell' Istituto Bolognese, or Academy of Sciences of the Institute of Bologna. In the following, it will sometimes be referred to as the Academy of Bologna for simplicity. Thus, through the intermediary of a wealthy patron with strong political influence, who also possessed an important scientific and cultural patrimony invested in the creation of an institute, the Inquieti's project, which was to implement the experimental program initially proposed by the Florentine academy Del Cimento, found its institutional consecration in the reunion of the Institute, the creation of Marsigli and the Accademia degli Inquieti (Boschiero 2004).

Six members of the Bologna Senate were entrusted with the execution of the project. The institute, named Institute of Sciences and Arts of Bologna, was spatially organized in the Palace, so that the different sciences had their own sectors. The palace was divided into departments, among which the instruments and funds necessary for their maintenance were distributed. Each department was staffed by the best professors in the discipline it represented.

> Soon there were seen by their [the Professors'] care, in the Palace devoted to this new Establishment, Chambers quite proper & well heard, either for History of Nature, or for Experimental Physics, or for Military Architecture, or finally for the other Sciences spoken of, all furnished with suitable Instruments. Soon, they had each their Professors & their particular Masters; so that living all together in this Palace, as in the place of their Confederations they have nevertheless each one their department, from which they pass each other the baton, to form this beautiful harmony which is between them (de Limiers 1723, pp. 15–16).

The Institute thus provided a tool for scholars to conduct their research in a coordinated manner, the latter belonging to the Academy, which pre-existed and did not merge with it. The Institute and the Academy had their own rules and regulations and were administratively distinct entities. The professors of the University appointed to the Institute were all academicians, and it was from among these professors that "the first heads of the academy are chosen, namely a president, a vice-president and a secretary" (ibid., p. 18). The academy, in a classical way for the time, was made up of four classes: (i) the ordinaries, those who "practise, work, and reason in conferences, either public or particular", from among whom were chosen the professors of the institute, (ii) the honoraries, who had no workload, (iii) the numeraries, who were brought in to replace the ordinaries when their posts became vacant, and (iv) the pupils, placed under the tutelage of an ordinary, the ordinaries being obliged to have at least one pupil each. Six subjects were present at the Academy: physics, mathematics, medicine, anatomy, chemistry and natural history. There was one professor and one substitute per subject, as well as a president, a secretary and a librarian. For the first term, the president of the Institute was a professor of natural history, the secretary a professor of medicine and anatomy, and the librarian a public reader in hydrometry. The faculty of the Institute also included an astronomer and professor of mathematics (Eustachio Manfredi), a mathematician, a professor of experimental physics and a chemist, the last two having the title of public reader in philosophy and medicine. This group of a few dozen professors (a little more if we include the substitutes) constituted the staff of the Institute, in much smaller number than that of the Academy. At the Academy, each of the six disciplines was placed under the responsibility of a pair of ordinary academicians, "that is to say, the professor & the substitute, who choose, on the same subject, the points susceptible of various considerations, & who, after having digested them well & examined them, put them on the carpet in common conferences that they make at different times, one after the other, where they teach by the prepared pieces, joining always the demonstration of the fact to the theory" (ibid., p. 21). Thus, the members of the Institute formed an elite core of the Academy, from which were chosen its president, vice-president and secretary. The institute, because of its equipment, constituted an operational center of research, which this core animated in connection with the Academy, and it was also a place of training and conferences. It was placed under the direction of an office, composed of 10 senators (Lalande 1790, p. 63). In addition, the "Clementine Academy" of fine arts, dealing with painting, sculpture and architecture, created by Pope Clement XI in 1712, joined the ensemble, occupying the second floor of the Palazzo Poggi. Here again, the two academies had separate rules, but were supposed to be able to help each other, for example, if the Academy of Sciences needed help with drawings or plans, or the Academy of Arts needed help with certain technical aspects of its work. The Academy of Sciences of the Institute occupied the second floor, and the Observatory the third. Jérôme Lalande detailed the system, such as it was in the middle of the 18th century:

> There is a room for the assemblies of the academy of sciences, a library, a very well set up observatory, a large natural history and physics cabinet; rooms for the navy, for military art, for antiquities, for chemistry, anatomy, childbirth, for painting and for sculpture; professors skilled in each of these parts, give lessons on the days marked, and even with very mediocre salaries.
>
> [...]
>
> The observatory, the Specola, is a large tower, very high and very convenient, equipped with good instruments of the modern kind: there is a wall-mounted quadrant, a meridian telescope or instrument of the passages, & several other considerable instruments (Lalande 1790, pp. 63–64).

As Fontenelle noted, the fact that these institutions remained administratively distinct suggests that Marsigli encountered political difficulties in achieving his goal:

> No doubt he had difficulties to overcome on the part of the older societies, different interests to reconcile together, and even caprices to overcome; but no trace of this remains, and it is as much lost for his glory, unless we take into account that no trace remains. He subordinated his Institute to the University, and linked it to the two Academies. From this new arrangement, made with all the necessary skill and care, it certainly follows that Physics and Mathematics now have considerable help and advantages in Bologna which they never had before, and whose fruit must be communicated by a happy contagion (Fontenelle 1766, p. 413).

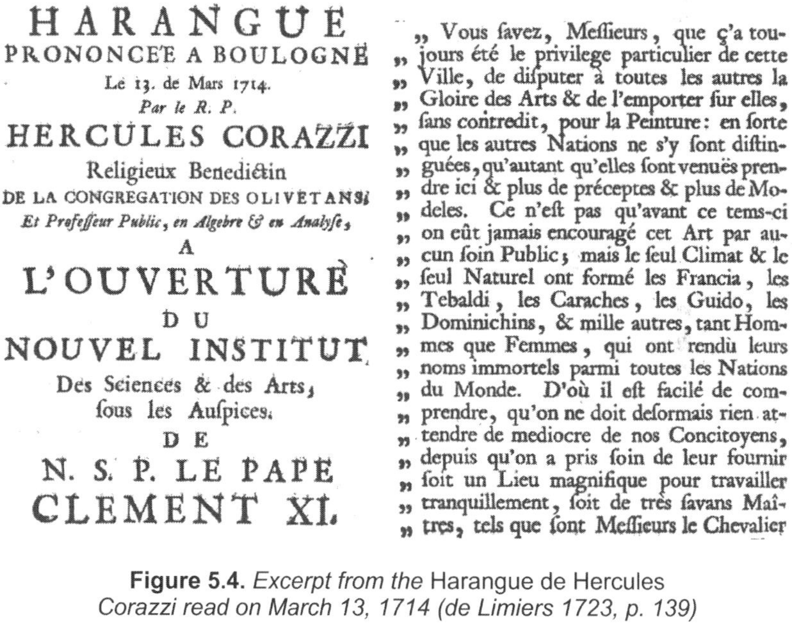

HARANGUE
PRONONCÉE A BOULOGNE

Le 13. de Mars 1714.
Par le R. P.

HERCULES CORAZZI

Religieux Benedictin
DE LA CONGREGATION DES OLIVETANS,
Et Professeur Public, en Algebre & en Analyse,

A

L'OUVERTURE
DU
NOUVEL INSTITUT
Des Sciences & des Arts,
sous les Auspices,
DE
N. S. P. LE PAPE
CLEMENT XI.

„ Vous savez, Messieurs, que ç'a toujours été le privilege particulier de cette Ville, de disputer à toutes les autres la Gloire des Arts & de l'emporter sur elles, sans contredit, pour la Peinture: en sorte que les autres Nations ne s'y sont distinguées, qu'autant qu'elles sont venues prendre ici & plus de préceptes & plus de Modeles. Ce n'est pas qu'avant ce tems-ci on eût jamais encouragé cet Art par aucun soin Public; mais le seul Climat & le seul Naturel ont formé les Francia, les Tebaldi, les Caraches, les Guido, les Dominichins, & mille autres, tant Hommes que Femmes, qui ont rendù leurs noms immortels parmi toutes les Nations du Monde. D'où il est facilé de comprendre, qu'on ne doit desormais rien attendre de mediocre de nos Concitoyens, depuis qu'on a pris soin de leur fournir soit un Lieu magnifique pour travailler tranquillement, soit de très savans Maîtres, tels que sont Messieurs le Chevalier

Figure 5.4. *Excerpt from the* Harangue de Hercules Corazzi *read on March 13, 1714 (de Limiers 1723, p. 139)*

The new academy was solemnly inaugurated on March 13, 1714, in the presence of political and religious authorities (de Limiers 1723, pp. 23–41). The first President of the Institute made the opening speech. Then, Father Don Hercules Corazzi, a mathematician, described the organization and the usefulness of the new Institute. He explained that the quality of the instruments with which the Institute was equipped was unprecedented. He spoke about experimental physics: pneumatic machines, thermometers, barometers, microscopes, hydrostatic machines, balances, etc... Then, he moved to chemistry and astronomy, and praised the instruments with which the Astronomical Observatory was provided. We note, in the harangue of Corazzi, a passage where he mentions the women, certainly not concerning the scientific subjects, since it was about painting, but the fact deserves to be noted:

> It is not that before this time this Art was ever encouraged by any Public care, but the Climate & Nature alone formed the Francia, the Tebaldi, the Caraches, the Guido, the Dominichins, & a thousand others, both Men and Women, who have made their names immortal among all the Nations of the World (ibid., p. 139).

Several academicians took the floor, of which Manfredi, who presented an essay about his tables calculating the solar eclipses, described the use that could be made of the tables for geography. It was learned that a room for anatomy was still missing, and another for botany. Marsigli was busy collecting plants and flowers in the Netherlands to install a botanical garden. As for anatomy, Marsigli had not collected anything on this subject, but the Senate was ready to do what was necessary to make up for this lack.

In spite of this promising start, the institute quickly floundered, not fulfilling the hopes that Marsigli had placed in it to become a research center of the highest European level. Marsigli himself had a real European dimension, being a member of the Royal Society since 1691, and being appointed to the Académie Royale des Sciences in 1715. After the foundation of the Academy of Bologna, he was absent for several years and traveled to England, where he met Newton in 1722, and to the Netherlands, where two of his books were to be published. On his return to Bologna in 1723, he noted that the Institute was only a place of intellectual entertainment, and not an academic center of European dimension, equal to the Parisian and London centers (Dragoni 1993, p. 234). He then entered into conflict with the Senate of Bologna, which oversaw the Institute, and reproached it for its lack of interest in the scientific project. Among the remedies he proposed were (i) the creation of a permanent librarian post, and the creation of a new room to rationalize the organization of the library, (ii) the opening of professorships at the University of Bologna to the non-Bolognese, (iii) the raising of professors' salaries. The book of grievances submitted by Marsigli ended with a request addressed to the pope, in case the Senate of Bologna would refuse his recommendations. The pope's response was favorable and, on March 24, 1727, Marsigli completed his 1712 donation with that of his natural history cabinet of the Dutch Indies, and of a certain number of books obtained in payment for the publication of his books in Amsterdam. In the meantime, some improvements had been made, including the construction of the astronomical tower in 1725. Nevertheless, the situation did not improve, because of bureaucratic red tape and quarrels with the enemies that Marsigli could not avoid during the accomplishment of his great project. Dissatisfied with the behavior of the politicians and some professors of the University, Marsigli announced in a letter of July 10, 1728, to the Pope that he was leaving the city and disengaging himself from the Institute. He died two years later (ibid., p. 236). At the same time, as a sign of the undeniable international influence of the new structure, several members of the Academy of Bologna, all former Inquieti, were appointed to the Royal Society: the mathematician Orsi in 1716, the historian Muratori in 1717 (Hall, p. 74), the anatomist Morgagni in 1722, the astronomer Manfredi and the professor of experimental physics Iacopo Bartolomeo Beccari in 1728, the physicist Francesco Algarotti in 1736, the astronomer Eustachio Zanotti in 1740 and the mathematician François-Marie Zanotti in 1741, followed by six others in the second half of the 18th century (Cavazza 2002, p. 16). Among them, some were also appointed to the Académie Royale des Sciences: Manfredi in 1726, Morgagni in 1731 and F.-M. Zanotti in 1750 (Hall 1982, pp. 79–80). In his fight for the success of the Institute, Marsigli was supported by an eminent representative of the church, Prospero Lorenzo Lambertini, whom we have already mentioned, a native of Bologna, archbishop of that city from 1730, then pope under the name of Benedict XIV in 1740. The humanism of Lambertini, his interest for the heliocentrism and Newtonianism, his support for the access of women to knowledge and his love of Bologna made him an essential ally. It was through him that Marsigli obtained the support of the pope in 1727. A year after becoming Benedict XIV, in 1740, Lambertini continued Marsigli's work with a series of large-scale actions intended to strengthen the Institute (Dragoni 1993, pp. 231–232): construction of a new wing of the palace and reorganization of the library (1741), equipping the Observatory with instruments purchased from the manufacturers George Graham and Jonathan Sisson in London (1741), donations of various collections of natural history (1742, 1743), realization by Ercole Lelli of anatomical statues in wax for the museum of anatomy (1742–1751), publication of the complete works of Galileo for the centenary of his death (1744), supply of instruments and experiments for Galilean and Newtonian physics by Dutch instrument makers, under the supervision of Pieter van Musschenbroek and Willem Jacob's Gravesande (1744), installation of a cabinet of physics which was not only a museum but also a laboratory, thus institutionalizing the place of experimental physics (1745), publication of three volumes of the journal of the Academy, the Commentarii Academiae scientiarum Instituti Bononiensis (1745–1547), only one volume having been published until then (in 1731) and five others to be published later in the 18th century, renovation of the laboratories, notably of chemistry (1749), donation of the 12,000 volumes of Cardinal Monti's library and of his collection of more than 400 paintings including portraits of Copernicus, Bacon, Galileo, Boyle and Newton (1754), donation of

25,000 volumes from the library of Benedict XIV, including 400 codices and manuscripts (1755), opening to the public of the library of the institute (1756) and the creation of a chair of obstetrics (1758).

Figure 5.5. *Cover page of the first volume of the* Commentaries *of the Academy of Bologna (*Commentarii de Boniensi Scientarium et Artium Instituto Atque Academia *1731)*

The correspondence between Joseph-Nicolas Delisle, from his return from St. Petersburg, and Eustachio Zanotti gave a perspective on the works carried out in Bologna in astronomy and in physics. In a letter of June 21, 1748, Delisle declared to Zanotti to want to resume the interrupted correspondence with the University of Bologna. He sent him a copy of his foreword on the next eclipse of the sun of July 25 of the same year. In this foreword, Delisle described his total artificial solar eclipse of 1715 showing the formation of a luminous ring (by diffraction), which Stancari, who died in 1709, also appears to have done:

> You can see there, Sir, among other things, the use that I make of the artificial total eclipse of the sun that I invented in 1715 to explain the cause of the luminous ring that appears around the moon in the total eclipses of the sun, without knowing that Mr. Stancari had had the same thought. As I invite astronomers to imitate this artificial eclipse in order to see for themselves all the singular phenomena that accompany it, not only in the total eclipse but also in the annular eclipse or even in the partial one; I beg you to take the trouble to search the papers of the Institute to see if there are any details of Mr. Stancari's experiments on the artificial eclipse that he has imagined; if not, I hope that the mathematicians of the Institute will take the trouble to make new experiments according to the method that I have indicated, in order to compare them with those that I propose to make as soon as possible.

In his answer of August 18 of the same year, Zanotti says he was unable to observe correctly the eclipse or have time to transmit the warning to the other Italian astronomers. He asked the secretary of the Academy if he had the notebooks of Stancari's experiment, but the secretary said to him that we could not draw from it more than what was reported in the *Commentarii* of the Academy. No one, after Stancari's death, took over on this question. On December 6, 1748, Delisle answered Zanotti, his letter covering mainly the solar eclipse and the various observations which were made of it. He said to have found in the 2nd volume of the *Commentarii* (thus of 1745) all the "beautiful observations" of comets, as well as of the eclipses of the moon and of the sun. He asked his interlocutor if he did not think of publishing all the observations made since the beginning of the Observatory of Bologna, as Giovanni Giacomo Marinoni had suggested to him, his correspondent in Vienna. Delisle was particularly interested in the eclipses of fixed stars by the moon. He asked Zanotti to communicate to him all the observations of eclipses of Jupiter's satellites made in Bologna over a 10-year period. In his answer of January 10, 1749, Zanotti sent to Delisle several observations of the eclipse made in Italy, in particular in Verona and Turin. He indicated to his correspondent that he did not plan to publish the integral of the observations made in Bologna since the creation of the Observatory, the measurements made before 1742, the year of the installation of the instruments bought in London, were too imprecise. It is known that during the 1730s, Manfredi and Zanotti lobbied for financial support from Rome to improve the quality of their instruments. A wall-mounted quadrant, a mobile quadrant and a transit telescope were ordered in 1738 by Georges Graham and his assistant Jonathan Sisson in London. Zanotti took particular care to arrange the meridian room, where he installed the transit telescope in August 1742, and spent another year calibrating the instruments (Findlen 2016, p. 47). Thus, the Bologna Observatory only reached the best European standard after 1742, 50 years after the first observations of Manfredi and Stancari, then the young Inquieti. The continuation of the correspondence, between 1750 and 1753, relates to the solar eclipse, the satellites of Jupiter, the measurements made in correspondence with those of Lacaille at the Cape of Good Hope in 1751 and 1752 to determine the parallaxes of the moon of Venus and Mars, and the transit of Mercury in 1753. Delisle asked Zanotti for the series of solstitial heights of the sun measured in Bologna to verify that the variations of the obliquity of the ecliptic followed the variations predicted by Newton (precession of the equinoxes), which required an accuracy of a few seconds of arc.

5.3.2. *The Academy and the Observatory of Berlin*

On July 11, 1700, Frederick III of Brandenburg, who would become King of Prussia the following January under the name of Frederick I of Prussia, signed in Cologne the letters patent instituting the new Society of Sciences in Berlin (Bartholmess 1850), the Royal Academy of Sciences of Prussia, which we will often call for simplicity the Berlin Academy. The next day, Gottfried Wilhelm Leibniz was appointed the Society's President for life. Before his appointment, Leibniz had submitted the rules of the Academy to Frederick. The objective assigned to the Society was stated in the founding documents to be

> in accordance with the designs that had animated the academies of Paris, London and Florence [...] The Academy of Berlin also receives the mission of collecting the knowledge scattered in the middle of the world, and acquired by the human mind; to put it in order; to form a precise and regular whole; then, that of increasing and multiplying it; finally, that of learning how to make a sure and legitimate use of it.

The goal was, with Galileo, Bacon and Descartes, to recognize the obvious necessity to observe nature, to make experiments, to describe the facts, just like for the Academy of Bologna, and those of Paris and London. With the Academy of Berlin, "a true academy, one where analysis and method prevail over synthesis and system, over everything that is too dogmatic, is born of the need to associate to better know the phenomena and their laws". This approach implied the setting up of an assembly of men struck "by the importance of making in common varied and repeated observations, compared and combined, to which the isolated genius would seek in vain to suffice". In view of the multitude of objects to be studied, and the difficulty of tracing the causes of phenomena, it was necessary to "divide one's work, and it is in the academies that this division takes place". The sharing and distribution of studies, the association of efforts and trials, the exchange and verification of discoveries, this is the task of scientific bodies, both diverse and simple; this was the vocation addressed by Leibniz to the Berlin Society. The resolutely applied character of a science useful to society was particularly marked in the rules of the Berlin Society, as Leibniz stated, in a formulation more politically oriented than scholarly:

> It is not a question of satisfying pure curiosity, nor even the natural desire to know. It is not a matter of inventing things that have no immediate practical influence. This institution must think of science and useful application at the same time, by imagining objects which can both honor its majestic founder and benefit the world. Let it combine practice with theory; let it perfect, alongside the arts and sciences, and by means of them, everything that interests the country and the people, agriculture, industry, commerce and even food!

Following Frederick's wish, Leibniz added a patriotic function to the practical function of science: improving life and ensuring the prosperity of the nation. This society, said Leibniz, had to "be penetrated by German sentiments, and zealous for the glory of Germany". The German language had to be perfected, and the study of German history had to be deepened. The society also had to propagate the true faith (Christian reformed) and the evangelical virtues. But Leibniz stipulated that the Society could employ people of other religions, or of other nations. Leibniz wished that the churchmen members of the Society were also men of knowledge and understanding. The Academy's literature class needed to promote the study of Asian literature (China, India, etc.). The aim was to "bring the Academy into close contact with the missionaries and scholars of Asia". Leibniz did not institute a philosophy class, partly because there were no philosophers of note in Berlin, but also and above all because he wanted to ban pure speculation from the program. The practices put forward, as in London, Paris or Bologna, were "observation and science of the physical world, from mineralogy to astronomy, including human anatomy and physiology" (ibid., p. 31).

We have seen that the development of astronomy constituted one of the queen's motivations to set up the Academy, this offering in addition an opportunity of financing by the sale of the calendars. Leibniz attracted Gottfried Kirch to Berlin in 1700. Kirch had to draw up calendars in order to raise the necessary funds for the construction of the Observatory and the premises that the members of the Society would occupy. Leibniz proposed to the King the publication of various almanacs ("Calendars of the Empire, the Court, the State, the Churches, the Post Office, the Courts, the Police, etc."), mainly for the use of country people, "a kind of popular directory which could become the directory of the common people, a kind of handbook of simple morals, of useful knowledge, of practical hygiene". But the construction of the Observatory progressed very slowly at first. Leibniz had many ideas to raise funds. He thought of organizing lotteries, sales, of asking for the privilege of paper manufacturing for the academy, or a monopoly on bookstores, on all articles of instruction and education, or even the privilege of levying a tax on books judged bad. But these suggestions were not heard by the king. Finally, in 1707, the Society obtained the exclusive right to cultivate mulberry trees and silkworms, a request supported by the queen. As early as 1700, Leibniz had set up a Board of Directors to defend the Company's interests in his absence. Between 1700 and 1710, the Society chose nearly 80 members from among the most renowned scholars of the time. Among them were Jacques and Jean Bernoulli, Pierre Varignon, Domenico Guglielmini, Nicolas Hartsoeker, Ole Christensen Rœmer and the German astronomer Henri Hoffmann, who succeeded Kirch as director of the Observatory in 1710. But the selection process of the members was contested. The members of the clergy and the doctors were too numerous, and they coopted themselves. The Society, during the first 10 years, functioned at very low speed. The War of Spanish Succession (1701–1714) absorbed the resources of the state. Moreover, Queen Sophie Charlotte, Leibniz's main supporter, died in 1705, which considerably reduced the credit of Leibniz, President of the Society, at court. A first volume of the memoirs of the Academy, the *Miscellanea Berolinensia*, appeared in 1710. Leibniz, who presided over the choice of the memoirs published in this volume, appeared at the head of the four classes constituting the Academy: (i) physics, (ii) mathematics, (iii) philology, and (iv) history. The dedicatory epistle he wrote for Frederick, master and founder of the Society, was very eulogistic, but it irritated some courtiers, amplifying the disgrace in which the death of the queen made him fall. The buildings of the Observatory were almost completed in 1710, and the king ordered that the Society be installed there. The inauguration ceremony was set for the following January 19, but Frederick not only did not consult Leibniz about the definitive institution of the Academy but also did not invite him to the inauguration. Leibniz was replaced as president of the society by the Baron of Printzen, minister of the interior, and kept only the honorary title. It was foreseen that at the death of Leibniz, Printzen would be the only one to govern the society as Honorary President.

Figure 5.6. *Cover pages of the first volume of the* Memoirs of the Académie Royale des Sciences and Belles-Lettres of Berlin *(Commentarii de Boniensi Scientarium et Artium Instituto Atque Academia 1710)*

The day before the inauguration, without informing Leibniz, Printzen proceeded to the definitive constitution of the classes of the Academy, and to the distribution of the members among these classes. He maintained the four departments that Leibniz had instituted: (i) physics and medicine, (ii) mathematics, (iii) philology and (iv) national history. Each department chose a moderator. For the first exercise, these moderators were, respectively: (i) Krug de Nida (the king's first physician), (ii) Jacques Cuneau (head of the national archives), (iii) Jean-Charles Schott (the king's librarian) and (iv) Daniel Ernst Jablonski (bishop, first preacher to the king, who also became the vice-president of the Society), all extremely close to the king, who was clearly the master of the Academy. These four directors referred to Printzen, without going through Leibniz. On January 19, 1711, for the inauguration in the new building of the Observatory, the decor was austere: "A long table in the middle of a vast room, next to this table a chair, then a few benches, that is all the furniture of the Academy at its birth" (ibid., p. 57). Jablonski, in his speech (Histoire de l'Académie Royale de Berlin 1752, pp. 36–46), which at no point quoted Leibniz, called for the constitution of universal academies, for "scientific synods, literary councils where each people would be represented by its most illustrious scholars, meetings, either periodic or permanent, finally something analogous to what was organized later under the title of scientific congresses". On the means attributed to the Academy, he expressed himself thus:

> We are in this respect in a different case from that of other Societies. The Members of the English Society are obliged to contribute to the expenses at their own expense: the French Society, on the contrary, is magnificently maintained by the Royal Liberties. Here it is neither the one nor the other. We have no income; but we are not charged with any expenses; we possess, thanks to the Royal Protection, all that is necessary to excite our activity, & encourage our work, to acquire Mathematical Instruments, Books, & to provide ourselves in general with other necessary things (ibid., p. 45).

It should be noted that the primitive statutes of the Berlin Academy did not formally exclude women, any more than those of the Academies of Paris or London. Leibniz, in his 1700 draft of the Academy's rules, wrote that a scientific academy should promote good taste and a solid understanding of God's work not only among the German nobility, "but among other people of high standing (as well as among women)" (Schiebinger 1987, p. 196). Thus, Leibniz mentioned women in his draft of the regulations, but this overture was not acted upon. A few weeks after Gottfried Kirch's death in July 1710, his wife Maria Winkelmann wrote a letter to Jablonski, asking him for her and her son Christfried, the position of "Assistant Astronomer". Her two arguments were that she was qualified as an observer (she discovered a comet, recorded observations and prepared ephemerides with her husband), and that on the other hand she had been working for 10 years for the Observatory, and that she had even provided alone the production of the calendars when her husband was ill. She did not ask for this position as an honorary position, but rather as a source of income for herself and her four children. Leibniz recognized Winkelmann's scientific abilities and presented her in 1709 to the Prussian Court, where she detailed her observations of sunspots. Here is what he wrote in a letter of introduction written for her:

> There is [in Berlin] a most learned woman who could pass as a rarity. Her achievement is not in literature or rhetoric but in the most profound doctrines of astronomy. […] I do not believe that this woman easily finds her equal in the science in which she excels. […] She favors the Copernican system (the idea that the sun is at rest) like all the learned astronomers of our time. And it is a pleasure to hear her defend that system through the Holy Scripture in which she is also very learned. She observes with the best observers, she knows how to handle marvelously the quadrant and telescope (quoted in Schiebinger 1987, p. 183).

The presentation made at the Court seemed to convince Frederick and, moreover, on July 17, 1709, Winkelmann wrote to Leibniz having been congratulated by the Danish Ambassador visiting the Observatory for the help she gave to her husband.

At that time, it seemed to be accepted that she could play a representative role with respect to distinguished external visitors. Jablonski did not deny Winkelmann's qualities. In a letter to Leibniz in September 1710, he indicated his hostility to the possible appointment of Winkelmann to replace her husband:

> You should be aware that this approaching decision could serve as a precedent. We are tentatively of the opinion that this case must be judged not only on its present merits, but also as it could be judged for all time, for what we concede to her could serve as an example in the future.
>
> [...]
>
> That she be kept on in an official capacity to work on the calendar or to continue with observations simply will not do. Already during her husband's lifetime, the society was burdened with ridicule because its calendar was prepared by a woman. If she were now to be kept on in such a capacity, mouths would gape even wider (quoted in Schiebinger 1987, pp. 186–187).

It was therefore important not to set a precedent with Maria Winkelmann. Leibniz supported Winkelmann in his approach, but he no longer had any influence in the Academy. At one of the last assemblies he attended, on March 18, 1711, only two months after Printzen's inauguration, Leibniz defended at the very least the granting of six months' accommodation and salary to Winkelmann as a widow. The Academy guaranteed Winkelmann the right to stay in her house but refused the principle of a salary. In the same year, the Academy awarded her a medal to make a good impression. After Leibniz's final departure, she went directly to the king. In 1712, after a year and a half of insistent requests, Winklemann received a definitive refusal. Regarding her request for an appointment as Assistant Astronomer, the Council's decision was final:

> Frau Kirch's request is in many ways unseemly and inadmissable. We must try and persuade her to be content and to withdraw of her own accord; otherwise we must definitely say no (quoted in Schiebinger 1987, p. 188).

The reason for this refusal was not explicitly given to her, but she naturally interpreted it as being due to her sex, a sanction all the more humiliating as, in the astronomical household that she formed with her husband, the two spouses lived in equality of investment and discovery in their astronomer profession. Very hard hit, she wrote: "Now, I go through a severe desert, and because ... water is scarce ... the taste is bitter" (ibid.). In the preface to one of her works, Winkelmann expressed herself as follows on the ability of women to practice astronomy: the "female sex as well as the male sex possesses talents of mind and spirit". A woman can become as "skilled as a man at observing and understanding the skies". Leibniz died in November 1716, and Winkelmann, consumed by bitterness, in December 1720. Her son Christfried then took over, and we have seen that after Christfried's death in 1740, his sisters benefitted from a salary to continue the production of calendars. This favor had been refused to their mother 30 years earlier.

The death of Frederick I in 1713, and his succession assured by Frederick William I, son of Sophie Charlotte, which lasted 27 years, marking the beginning of a long crossing of the desert for the Academy, the new king despising the arts and sciences. The Academy came very close to being dissolved. In 1733, Daniel Ernst Jablonski, known for his rigor and his love of science, was appointed President of the Academy. The political direction was given to Adam Otto de Viereck, a minister who became "Protector" of the Academy, a role that had been assumed by Frederick I, but not by his successor (Bartholmess 1850, pp. 120–121). In 1735, Jablonski and Viereck obtained that the king detach from the royal library all the works relating to mathematics and physical sciences to join them to the library, until then poor, of the Academy. Moreover, they revived the publication of memoirs, of which a first volume had appeared in 1710 under the aegis of Leibniz. The second, third and fourth volumes were published in 1723, 1727 and 1734, respectively, and three more volumes were published in the following 10 years. The death of Frederick William in 1740, and the accession of Frederick II, known as Frederick the Great, marked the true resurrection of the Academy. The Silesian wars, between 1740 and 1745, nevertheless delayed its restart, while bringing to the Academy new incomes, with the extension of the privilege of the calendars to the new borders of the kingdom, Christine Kirch being in charge of producing these calendars.

The Academy was re-founded in 1746 under the leadership of Field Marshal Samuel Graf Von Schmettau (ibid., p. 147), a military man with extensive scientific knowledge in the fields of geography and engineering. Schmettau was compared in profile to Marsigli, the founder of the Academy of Bologna. In the manner of Marsigli, he could not suddenly impose a change of structure and purpose, so he proceeded step by step by reorganizing the old academy and adding to it a component of science and literature lovers from the court and the state apparatus, among whom Schmettau and Viereck, for example, counted themselves. The whole thing was called the "Literary Society", and its aim was to cultivate both the "useful" and the "interesting", as the introduction to the society's rules attests:

> Some inhabitants of Berlin, who have a taste for science and literature, wishing to extend their knowledge and make themselves more and more useful to the public, believed that the best way to achieve their goal was to form a literary society among themselves […]. The

principal aim of this Society being to cultivate what is interesting and useful in the different parts of philosophy, mathematics, natural, civil and literary history, as well as in criticism, they will not stop at questions which, instead of instructing and perfecting the mind, could only serve to amuse it uselessly (ibid., p. 150).

The new Society was close to the one created by Leibniz in that it focused on the experimental and practical side of science. But it differed from it by two innovations, in a frank break with the options of Leibniz: the study of speculative philosophy and the use of the French language, thus giving to the whole a more literary and European turn. At first, the old Academy, whose members formed a subset of the new one, and met at the Observatory, remained as such. It published a last volume of *Mémoires* in 1743. The new Society had its own rules, and a room in the King's castle for its meetings. After a few years, Frederick ordered the organic reunion of the two Societies as "a perfected continuation of the Society founded by Leibniz" (ibid., p. 152). The Observatory's Assembly (the old Academy) brought the traditions of the past and their continuation; the Assembly of the Castle (the new Society) represented the contemporary aspirations towards the future and perfection. Schmettau and a commission formed for the occasion created, by merging the two societies, the Royal Prussian Academy of Sciences and Letters. The new Academy was divided into four classes, like the old one, but the outlines of the classes changed: (i) physics or experimental philosophy, (ii) mathematics, (iii) speculative philosophy and (iv) belles-lettres or philology, each under the responsibility of a curator, the four curators presiding alternately per term. Two years later, on March 3, 1746, Pierre Louis Moreau de Maupertuis, who had come to live in Prussia, became the permanent President of the Academy, and Frederick II officially took the title of Protector. The Germanic tendency was abandoned in favor of a cosmopolitan dimension. Utility became more scientific than social, the latter being a prerogative of the government. The evangelical character was erased, and speculative philosophy was reintegrated, including metaphysics, morals, natural law and the history of philosophy, which made the Berlin Academy the only one in Europe where philosophy was studied at this level. Maupertuis justified it as follows:

> "Experimental philosophy", said Maupertuis, "physics, this word taken in its broadest meaning, examines bodies as they are, with all their sensible properties. Mathematics considers them stripped of most of these properties. Therefore, every academy of sciences has a mathematics class next to a physics class. But, next to bodies, are there not objects that have no corporeal properties? What right does an academy of science have to neglect this order of objects? An academy, on pain of being incomplete, must therefore have a class of speculative philosophy, that is to say, a class whose object is the

nature of the supreme being, the nature of the first causes, the nature of the human mind and of everything that belongs to the mind. The nature of the bodies themselves, as represented by our perceptions, is within its scope" (ibid., p. 170).

The memoirs were now published in French. The works were read at the Academy in the language chosen by the academician: German, Latin or French, a translator being responsible for translating them into French for the *Mémoires*. In a classical way, the Berlin Academy, as stipulated in its regulations (Histoire de l'Académie Royale des Sciences et des Belles Lettres 1752, pp. 98–102), had ordinary members: members, foreign associates, veterans and honorary members. In its 1752 version, the composition of the Academy included 50 members in addition to the president Maupertuis: 4 curators, 16 honorary members, 2 veterans and 28 members (13 in experimental philosophy, 3 in mathematics, 6 in speculative philosophy, 6 in belles lettres). The influence of experimental philosophy remained significant. The number of foreign associate members was 136, covering all the major Academies of Europe. The election of members was done by a plurality of votes of all the academicians present. Each academician could participate in the work of all classes. Every resident academician had to read two memoirs per year in session, and every foreign associate had to send one. The permanent secretary kept the registers, maintained correspondence and attended all weekly meetings. The Academy had its own infrastructure: observatory, library, amphitheater of anatomy, plant garden, museum of natural history, cabinet of machines, and its own personnel: translator, bookseller, printer, curators and inspectors in charge of the collections and the establishments. To enable the Academy to finance its personnel, its infrastructure and its operations, "Frederick granted it, in addition to the land and buildings, fairly extensive plantations of mulberry trees, the privilege of publishing civil laws and geographical maps, and the monopoly of the composition and sale of almanacs" (Bartholmess 1850, p. 179).

The correspondence between Joseph-Nicolas Delisle and Johann Wilhelm Wagner, who took over the direction of the Berlin Observatory following Kirch in 1740, and kept it for five years until his death, was instructive on the poor state of the observatory and its equipment. In a letter of July 1741, Wagner declared to Delisle not to know what Maupertuis, who visited the Observatory in December 1740, thought of its state. This state, he told him, was the one in which Delisle saw the Observatory 15 years earlier, when he went there to visit Kirch during his trip to St. Petersburg. It is understood that the state of the Observatory was worrying, but that the Academy did not judge opportune to make the required improvements. Wagner said that they could see more clearly at the end of the war. He indicated to his interlocutor that he was occupied most of his time to prepare the calendars. Here is how, in a letter of August 5, 1742, Wagner described to Delisle the precise state of the Berlin Observatory:

> You have, Sir, seen the observatory (16 years ago) and know that it deserves to be better established and supplied with better instruments; there are some, particularly quadrants, large and small, but none of which is sufficiently accurate or suitable to be used to make exact observations; the building of the observatory itself does not have the convenience due to place the instruments here and there, as I found it lately to my great displeasure, during the appearance of the comet; there is also a lack of some exquisite approach telescopes (with micrometers) of a suitable and manageable size, we do have three, a little big and quite good, 18, 20 and 25 feet long, one of which is made [by] Campani in Rome, but there has been, for several years, no place where they can be placed that allows them to be used at ease [...] there are other circumstances, which prevent a better establishment, namely, first and mainly the fund of the society is not sufficient; secondly, we have not found it appropriate to move anything at the observatory and to make useless expenses since two years ago the rumor spread that His Majesty our King had the great intention to build a pleasure castle with a large garden in place of the stable and the observatory and the surroundings, and that in this case these large buildings with our residence and several other houses would be demolished.

In a letter to Delisle dated February 1, 1744, Leonard Euler, former professor at the Academy of St. Petersburg and director of the mathematics class of the Academy of Berlin, insisted on the little means of the Academy, and on the low level of the remuneration of the director of the Observatory, which did not enable good astronomers to be attracted:

> I was quite mistaken when I thought that the new Academy would be put on the same footing as the one in Paris [...] There are four classes [...] which are filled with members from this city who for the most part have either no pension or very little: for we have no other income than that which we get from the sale of almanacs, which does not matter much [...] Among these members I am the only one to whom the King gives a considerable pension from his own fund. However, we are taking all possible pains to increase the funds, and we have already proposed to put the observatory back on a good footing; but as it would be necessary to bring in excellent instruments from England for this purpose, and as good Astronomers could not be found for such a small salary, which we are in a position to give, I do not know yet, how much I can hope for in this enterprise.

5.4. Astronomical households, institutions and gender in Bologna and Berlin

For almost half a century, no scholar of the Bologna Academy was a foreign associate of the Academy of Berlin. This was due to a quarrel between Maupertuis and Francesco-Maria Zanotti, secretary of the Academy of Bologna for more than 40 years, who criticized Maupertuis' "Essay on Moral Philosophy" (Denina 1798, p. 563). Francesco Maria Zanotti, Eustachio Zanotti and the anatomist Caldani were appointed foreign members of the Berlin Academy in 1760, the year following Maupertuis' death. Two major figures of the Italian scientific scene, the marquis Scipion Maffei of Verona, and the cardinal Ange Marie Querini, Bishop of Brescia, who inspired Benedict XIV in his actions of consolidation and renewal of the Academy of Bologna in the decade of 1740, by the gifts that they made to their cities of remarkable artistic or scientific facilities, were among the foreign associates of the Academy of Berlin in its composition of 1752. The four Italian scholars who were members of the Academy of Berlin in its first version of 1700 were all from Padua, one of them, Domenico Guglielmini, having also taught in Bologna. The marquis Giovanni Poleni, professor of mathematics in Padua, whom Celsius met during his European trip, was also a long time member of the Academy of Berlin. There was thus a certain level of interweaving between the two academies, even if it was only created late.

But beyond these few individual links, we cannot fail to be struck by certain similarities in the way these two Academies were created, on the basis of pre-existing Academies: the Accademia del Cimento of Florence in Italy and the Accademia Naturae Curiosorum in Prussia. In both cases, the members of the old academies were absorbed into the new ones. In Bologna, a young scholar passionate about astronomy, Eustachio Manfredi, built a first scientific circle, the Accademia degli Inquieti, on the initial experimental project of the Academy of Florence. Then, a military man with a passion for natural sciences and many political connections, Luigi Ferdinando Marsigli, created an institute to house his rich collection of scientific curiosities, later merging it with the Accademia degli Inquieti, and adding an art academy, with a clear ambition for universality. The practice of arts, inherited from the Renaissance, was in fact in Italy a natural extension of science. After Marsigli's rejection and death, the archbishop of Bologna, who had been enthroned as pope, became the de facto political head of the institution, and continued Marsigli's work by endowing it with absolutely exceptional means. In Berlin, the first step was taken by the famous Gottfried Wilhelm Leibniz, politically supported by the Queen of Prussia, who created an academy of experimental sciences from the old Academia Naturae Curiosorum, which was identical in its objectives to that of Bologna. Then, after the rejection of Leibniz, deprived of political support by the death of the queen, the military man Samuel Graf Von Schmettau, under the aegis of Frederick the Great, proceeded by enlarging the old academy, creating an academy of sciences and fine arts, also with a universal vocation, but there in the direction of philosophy and literature, and not of arts as in Italy. In both cases, it took almost half a century for the institution to fully respond to the role assigned to it by its founders. Both projects were similarly accompanied by the construction of an astronomical observatory, which was a structuring element: in Berlin because the Observatory building was to house the Academy, and also because the financial survival of the Academy depended on the sale of calendars, and in Bologna because astronomy enjoyed a great prestige there, in particular towards certain progressive religious authorities of the time, with whom this discipline, with its applications to geography, constituted an attractive showcase. In both cases, the means allocated to the instruments of astronomical observation, beyond the facade that this activity constituted, were limited, the astronomers working until the middle of the 18th century only with insufficient means, not to say rudimentary, just as we saw that it was the case at the Observatory of the Royal Society of Uppsala. More generally, the salaries of the academicians, either coming from the Academy or paid by the University where they taught, were most often insufficiently attractive, generating problems of high-level recruitment, and demotivation. These low salaries explain in particular the formation of the astronomical households, the family grouping together to optimize its production in its field of competence and to minimize its global expenditure, as in the case of the Kirchs or the Manfredis, and to a lesser extent the Delisles.

Concerning the condition of women, in astronomical households and in relation to the institution to which the man of the household belonged, the situation was clearly more favorable in Italy than in Germany, where the Academy of Berlin remained totally closed to women. Some Italian women, even in small numbers, managed to gain visibility and recognition from the institutions. In both the Kirch and the Manfredi families, men and women in the household seemed to function in an egalitarian way, in an atmosphere of active solidarity. Besides the interest in astronomy, the members of these households were linked by shared convictions or cultures (religious for the Kirch family, humanistic for the Manfredi family). Moreover, the men did not fail to thank the women explicitly in their publications for their help, despite Gottfried Kirch's initial hesitation, materializing the agreement that reigned in the household. In the case of the Manfredis, the literary publications of the sisters, although unsigned, circulated among the Bolognese bourgeoisie, who knew the authors, which constituted a recognition for these women, all the more so as they thus participated in the household economy by supplementing the modest salary of their professor brother. In the case of the Kirchs, Winkelmann benefited from a de facto recognition of her scientific value, including within the institution, which rejected her in the shadow of her husband for reasons of convenience and image, on the pretext that her recruitment would create a precedent, inciting other women to want to invest in the institution. Both Winkelmann and the Manfredi sisters were supported by men of great intellectual or political authority, such as Leibniz in the case of the former, or the future pope Lambertini in the case of the latter, but this positive judgment on the capacity of an individual woman to exert a scientific activity, shared in an almost unanimous way, did not constitute the root of the question. The place that the society of the time assigned to women was at home, where their role was to have children and to educate them. This question of submission to the rules of propriety was particularly well illustrated by the prohibition on Kirch women being present at the Observatory during visits by outside officials. The case of Angélique Delisle is particular, since she was the official representative of her brother, receiving the salary that he would have received if he had remained, at the scientific institution on which he depended. It seems that she had no academic activity as such, and it was as a steward of her brother's affairs that she was seen at the Academy, and certainly respected as such.

In Italy, the tradition of Renaissance Italian city-states, encouraging exceptional women to contribute their knowledge to the good and influence of the state, made the situation for women somewhat different (Frize 2013). Many Italian academies, including that of the Inquieti, began as discussion clubs, meeting in each other's salons, where men and women discussed without distinction of gender topics of philosophy or literature, even science, or played games. From the beginning of the 17th century, women were associated with the Accademia degli Incogniti in Venice or Rome. A woman called Margherita Sarrochi created around this time an Academy in her Roman house, inviting Galileo during his stay in that city in 1611. Some scientific academies, however, such as the Accademia dei Lincei, founded in 1603 in Rome on the model of a religious brotherhood, or the Neapolitan Academy of Sciences, founded in 1732, did not accept women. The Academy of Bologna was one of the (somewhat) feminized academies. It first had the Manfredi sisters, as part of the Inquieti, and then several other women scientists in the 18th century. The physicist Laura Bassi was one of them. On April 17, 1732, at the age of 20, she presented a thesis in philosophy at the University of Bologna, and was elected member of the Academy of Bologna on May 20. She was, with the mathematician and physicist Faustina Pignatelli Carafa, elected the same year as her, the mathematician Maria Gaetana Agnesi, elected in 1750, and the anatomist Anna Morandini Manzolini, elected after the death of her husband in 1755, one of the few Italian women integrated into the Academy of Bologna in the 18th century. Another famous woman, the French Émilie du Châtelet, author of a translation of Newton's work and herself a high-level philosopher (see, for example, Ray 2013), was associated with the Academy as a foreign member in 1746. About Pignatelli's election, it was not without difficulty: "Soon after being admitted, Pignatelli anonymously published her solutions to several mathematical problems in the *Acta Eruditorum*. The Bologna academicians felt obliged to admit Pignatelli because of this virtuoso display of erudition, 'having the most certain testimonials of this lady's great and marvelous worth in mathematics, and especially algebra,' but they did so grudgingly, swearing to each other that they would 'not accept any other woman into the academy'" (Findlen 2016, p. 46). Here, as in Berlin, the fear of setting a precedent by recruiting a woman was not absent, though not prohibitive. Agnesi, a protégé of Benedict XIV, like Bassi, was offered a professorship at the University of Bologna in 1749, but she refused it and withdrew to devote herself to her research and to charitable works. Laura Bassi accepted the professorship offered to her in 1732, but her duties were ill-defined; she was not invited to the meetings, nor was she allowed to teach publicly in the university premises, and the university authority tried to confine her to literature and the writing of poems, far from Newtonian physics, of which she was nevertheless a highly competent follower (Pigeon 2014, p. 275). Married to a professor of medicine and physics at the university, she played her role as a mother by having nine children. She sporadically taught anatomy. Tired of a position of which she had only the title, without being able to play a true role as professor within the university, she decided in 1744 to found a school at home with her husband and to install a laboratory there. For 30 years, she devoted herself, in her house, to teaching and research, notably in mechanics and electricity. In 1758, another woman, Anna Morandi Manzolini, whose reputation in wax modeling of various body parts spread throughout Europe, obtained a chair of anatomy at the University of Bologna (*Portraits de médecins* 2017), and thus became the second woman professor after Laura Bassi. These two women, also members of the Academy of Bologna, published scientific articles under their names in the memoirs of the Academy, unlike the Manfredi sisters. They were therefore recognized as equal to men, even if their life at the university was more complicated and less rewarding than that of their male counterparts. Bassi had no money problems, inheriting a comfortable paternal fortune. As for Manzolini, she had financial problems when she fell ill in 1765, but she was helped by a wealthy senator who bought her collection of instruments and books, and gave her shelter in his palace. These few women recognized by the institution benefited from political protection at the highest level, including from Pope Benedict XIV, a progressive who advocated for a greater opening of the Academy of Bologna to foreigners and women.

The presence of women in the Academies, including those dealing with science, was quite specific to Italy, the other Academies in Europe being exclusively male. But can we consider that women scientists in Italy enjoyed true equality with men as women? The answer is clearly no. We have underlined the ambivalence of the judgment of Fontenelle, who respected these women only because, according to him, they did not have the defects of superficiality that he lent to women in general, in a way because these women were in fact men, which made them respectable. The case of the Accademia dei Ricovrati (literally "Sheltered Academy") of Padua, and the debate that was organized there in 1723, is revealing in this respect. The Accademia dei Ricovrati was founded in 1599, and included Galileo among its founders. By the end of the 17th century, it was one of the Italian academies that had women members, but only as honorary ones: they could not vote, hold administrative positions, or address the Assembly, except by invitation, on rare occasions (Frize 2013, p. 19). Of the 26 women admitted to the Ricovrati before 1723, only four were Italian. The others were French, and did not attend the meetings. One of the members was the Venetian philosopher and mathematician Elena Cornaro Piscopia, the first woman in the world to obtain a doctorate, in 1678, at the age of 32. Among the French women, there was the philologist Anne Dacier, elected in 1679, and the writer Madeleine de Scudery, elected in 1685. On June 16, 1723, a debate was organized in Padua on the question: "Should Women Be Admitted to the Study of the Sciences and the Noble Arts?" (Messbarger 2002, pp. 21–48). The debate was introduced by Antonio Vallisneri, "Prince" (President) of the Academy since the previous year, a philosopher and professor of medicine who was a follower of experimental science, in the tradition of Galileo. Vallisneri was committed to a modernization of the institution in accordance with the values of the Enlightenment, putting in particular the issue of education of women at the center of the group's concerns. Guglielmo Camposanpiero, aristocrat of Padua and librarian of the university, defended the instruction of the women in sciences and arts, and Giovanni Antonio Volpi, professor of Greek and Latin at the University of Padua, was opposed to it. The debate was held in the evening, in a room of the prefectural palace, between the two protagonists, under the arbitration of Vallisneri. The debate in the tribune was between men, the women being confined to the assistance, considering the conventions of the time preventing the women from intervening publicly.

In his defense of access to education for women, Camposanpiero first insisted that "the ability of women to master the arts and sciences" was a proven fact, even within the Academy. The educability of women was widely accepted, even by misogynists who believed that despite this ability, women should not be educated. For him, the point of the debate was not whether women had the intellectual capacity to be educated, but rather what effect the mass education of women would have on established social structures. He did assert that it was the selfishness of men that was at the root of a policy of keeping women ignorant, but this criticism of men, a constituent element of Enlightenment ethics, led only to traditional ideas about the benefits of women's education: a better exercise of responsibilities at home, an optimized management of the family, a good moral education for children. Thus, Camposanpiero came to confuse the public good with the domestic confinement of women. He subscribed to the widespread opinions of the time (as we have seen with Fontenelle), according to which it was necessary to educate women to correct their corrupted nature, poured in frivolous or immoral preoccupations. He said in particular: "How great an advantage to the whole world would be knowledgeable women, who would delight themselves with anything but vanity! They would not look for ornate and affected attire in men, nor certainly weak and insipid grace. Rather, they would enjoy seeing men adorned in those rare and useful doctrines, and schooled in beautiful and eclectic arts" (ibid., p. 35). With this clever discourse, he satisfied his male audience by making the satisfaction of men and the moral integrity of women the goal of women's education. At the same time, by making men puppets of fashion and by advocating the education of women, he flattered his female listeners of the good urban bourgeoisie, most of them better educated than the average woman of the time.

Figure 5.7. *Coat of arms of the Accademia dei Ricovrati representing the image of a cave with an opening at both ends and sheltered by an olive tree, with the motto borrowed from Boethius: "Bipatens animis asylum" ("a sanctuary of the spirit open at both ends") (see Wikipedia, Accademia dei Ricovrati)*

His opponent, Volpi, started off head on, but skillfully, by affirming that not only was the woman different from the man, but above all that women of different social and economic levels were different among themselves. This formulation allowed him to differentiate the educated women in his audience, whom he spared, from the "masses of ordinary women". Volpi's central argument was that women's submission to men was the very foundation of their happiness. Prohibiting women's knowledge of the arts and sciences was useful not only to the Republic but also to women themselves. Nature had entrusted men with the task of shaping history and directing social progress, and women, intellectually deficient in comparison with them, were thus protected from the anxiety inherent in any active participation in the world. Second, he made historical arguments. Women have always been subjected to men, from civilization to civilization, proving that women were unable to leave their homes. Then followed a development on the biological and physiological inequalities between men and women, a link being explicitly established, concerning the woman, between weakness of the body and weakness of the spirit. Finally, at the top of his argument, Volpi placed the reproductive function of the woman: stating that nature had produced women for the sole purpose of having children. The nine months of motherhood presented as a "punishment for original sin", a period of suffering, "with atrocious pain and evident risk of death", to be followed by breastfeeding and the care of infancy, "burdens all of which are vexing if not intolerable, and in regard to which the union of Man and of Woman, *Matri*mony, was named for her rather than for him" (ibid., p. 38). But the ultimate argument Volpi made was that of the risk of the end of the human race with the education of women:

> Once she is instructed, a woman's lust for knowledge will become insatiable; educated women will thwart the public good and jeopardize the survival of the species by refusing to marry and to bear children; educated, married women will rebel against the authority of their husbands. The woman philosopher will plague her husband with academic questions when he returns from a day made difficult by the demands of the world (ibid., p. 39).

In Vallisneri's judgment at the end of the debate (ibid., p. 40), he called for the admission to the sciences and arts of women who were in love with them, "predisposed to virtue and to glory by a noble and occult genius, in whose veins flow clear and illustrious blood, and in whom excites and sparkles a spirit outside of the norm, surpassing that which is common to the masses, exactly that which [he] recognizes" in the women of his audience, whom he sought to flatter. He thus affirmed, by promoting education only for an elite of "exceptional women", the general inferiority of women, and was in fact opposed to education for all women. Vallisneri's position was probably representative of the opinion of a very large proportion of educated and progressive men of the time defending women scientists. Volpi's positions, which he later amended, were refuted by women like the mathematician Maria Gaetana Agnesi and the Sienese painter and poet Aretafila Savini de Rossia, member of the Accademia d'Arcadia, who in 1723 wrote a plea in favor of women's studies. In the final account of the debate that Vallisneri had published in 1729, which included contributions from several women, Savini de Rossi spoke in the name of all women, systematically using the "we" form. She refuted the distinction between exceptional women and a mass of women. She wrote that, even if she had not been educated, she would be able to defend the cause of women. In her proposal for the education of women, she defended an essential transformation in the existing social order for the general benefit of women. She advocated for an egalitarian system. "Allowing women access to a formal education, each according to her constitution, her circumstances, and above all her talent, would not only be a great but an honest diversion for women." She opposed the class distinction introduced by Volpi between educated and common women: "Nature does not demonstrate greater partiality toward the poor man, the rich man, the nobleman, or the plebeian" (ibid., p. 45). She anchored her advocacy, not in global history, but in the microcosm of individual experience. She linked the public good directly to women's education, and contradicted Volpi's claim that educated women would refuse to bear children. But she did not question the social and political model responsible for confining women to the domestic circle. Although she argued that education would lead women to truthful and virtuous lives, the honest woman, for Savini de Rossi as for Volpi, was the one who actually fulfilled the demands of domestic life. A woman's virtue was measured by the extent and mastery of her care for her family and her home, and women's well-being was defined in this domestic context, being therefore essentially contingent. Here is how Messbarger concluded:

> In citing in her argument the martellian verse: "A daughter serves her father; a wife serves her husband; a widow her decorum, and they die having only served," Savini De Rossi points to women's perpetual state of subjugation not to inveigh against men's abuse of their daughters and their wives and to advocate an alternative, but instead to counter her opponent's claims that an education will lead women to abandon marriage for a life of freedom. Although she considers this condition "tragic," and the laws of the state that perpetuate women's debasement harsh, she would appear to accept that woman "is never free" (ibid., p. 47).

5.5. Conclusion

The debate in Padua, and the reactions it provoked, show the nature of the freedom then given to some women to participate in scientific institutions in Bologna, and elsewhere in Italy. The condition of these women scientists, which was part of the Italian tradition of "exceptional women" of the Renaissance and the early modern period, most often from aristocratic backgrounds, and which was put into perspective by the spirit of the Enlightenment, was at the same time that of a recognition of their own scientific value by educated men who respected them as such. It was also of an exhortation to use this knowledge primarily to run their households intelligently and to bring up their children morally well. A woman scientist had to be above all a good wife and an exemplary mother for her children. The integration in the University or in the Academy of Bologna of women like

Bassi, Pignatelli and Manzolini was not made without difficulty, remaining strictly limited to a small number of individuals, and perhaps it would not have taken place without the support of the archbishop Lambertini, a powerful man and benefactor of the Academy. It is very probable that the Manfredi sisters, who were members of the Accademia degli Inquieti from the beginning, a small scientific enterprise of family character, would not have been able to enter directly into an academy of the level of that of Bologna, whose level of institutionalization was quite different, and the gender conventions were more restrictive. Nevertheless, it was the heritage of these small academies of the 16th and 17th centuries that, in Italy, made possible the integration of some exceptional women into the Academy of Bologna, the absence of this heritage elsewhere in Europe leading to the unquestionable rejection of women, as shown by the example of the Kirchs at the Academy of Berlin.

Conclusion

The period covered by the facts recounted in this book, the first half of the 18th century, in addition to being part of a profound change in scientific practice, saw, as we have seen, the institutionalization of science throughout Europe, with the creation of large academies, mainly in St. Petersburg, Bologna, Berlin, Uppsala and Stockholm. The aim was to compete with the state academies created in the previous century in London and Paris, these seven academies alone gathering more than a third of the European astronomers of the time. The new academies were all equipped with an astronomical observatory, just like their predecessors. The correspondence of Joseph-Nicolas Delisle, presented in the chapters of this book, provides a lot of information on the difficulties encountered in these different observatories during the first half of the 18th century, which were due to several factors, the most significant of which was the lack of means, both in terms of remuneration of the astronomers and the purchase of equipment, especially in Bologna, Berlin and Uppsala, a lack of means partly resulting from a lack of political support for science in the first stages of development of the academies that were founded there. The service tasks introduced in the observatories, such as the preparation of calendars in Berlin, for example, or the cartographic activity, especially in St. Petersburg, which took up part of the astronomers' time, also contributed to slowing down the scientific development. Another obstacle was the power struggles in the academies, as in Uppsala where after the death of Celsius, the observatory was privatized to the benefit of a single man, or in St. Petersburg where the perilous exercise of the management of the new academy under the pressure of contradictory political injunctions led to a serious conflict between the administrator Schumacher and a number of the academicians. To a large extent, the service tasks that diverted the astronomers from their main mission were the consequence of the lack of means allocated to the pure scientific activity, the scientists having to finance their science on the income of these additional activities. As for the conflicts, they were also the consequence of the limited means allocated to science, which individuals or small pressure groups tried to recover for their profit.

Let us return briefly to the lack of political support experienced by some of the young academies (Chassefière 2021d). In Berlin, the period from 1713 to 1740, corresponding to the reign of Frederick William I, a king who was not interested in the arts and sciences, constituted a period in the wilderness for the nascent academy. The publication of the Academy's memoirs, revived in 1735 by Jablonski and Viereck, nevertheless led to the printing of three volumes between 1734 and 1744, as many as in the first 30 years of life of the Academy. Each one of these volumes contains astronomical observations worth "a beautiful fame to the most useful members then to the society, to the astronomers whose ephemerides and the calendars were, until 1740, the only assured resource of the academy" (Bartholmess 1850, p. 124). The observations were nevertheless carried out with deficient instruments, as described by Wagner in his correspondence with Delisle. After the accession of Frederick the Great, the observatory was threatened with destruction for a time. The three Silesian wars between 1740 and 1763 between Prussia and Austria monopolized the attention and the finances of the King of Prussia. It was only in 1764, five years after the death of Maupertuis, president of the Academy from 1745 to 1759, that Frederick the Great, anxious to develop the sciences, posed himself, "no longer only as *Protector*, but as supreme director, as high administrator of the Academy, as its *Curator*" (ibid., p. 205). In Bologna, it was the placement of the nascent academy under the direct tutelage of the city's Senate that thwarted the proper scientific development of the institution, which its leaders conceived as a mere showcase for the public and foreign visitors. It was finally thanks to the mediation of the archbishop of Bologna, Lambertini, who became pope in 1740 under the name of Benedict XIV, that the Academy obtained significant means, enabling in particular the observatory to be equipped with quality instruments, in replacement of a mediocre stock, as Zanotti highlighted in his correspondence with Delisle. It was necessary to wait until 1742 so that the new instruments were installed and ready to function. Lambertini's role was decisive, not only for astronomy but also for all the scientific and artistic disciplines represented at the Academy, with numerous financings and donations during the 18 years of his pontifical mandate. Uppsala lacked the presence of a powerful and enlightened benefactor, such as Frederick the Great in Prussia, or Benedict XIV in Italy, for astronomy to develop fully. Celsius nevertheless obtained at the beginning of the 1740s from the Senate of his University the construction and the equipment of an observatory, certainly modest, but operational and equipped with some good instruments. His death in 1745 interrupted the dynamics of equipment at the observatory launched with the Swedish manufacturer Ekström, probably because Hiorter was not a famous observer, and that he could not or did not wish to continue the equipment policy of his predecessor. He thus broke the dynamics of the Astronomical Observatory of Uppsala's rise in power. It was, as we saw, only in 1751, one year after Hiorter' death, that Wargentin succeeded in interesting the royal power thanks to the campaign of observations coordinated with that of Lacaille at the Cape of Good Hope, for which he obtained from the king a financing for a team of Swedish observers spread over the territory, and at the same time the recognition of astronomy at the highest level of the State. But Wargentin, unable to assert himself in Uppsala, set up an observatory in Stockholm, operational in 1753, within the framework of the newly created Swedish Academy of Sciences.

The low salaries of astronomers in the three observatories that we have just mentioned generated, on the one hand, a lack of attractiveness that did not allow for the recruitment of quality astronomers at an international level, as Euler said about the Berlin Observatory, and on the other hand, a reduction of the time devoted to astronomy due to the exercise of other remunerated activities, such as Manfredi's role as superintendent of water in Bologna, or the additional activities of Wargentin or Hiorter in Uppsala (Chassefière 2021d). We have seen that Delisle's salary as the head of the Imperial Observatory of St. Petersburg was very high: 1,800 rubles per year, or 9,000 francs. In the same period, Jacques Cassini earned 3,000 francs at the Paris Observatory (Hahn 1975, p. 507), corresponding somewhat to Edmond Halley's salary as director of the Royal Observatory of Greenwich from 1727 (Ditisheim 1925, p. 554). While astronomers in the Paris and Greenwich Observatories thus had decent salaries, the same was not true in Berlin, Uppsala and Bologna, where the annual salary was rather around the equivalent of 1,000 francs: 500 thalers in Berlin (des Vignoles 1741, p. 337), 2,100 dalers in Uppsala (Ekman 2016, pp. 72–73), and less than 800 lire in Bologna (salaries of the Institute's president and secretary, see Elena 1991, p. 514). In terms of means for the equipment, it was incontestably at the Observatoire Royal in Paris that the expenditure was the most significant, approaching 70,000 francs in the 17th century, completed by purchases on funds of the Academy up to approximately 20,000 francs in the following century (Wolf 1902). Most of the instruments bought at the time were used for distant measurement campaigns; either they were deteriorated or the astronomers having used them then recovered them for their private observatories, but they did not remain usable for a long time at the Observatory. Indeed, Colbert did not provide the Observatory with an annual rent allowing the maintenance and the regular renewal of the instrumental stock. The Royal Observatory of Greenwich had no recurrent funds for the acquisition and the maintenance of its instruments (*History of the Royal Observatory* 2021), and its first astronomer, John Flamsteed, had to have it equipped at his own expense. His successors in the 18th century, Halley then Bradley, received a total royal investment of £1,500, or about 20,000 francs, for the purchase of instruments (The Royal Observatory Greenwich 2021). In St. Petersburg, the substantial instrumental stock set up at the Imperial Observatory in the middle of the 1730s (Struve 1845, pp. 7–8), whose quality seemed equivalent to that of the equipment of the Paris and Greenwich Observatories, was entirely destroyed in the fire of 1747, and it was not until 1760 that new instruments were ordered from English manufacturers. In Berlin and Bologna, Delisle's correspondence tells us that the instruments remained of poor quality during the first half of the 18th century, improving nevertheless in Bologna from 1742 thanks to the financial support of the

Vatican which enabled the purchase of instruments from the English manufacturer George Graham, then in Berlin, in 1768 through the order of an instrument from another English manufacturer, John Bird (Turner 2002, p. 375). The Astronomical Observatory of Uppsala was equipped, at the construction of the new building in 1741, with a modest set of instruments purchased in Paris, and in London from George Graham (Ekman 2016, pp. 72–73). Thus, the Observatories of Bologna, Berlin and Uppsala were equipped with quality instruments only from the middle of the 18th century onwards (Chassefière 2021d).

The aurora borealis is an emblematic object of the experimental approach. Its reappearance at the beginning of the 18th century, after nearly a century of absence during the Maunder Minimum, was the subject of detailed accounts of the observations that were made throughout Europe. The richness in colors and shapes of the visual phenomenon, and the complex evolutionary dynamics of these shapes, gave rise to detailed descriptions, which were described as "Baconian", by the scientists of the time. The phenomenon was then judged of atmospheric nature, and it was the astronomers in charge of the meteorological statements in the observatories who were more particularly in charge of observing it, like Friedrich Christoph Mayer at the Imperial Observatory of St. Petersburg, or Jacques Philippe Maraldi at the Observatoire Royal. But most of the astronomers of the time, spending many nights observing the sky, were in fact the spectators of the aurora borealis, irregular phenomena whose occurrence cannot be predicted, particularly frequent in northern cities, such as St. Petersburg or Uppsala. The observation of the aurora borealis, for many astronomers of the time, was not the primary objective of the nights spent observing, centered on the observation of various kinds of eclipses and occultations, and also the trajectories of comets, which allowed for Newtonian scientists such as Delisle and Celsius to test Newton's theory of attraction in order to validate it. As we have said, the aurora borealis are almost absent from the correspondences maintained by the scientists of the time, which did not prevent them from accumulating observations, that their astronomical instruments, within reach when they appeared, allowed them to characterize angularly. The aurora borealis, whose answer to the explanation cannot be found in the refinement of the calculations of celestial dynamics, even if the system proposed by Mairan used an astronomical determinant, is clearly outside the field of the disputes which opposed Newtonians and Cartesians, or Newtonians and Wollfians, on the continent. On the contrary, there was a consensus among scholars of all persuasions that the precise observation of the aurora borealis, accompanied by a detailed account of its manifestations, was necessary in order to identify some recurrent characteristics that would allow for understanding its mechanisms. While certain scientists put forward proposals of systems, like Halley in London, Mayer in St. Petersburg, Maraldi, then Mairan, in Paris, Plantade in Montpellier, most of the aurora observers, like Celsius or Delisle, retained positions of pure observers, not taking sides for any system. The aurora borealis, like meteorology, are perfect examples of phenomena that were difficult to explain, and which were only understood in the 19th century, or even a little later for the aurora borealis, and to which the experimental philosophy of the Enlightenment, with the asserted need to accumulate observations and experiments over a long period of time before being able to build a system of physics, applied perfectly. They are, in a way, textbook cases, to which, for example, Fontenelle referred explicitly in his articles in *Histoire*. This emblematic nature, vis-à-vis the new philosophy, of the aurora borealis, combined with its sudden appearance in this period of change in scientific practice, were responsible for the enthusiasm that manifested itself with regard to the phenomenon throughout the first half of the 18th century.

There was no clear-cut boundary between Cartesian (or Wolffian) and Newtonian doctrines in this period of change in scientific thought. In this book, we have used the term Cartesian to refer to those who used vortex models of subtle matter whose origin is to be found in Descartes' work, although there were evolutions between his theory and that of his successors (see, for example, Schmit 2020). Mairan can thus be defined as Cartesian, even if he was concerned to include in his systems elements of Newtonian theory, without adhering to the idea of an attraction at a distance as an intrinsic property of bodies, an idea of which we have seen that Newton himself doubted. Anders Celsius was deeply influenced by Wolff's philosophy and his mathematical method, which did not prevent him from adhering to the Newtonian vision of the world, which he shared with Delisle, or with the members of the Lappish expedition to measure the shape of the Earth directed by Maupertuis in which he participated, and this in spite of the basic oppositions between Cartesian or Leibnizian rationalists and English empiricists. Fontenelle, a last-minute defender of Cartesian vortices, advocated the necessary anteriority of observation and experimentation on the construction of systems of physics, breaking by his historical approach with Descartes' approach consisting of affirming a priori a metaphysical system of the world (the vortices of subtle matter), Fontenelle's Cartesianism should rather be considered as an adherence to the occurrence of a time of access to reason that Descartes embodied (Mazauric 2007). Fontenelle or Mairan's experimental philosophy, which was also Wolff's, was in many ways closer to Newton's empirical approach than to Descartes' aprioristic metaphysics. Concerning the aurora borealis, there was a great porosity between the schools of thought. Thus, the Aristotelian explanation in terms of inflammation of terrestrial exhalations applied to the aurora borealis as well as to other luminous meteors, which Descartes took up in his *Météores*, was adopted by the Newtonian physicist Pieter van Musschenbroek. In the same way, Halley, a convinced Newtonian, took up Descartes' magnet theory to explain the aurora borealis without adhering to the explanation of gravity in terms of pressure exerted by subtle matter, an idea about which he expressed his deepest incomprehension (Halley 1687). The scientific advances of the time, as we can see, were not hindered by the spirit of the school of thought, quite the contrary. All ideas were good to take, in this period of intense expansion of the new scientific thought.

In an interesting way, the partisans of the Cartesian vortices at the Académie Royale des Sciences, Mairan and Fontenelle, seized upon the aurora borealis, an object we have just seen that was not very divisive by nature within the philosophical landscape of the time, as an opportunity to assert the Cartesian mechanism against the progression of Newtonianism. It was not so much by taking up the vision of the aurora borealis expressed by Descartes in his *Météores*, strongly inspired by the Aristotelian vision, and revived in particular by Wolff and Mayer in the German–Russian world, and Maraldi in France, but rather by proposing a large-scale cosmo-atmospheric mechanism involving the entire planetary and cometary environment of the sun, that Mairan affirmed the French specificity. The explanation in terms of magnetism proposed by Halley was itself inspired by Descartes' magnet theory, and could be considered more "Cartesian" than Mairan's, even if Halley was deeply Newtonian. But Mairan placed himself on a completely different plane, proposing a mechanism involving subtle matters, in the best Cartesian tradition, while affirming that he did not belong to any school, and while claiming the integration of Newtonian mechanics in his system when he calculated the size of the Earth's sphere of gravitational attraction using the inverse square law of distances. Mairan's system, which claimed the assimilation of the Newtonian theory of attraction, retaining the formulation of the law of attraction at a distance, while rejecting attraction at a distance deemed non-physical, constituted a form elaborated from the Cartesian mechanism, revived in the circumstances and adorned with the explicit refusal of the schools of thought by its author. The rejection by the Parisian Academy of François de Plantade's memoir emanating from its Montpellier branch, a memoir defending a system similar to that proposed by Halley, at the cost of a serious institutional crisis, illustrates in our opinion the will of the Academy to affirm a French Cartesian specificity in a period of renewed tension between the two great schools of thought on the continent. Moreover, the way opened by Mairan to a very great height of the atmosphere allowed, a few years later, Jacques Cassini to propose a revision of the Boyle-Mariotte law on high mountains, where the air is different (purer, less loaded with vapors and exhalations of all natures), enabling reconciliation between the heights of the atmosphere deduced from the parallax measurements made on the auroral arcs and those based on the measurements of barometric pressure made on the mountains around a value of about 2,000 km (Cassini 1733). Cassini, two years later, published a theoretical article in the *Mémoires de l'Académie* in which he asserted that the height of the atmosphere required reconciling the Earth's speed of rotation on itself with the Cartesian idea that this rotation was caused by the effect of frictional momentum by the aether on contact with the upper surface of the atmosphere was 40,000 km (Cassini 1735). This height was certainly greater than the value of 2,000 km deduced from observations of aurorae and pressure, but Fontenelle found, in his *Histoire* article of

1735 introducing that of Cassini's, possible explanations to this disagreement. The aurora borealis was thus an important part of the Cartesian counter-offensive at the end of the 1720s and the beginning of the following decade (Chassefière 2021b). Moreover, Mairan's system, which was never presented by its author as definitively proven, called for long series of observations intended to validate the link between the position of the Earth around the sun and the frequency of occurrence of aurorae, and was perfectly in line with experimental philosophy, as outlined by Fontenelle in his foreword of 1699. It is effective that the system proposed by Halley and Plantade did not lend itself to the idea of a long-term validation by accumulation of observations.

Let us return to the scientific question of the aurora borealis and to the evolution of ideas on the subject during the second half of the 18th century. The 1750s were marked by the rapid development of the observation of atmospheric electricity by metallic points and kites, which led to the demonstration of the electric character of stormy phenomena, and the attribution to electric fluid of the origin of flying fires (meteoroids entering the atmosphere) and the aurora borealis. These different phenomena were considered by many scientists of the time to be closely linked (see, for example, Bertholon 1787). From 1750 onwards, there was a general tendency to attribute many phenomena involving fire to electricity, and the aurora borealis was no exception to the rule. The missing links in the chain to synthesize the existing systems of the aurora borealis were (i) the link between electricity and light emission, and (ii) the link between electricity and magnetism. The Dutch physicist Jan Hendrik Van Swinden reported in his collection of memoirs on the analogy of electricity and magnetism (1785) detailed observations, over long periods of time, of the irregular variations of the magnetic needle during the aurora borealis discovered by Celsius and Hiorter a few decades earlier. It was based on two series of measurements: one made by Father Cotte between 1768 and 1779 and which related to 134 auroras, of which only 53 agitated the needle, the other made by himself between 1771 and 1781 relating to 284 aurorae, of which 122 agitated the needle, and 20 altered its daily variation. Swinden admitted that he had no explanation for the randomness of the connection between the two phenomena, which occurred at best in only half of aurora cases. Favoring the Mairan system, he imagined that the solar matter acted on the magnetic needle when it penetrated sufficiently deeply in the atmosphere, without however finding an explanation to the phenomenon.

He then examined the hypothesis of a possible electrical origin for the irregular agitation of the needle often observed in the presence of an aurora borealis. John Canton was one of the first, at the beginning of the 18th century, to notice that an electric current passing through an airless enclosure, or at very low pressure, produced a light comparable to that of the aurora borealis. In the middle of the century, Canton presented a memorandum to the Royal Society in which he defended the idea that the aurora comes from a discharge between clouds electrified positively and negatively, occurring through the upper atmosphere, with a lower electrical resistance (Canton 1753). Many physicists, following Canton, attributed to the aurora borealis an electrical origin, without having a precise explanation, and deduced from the agitation of the magnetic needle a link between electricity and magnetism. How could the agitation of the needle depend on the electricity in the air? Swinden cited three sources of induction in support of this hypothesis: "1. the electric constitution of the air during the time that this agitation is observed. 2. The similarity of these agitations with those which Electricity can produce, & which it sometimes produces: finally 3. the influence of Thunder" (van Swinden 1785, p. 202). He reviewed the three types of effects in light of all existing observations. Concerning the first, he showed that there was no evidence from existing measurements that the air was more electric when the needle was agitated. The second effect would be that the electricity agitated the needle as it agitated any light and mobile body in the vicinity. However, a brass needle was not agitated during the aurora period, whereas a steel needle was, which contradicted the hypothesis of an electrical origin. Moreover, experiments comparing needles in the open air and needles enclosed in boxes during thunderstorms showed that only the needles in the open air moved, suggesting an effect of air movement rather than electricity. The only effect that Swinden recognized as real was the direct effect of lightning on the needle, because it was a particular third effect, which could in some cases reverse the polarity by changing the substance of the needle itself. From his study, Swinden concluded that "the irregular movements of the Magnetic Needle do not depend on Electricity; but that it is very probable that they are all due to the action of the Aurora Borealis, though we do not know how this meteor acts" (ibid., p. 238).

An article from the end of the 18th century mentions a statement by Joseph Priestley, an English theologian, chemist and physicist, known for his work on electricity, reporting the opinion of Giambatista Beccaria, a physicist from Turin of great reputation, specialist in atmospheric electricity, as to the fluid responsible for the aurora borealis, which he identified with electric matter. According to Beccaria, "since a sudden stroke of lightning gives polarity to magnets [...] a regular and constant circulation of the whole mass of the fluid from north to south may be the original cause of magnetism in general" (*An Historical Miscellany* 1794, pp. 43–44). He supposed that this electric current could have different sources in the northern hemisphere of the Earth and that the aurora borealis was formed by this electric matter making its circulation towards the south, made visible in certain places by a particular state of the atmosphere where it moved, or by the fact that it got closer to the Earth's surface. Beccaria thus took again, identically, Halley's explanation but substituted magnetic matter with electric fluid. Both Halley's explanation and Beccaria's were nevertheless contradicted by the first observations of aurorae in the south by Westerners, namely George Forster (1777), who observed in 1773 several aurorae between 58° and 60° south latitude. These aurorae showed an ascending movement of light columns rising from the horizon, identical to that observed in the northern hemisphere. This observation suggested a movement of the matter responsible for the aurorae from the poles to the equator, and not from pole to pole, as Halley, then Dufay (1730), hypothesized by attributing to the terrestrial magnetic vortex the origin of the aurora. The author of the 1794 article imagined, in order to conform to this fact of observation, a sink of electricity in the tropical region, the frequent thunderstorms discharging the hot and humid lower atmosphere of its electricity into the Earth, much more than at medium and high latitudes. This sink required a contribution of electric matter from the upper tropical atmosphere, which was therefore necessarily recharged by the circulation of electrical matter from the poles to the equator at high altitude. In a somewhat similar scheme, although reversed, Benjamin Franklin proposed in 1779 an explanation of the aurora borealis which appealed to the circulation of air from the intertropical region towards high latitudes, this one transporting the electrically charged air towards the poles and accumulating by precipitation the electricity on the electrically insulating ice cap (Franklin 1779). The accumulation of electricity on the ice would lead to an electric discharge between the surface and the upper atmosphere, the electric fluid migrating, in altitude, from the poles to the equator by diverging strongly, thus creating above the polar regions the luminous concentration of the aurora.

The electric explanation of the aurorae thus took the place of the magnetic explanation from the middle of the 18th century, by a substitution of electric matter for magnetic matter. Mairan's system was then abandoned. The discovery of the influence of the aurora borealis on the compass needle had contributed to a return to Halley's system, which had been eclipsed for a while, giving a certain legitimacy, even if the connection between electricity and magnetism to explain the aurora had then led to a loss of interest in the aurora borealis as a phenomenon produced by the terrestrial magnet, and was in fact useful to the understanding of the magnetism of our planet. A century later, in the middle of the 19th century, the community had a certain number of theories, as many pieces of the puzzle had been discussed for a hundred years, but none of which was satisfactory. Jean-Baptiste Biot wrote then on this subject:

> In reviewing these various systems, one realizes that each of them refers especially to some particularity of the phenomenon, on which it is made and molded, so to speak, while the others are forgotten; so that there is truth in all of them, although none of them is totally true. From the little success of these attempts, it seems that one would walk with more safety if one took an absolutely opposite road, that is, if one considered each observed peculiarity as a condition given by nature; and if, after having ascertained its reality and weighed its importance, one made it a character of the unknown cause by which this phenomenon is produced (Biot 1820, p. 349).

Halley saw terrestrial magnetism as a central element of the phenomenon, structuring the auroral figure around the magnetic field of the Earth's magnet, invoking a subtle magnetic matter circulating in the pores of the Earth and rising to a great height in the ether. Mairan attributed to the phenomenon a solar origin, imagining that in the manner of the comets, the Earth regularly collected the subtle matter of the solar atmosphere, which came to be arranged in strata in its high atmosphere of subtle air. The phenomenon of aurorae occurred therefore in the atmosphere, but at a very high level, this being possible only because of the extreme subtlety of the solar matter. With Canton, and Franklin, the hypothesis that it was electric currents circulating in the high atmosphere, very rarefied, which were responsible for the auroral gleams, which would have the same origin as the light which was born from a current in a low-pressure enclosure, emerged from experiments made in a laboratory, instead of the conceptions of Halley and Mairan resulting rather from theoretical considerations, inherited from Descartes for what concerned magnetic and solar subtle matters. These different conceptions contained at the beginning elements that would allow, half a century later, Birkeland and Störmer (Birkeland 1908) to solve the enigma for good.

Concerning the great height of the atmosphere deduced from parallax measurements made on the auroral arcs, it was far from being unanimously accepted, in particular in the English community, where the scientists were more than reserved on the question. Cavendish, in a 1790 memoir, estimated the height of an aurora that occurred in 1786 at between 83 and 113 km, that is, the height of the center of the auroral arc, the structure that seemed to him to be the most reliable for this type of estimate, but he did so only to immediately question this estimate. He hypothesized that auroral arcs may consist of bands of material that were wide in height, but thin in depth, so that an observer below the arc saw a narrow, well-defined band, while an observer far from the arc's plane saw a wide, tenuous band, and a distant observer far from the arc's plane saw almost none. He concluded his paper by judging that the height estimates were doubtful, because the observed structures could be specific to each observer, an opinion already expressed by Halley in his seminal paper of 1717. Another English scientist, John Dalton, published in his *Meteorological Observations and Essays* (second edition, Dalton 1834) height estimates made on aurorae seen between May 1786 and May 1793 in two places approximately 75 km away in the NW–SE direction. The height of the arc during the aurora of February 15, 1793, was of the order of 250 km, with nevertheless a significant uncertainty (130–1,200 km). This uncertainty resulted from the relative proximity of the two sites, and from a possible error of 1–2° on a weak parallax. Thus, the proximity of the stations may seem to guarantee that it was indeed the same arc that was observed, but this was at the expense of the precision on the height, which was more than modest. Dalton wrote that he was finally convinced, after having been reluctant for a long time, like most of his English colleagues, of the great height of the aurora borealis. He noted the extreme diversity of existing estimates of the height of the aurora borealis:

> There are, at this day, some persons who hold the height of the aurora to be *one thousand miles*, others who hold that *one thousand feet* may be nearer the truth, and others who think that its height may, in particular instances, be the one or the other of these extremes, as well as all the intermediate heights (Dalton 1834, p. 227).

Different testimonies from the beginning of the 19th century, in particular from navigators, who placed the auroral light at the level of the clouds, indeed favored a very low height of the aurora. This hypothesis was notably relayed by scientists of the stature of François Arago or Alexander von Humboldt, who did not believe in Mairan's estimates, and rather took sides with Halley, who was the first to underline the intrinsic difficulty in estimating the height of aurorae. Humboldt notably took up the comparison with the rainbow, whose position is specific to the eye of the observer (Humboldt 1855, p. 219). As we can see, the history of the aurora borealis and its observation was eventful, and all the more so since the phenomenon remained mysterious for nearly a century and a half after the first hypotheses on its origin were made in the first half of the 18th century. It is remarkable that the main components of the phenomenon (solar, magnetic, electric) were identified in the few decades following the resumption of the aurora borealis in 1716, even if the knowledge at the time did not make it possible to make the link between them. Furthermore, the question of the height of the aurora played a decisive role in the evolution of the representation of our atmosphere in the first half of the 18th century, giving rise to new frameworks of thought which made it possible to better understand its vertical structuring, preparing the advances of the next two centuries.

Acknowledgments

I would like to thank Christophe Schmit for his proofreading of a first version of this book and his numerous critical pieces of advice that improved it very significantly. My thanks also go to Isabelle Lémonon-Waxin for her information on the question of gender and her critical proofreading of a first version of Chapter 5.

Appendix

About the Polar Lights, by Friedrich Christoph Mayer

English translation of Friedrich Christoph Mayer's memoir[1] by François Mottais[2].

i) Since the appearance of lights in our northern regions is so frequent that even the common people are no longer amazed by it, it was easy for me to collect in a short time a large number of observations, which I provide in this writing with my own thoughts about the origin of polar lights, thoughts that I have decided, as they are new, to submit to the public judgment of scientists so that I may learn from somebody else what these thoughts are worth.

ii) I have decided to treat my subject as follows: first, to give a brief description of each of the phenomena collected from all the observations I have had to date; then to draw some conclusions from these observations to try to unravel the secret of the polar lights using the principles of physical science to support my thesis.

iii) Because they appear in two different ways, two kinds of polar lights can be identified. The first kind is common and more frequent in our regions: shaped as an arc, it is located in the north, sends towards the top rays that look like light beams and shines with a quiet light. The second kind, which is rarer, turns towards the sky of which it occupies every region, is fragmented into different parts where most commonly fixed beams are present, which are either very short or non-existent. The splendor of this kind of polar lights, as is commonly thought, is that it waves or trembles. As will be discussed later, it will be clearly proven that the two kinds are different in themselves, but that the appearance is related only to the place.

iv) The following phenomena are linked to the first kind of polar lights:

---

1 Mayer 1726, n. 25.
2 From the THEMAM/ARSCAN team at Université Paris Ouest Nanterre la Défense.

1) a shining arc turned towards the north, set in such a way that its concave part looks towards the horizon and the convex part towards the top;

2) the upper part of the arc always occupies, for us, exactly north;

3) the legs of the arc usually touch the horizon;

4) the higher the arc, the more extended the legs;

5) never have I observed an arc higher than 40 degrees and legs wider than a half-circle;

6) the lower the arc, the more proportionate and regular its roundness; usually, the higher the arc, the more imperfect it usually becomes because of the cracks, gaps, prominences, curves, etc. that we can find in it. Sometimes, without its height getting affected, an arc that lacks harmony gradually sets back into a smooth shape and vice versa;

7) it often happens that newly created arcs are usually higher or lower than an older arc, but yet uniting with it. This phenomenon makes the arc appear to be higher or lower;

8) however, an arc sometimes becomes higher or lower without other observable arcs uniting with it;

9) the space between the arc and the horizon that is enclosed is always shrouded in darkness; however, it transmits the rays of the stars in a particularly radiant way, unless it is full of black clouds;

10) the inside edge of the arc is usually dark and the outside edge luminous, but it is impossible to distinguish the meeting point of light and darkness as they gradually merge into each other;

11) light beams and sometimes very broad luminous trails are set on the outside margin of the arc and they shine a bit less than the arc itself. But it happens very often that this part of the arc where the beams exit shines more than the rest;

12) all spread straight towards the top that they often attain or sometimes seem to pierce;

13) every beam has approximately the same width at its base and at its top;

14) every beam has approximately the same shine except for that often the ones closer to arc sometimes shine more than the ones closer to the top;

15) the duration of a beam is short and uncertain; it usually is about 10 seconds of time;

16) their appearance and disappearance are more often spontaneous, but sometimes they appear next to the arc before appearing next to the top and sometimes in the reverse order; thus, they disappear in a way or another;

17) sometimes, but rarely, they are not complete and seem to be in pieces, as if they are scattered;

18) usually they are set in motion eastwards or westwards; sometimes rather slowly and sometimes rather quickly; sometimes one beam is quicker than another in such a way that they follow one another or get ahead of one another in their progression. But meanwhile neither the straightness nor the impetus of the beams towards the top is changed;

19) they are sometimes present in a small number or even absent despite the presence of a luminous arc; sometimes they are present in such a huge number that the arc looks like a comb;

20) it often happens that many beams appear at the same time and, on the contrary, disappear at the same time;

21) stars appear everywhere, except at the dark inside margin of the arc, which, if present, will cover the stars;

22) it often happens that some clouds appear, higher than the ones that the polar lights leave behind them; the phenomenon becomes evident because of the lights and different movements of these clouds: it is indeed impossible that cloud observed in the same area of the sky are carried and illuminated in a different way, unless they are very distant from the Earth because of their height[3].

v) The following phenomena are linked to the second kind of polar lights:

1) At first sight, the sky seems to burst into flames, a trembling, radiant light appearing in the approximate same way that a trembling light does when it roams in a pile of hot coals; but this radiance is sometimes very visible, sometimes less, sometimes hardly.

2) Upon further inspection, we can note small shiny clouds scattered here and there. This observation is sometimes easy and sometimes hard, depending on whether these clouds are dense or thin and, similarly, on whether they burst into flames slowly or quickly.

3) As I happened to observe a small cloud of this kind that was linking to the top and at once bursting into flames and that, conversely, was soon dying out, I observed that this small cloud, immediately after, as it had become darker, was

---

[3] So he considers that colors and movements peculiar to these clouds are the sign that they are located higher, so further from the observer than the other. This inference is due to its reason.

going away and was, little by little, leaving under itself a cloud that was darkening and finally covering the first one.

4) In the same way, I later observed and understood that, regarding some other small clouds, meanwhile they burst into flames and die out; they do not change their location nor the shape or just a little bit.

5) Once the star bursts into flames, they stay this way for approximately the same time that the beams of the other kind of polar lights last (see §IV, no. 15).

6) The small clouds that are far from the top sometimes emit beams of light, but short and thin; the closer they are from the top, the shorter they are.

7) On the basis of other people' reports, I know that sometimes dark bolts are sprinkled with such radiances.

8) I rarely observed this kind of polar lights alone: almost always there were, linked to these polar lights, polar lights of the first kind, as previously described; one precedes or follows the other.

9) The light of these small clouds is more important than the one usually present in an arc.

10) This light is usually stronger on the side of the horizon than on the top, so that the horizon sometimes shines from everywhere, approximately the same way that polar lights do (see §IV).

vi) The following phenomena are related to the two kinds of polar lights:

1) The two kinds of polar lights usually appear between the autumnal equinox and the spring equinox: I indeed never saw any during the summer.

2) More often the clouds follow the light, being at first disposed in the shape of a shell, then growing up and accumulating little by little.

3) The two kinds of polar lights tend to usually appear in a clear or half-clear sky; yet I observed once polar lights in dense clouds covering up the skies in every part, except for some cracks here and there.

4) Their radiance is pale or whitish.

5) Their light usually goes, under the effect of a soft wind, towards any region of the sky that is, however, always the region towards which the wind, in the lower part of the air, blows. Up to now, never have I experienced any example contrary to this phenomenon.

6) I often observed that a mild weather precedes or goes along with the polar lights and that ice-cold weather follows them.

7) If it occurs, which is frequent, that polar lights appear for a certain number of nights in a row, the successive lights afterward become weaker until they are barely visible.

8) The duration of polar lights of both kinds is highly variable; it often lasts for the whole night, sometimes a few hours, depending on whether the air is calm or disturbed.

vii) Based on the analysis of the above-mentioned phenomena, I conclude that these two kinds differ only in appearance. Most of their characteristics are indeed common (see §VI) and, regarding those that we might think of as peculiar to one kind of polar lights, I will explain later that this is due solely to the diversity of places, the diversity which is itself observed (see §III, §IV, no. 1, 2 and 3, §V, no. 1).

viii) Regarding the light beams, I assert that they are generated by the reflection of the rays cast by a light matter that is not present at the place where we see the beams. All that we see, we see under the effect of direct refracted or reflected rays: if the beams were seen under the effect of direct rays, it would be necessary that their light matter spread on a vast space 1) be present for any observer on vertical plans (because of the observation mentioned in §IV, no. 12), but these plans, on other places of observation, are bent towards the horizon, so the matter would be linked in the same time to vertical plans and to inclined plans, which is impossible[4]. 2) It would be necessary that this matter, in the places far from the observer, be wider and in the closer places proportionally tighter: indeed, in different places, the widths could not appear equal in every place (see §IV, no. 13) and, regarding different observers, this variability of width should be variable at the same time (which is established by optical laws) and the matter should be disposed depending on the locations where the different observers are; however, this is impossible, so a vision of the beams by a direct line is not possible. If, instead of different locations, we group a certain number of phenomena observed by the same observer at different times, we must draw the same conclusion, because there is no reason why all the fires would always be disposed in a constant direction towards the same observer. If the vision was due to refraction, there should be a light matter beyond the region of the steams, causing the refraction, light matter which would sometimes burst into flames and sometimes die out (see §IV, no. 3 and 4), which would be troubled by the winds (see §VI, no. 5), which would leave behind her thick steams (see §V, no. 3; §6 no. 2) and which would sparkle (see §V, no. 7); and all of this does not seem to apply to this region. I add that the Sun, the moon and brighter stars should sometimes emit such beams by reflection, which yet never occur in this way and

---

4 Note that for the tilt to be significant, the polar lights must take place quite far to the north (or south), which implies that it takes place at great height so that it is visible despite the roundness of the Earth.

could never occur by virtue of the optical principles, or can be demonstrated by similar cases. The vision by refraction is thus also excluded and only the vision by reflection remains, which is yet to be proven.

ix) I speak about the location of the light matter, since there is in this region of the air a location where the clouds usually stop. Indeed, the dark, lower edge of the arc, to which the bright edge is always contiguous, has to be regarded as a cloud or as steam (see §IV, no. 10), mostly because it is troubled by the winds (see §IV, no. 5), because the light itself usually goes into the clouds (see §VI, no. 2) or rather appears in the form of a small cloud or of luminous steam (see §V, no. 2) and becomes thicker towards the horizon (in virtue of what is mentioned in §5, no. 1): it is obvious that this matter has all the characteristics of the clouds and thus occupies the same position and is equidistant from the Earth.

x) In order to be able to draw more conclusions, I will introduce a number of auxiliary principles from physics, the first of them being as follows: in a calm air, the clouds are suspended in such a way that their interior surface is flat and is in every place equidistant from the Earth. Indeed, since the peculiar weight of the air is the same for objects equidistant from the center of the Earth and that the clouds, formed by the same kind of steams, stop where their peculiar weight matches with the weight of the air, it is obvious that the clouds stop in the place that is in every part equidistant from the center of the Earth and that they spread in a flat shape which to our eyes is parallel to the surface of the Earth, and this is more accurate as the air is quieter.

xi) An experiment gives credence to this principle: it is established that the closer the clouds are to the horizon, the flatter their base is, which is always parallel to the horizon. This is because the flat and inferior part of the clouds, located high in the sky, is seen from one side and thus appears as a line.

xii) Moreover, it is established by reasoning and experiment that some clouds are suspended on top of others. This is because the flat part of one cloud is farther from the Earth than the flat part of another cloud: this can be clearly observed when polar lights appear (see §4, no. 22).

xiii) Here is another principle: the farther they are from the observer's eye, the closer to the horizon appear the objects that in the air have the same height; conversely, the closer they appear from the horizon, the farther the objects at the same height are from the observer's eye. These principles come from the optical science, in which they are carefully demonstrated.

xiv) From the previous principle, we can also understand the following: when a line is suspended in the air, set down from the top at its height, spread following the parallel of the globe and equidistant from the Earth in each part, it appears in the form of an arc, like the arc of polar lights (see §IV, no. 1, 2, 3 and 4). Conversely, when a line equidistant from the Earth appears in the form of an arc with its highest point pointing exactly towards the north, the line spreads in the air along a line parallel to the Earth. This is easily understood by experts in optical science, so that it is possible, without a doubt, to abstain from a demonstration.

xv) Those elements being set, I assert that the matter of the polar lights is located in the air following a line almost parallel to the surface of the Earth. Indeed, this matter is located in the cloud region (see §IX), so it is in every place equidistant from the Earth (see §10). Furthermore, it appears in the form of an arc, the highest point of which points towards the north (see §IV, no. 1, 2, 3 and 4), so it follows from §XIV that it occupies in the air a trail that spreads along a line parallel to the Earth. But I add that this line is almost parallel because the curve of the boreal arc is not always exact (see §IV, no. 6).

xvi) Furthermore, we know from the optical sciences that, as we think, the body having an irregular, inconstant and unequal shape has a shape that is more regular, constant and equal than when it is far from the eye. Therefore, we understand that the closer the boreal arc is to the horizon, the more exact it appears (see §IV, no. 6). Indeed, it is farther in this case (see §XIII).

xvii) If several bodies are scattered and disjointed in any way on a flat surface, they will also appear disjointed to the eye if it[5] is set perpendicular to this flat surface; but the more the eye looks at this flat surface obliquely, the closer the objects seem to group together, or rather they will end up looking like they form only one body. Similarly, the light or color of each object will increase and become brighter. I suppose again that these principles are totally sure in virtue of optical science.

xviii) The flat surfaces of the clouds that link to themselves next to the top are viewed perpendicularly, but those which get closer to the horizon are viewed from a distance and obliquely. This principle is easier to understood from §X and XI.

xix) Based on these elements, I make the next proposal about the boreal light: the matter of this light is scattered and disjointed in the small luminous clouds; but the first kind of light appears continuous (see §V[6] no. 1) and more radiant (see §V, no. 9), because in this case matter is seen from a distance and obliquely. In fact, since this matter is spread on a flat surface that is parallel to the Earth (see §IX and X), it appears on the top according to the way it is spread (as shown in §XVIII and XVII), but it appears disjointed (see §V, no. 2) and is therefore also disjointed. Hence, the matter of the two kinds of polar lights is scattered in the small luminous clouds. But the first kind always goes towards the horizon (see §IV, no. 5), to the point that it is seen obliquely (see §XVIII), so it is necessary that it appears continuous and more radiant (see §XVII). QED.

---

5 The subject should be the eye.
6 The number is erased; it should be V.

xx) This being said, we understand how the phenomena that distinguish the two kinds of polar lights should be explained.

xxi) Now, before explaining the phenomenon of the beams, I shall start by presenting two principles. The first one is as follows: if a luminous dot is set on a flat plan a little bit wavy and constant on its polished parts, it will irradiate over and it will reflect its beams towards the looking eye; then to the eye will appear a line or a light beam that will extend straight towards the eye from the luminous dot. According to the optical laws, this proposal can be demonstrated; but a common experimentation can also prove it: in fact, the moon or any luminous star irradiating the surface of a quiet water sends back to the eye a beam similar to the one I described.

xxii) On this proposal is based another one: what has to be the inferior part of the clouds or the steams, the luminous surface, in turn reflects towards the bottom, in the form of a luminous beam, the dots which light it from below. In fact, the flat and inferior part of the clouds is wavy and flat[7], and results from the polished water or ice particles (see §X). But I also learnt by experience the veracity of this proposal: indeed, at night, not so long ago, whereas a fire broke out, two or three versts[8] away from my home in Saint-Petersburg, woken up by the usual noise I ran to the window along with a friend in order to observe and as we soon lifted our eyes to the sky, we both saw a radiant white beam spreading from the fire area towards our eyes or towards the top of the sky, reflected by the clouds from which some snow was falling, and the beam at times was growing up and at times was shrinking, depending on the increase of the fire.

xxiii) This proposal is confirmed by another phenomenon that is not as rare in Saint-Petersburg as it is in Germany or France. It sometimes happens, that is a certain fact, that during winter the Sun, stopped as it rises or sets, projects, as it is commonly thought, a luminous shiny tail, in a vertical way towards the top, which is described in the commentaries of the French Academy of sciences (year 1703, p. 78); this tail is in all aspects similar to the one associated with the fire mentioned previously (see §XXII). This tail is issued from solar rays set in motion towards the top on the flat surface of the steams; the rays result from the ice particles and are reflected towards the observers' eyes, much like the luminous beams that are issued from the surface of water (see §XXII).

xxiv) Based on these proposals, I assert that the beams of the polar lights are issued from the reflection of the light that is present in the small luminous clouds (see §XIX), while these small clouds project towards the top beams on the flat

---

7 *Sic.*

8 A verst is about 0.65 mile.

surface of the very tenuous steams that overlook them, by which light is later reflected in the form of beams. The beams are created by reflection (see §VIII), so they are issued from the small luminous clouds (see §XIX). In fact, at that time, no other light is present in the air and the radiance of the beams lasts as long as the burning of small clouds (see §IV, no. 15 and §V, no. 5). The flat surfaces of steams, because of their capacity for reflection, are in top of the small clouds, which we understand from the phenomena mentioned in §IV, no. 4, 10, 11 and 12 as well as in §XII. Consequently, they form the beams in the way explained in §§XXI, XXII and XXIII. The phenomena mentioned in §IV, no. 11, 12, 13 and 14 and in §V, no. 6 are explained by this proposition.

xxv) If we also assume that these flat surfaces, as overlooking as the small clouds are luminous, are troubled by a soft wind, then we will have explanations regarding the phenomena described in §IV, no. 16, 17, 18, 19 and 20.

xxvi) In the same way, since there are steams on top of luminous matter that illuminates them but none below it, the way in which the phenomena are described in §IV, no. 9 becomes obvious.

xxvii) If a liquid contained in any vase is moved, in its superior part, towards a zone, it happens that the inferior part of the liquid, in the bottom, is moved towards the opposite zone. The veracity of this principle is easily checked by reasoning and experiment. Thus, it is easy to draw conclusions from this about the way in which the phenomenon mentioned previously in §VI, no. 5, works.

xxviii) The explanations I gave up to now have, I suppose, a degree of certitude; but there are still other phenomena whose explanations I do not promise to be equally certain. I will tell a few words about my opinion towards them, but I will beforehand recall a few auxiliary principles.

xxix) I assume that the fact that the air is from every part weighted down by flammable exhalations and very tenuous steams is established, because of the bolts, of the perpetual exhalations of the Earth (sulfur, niter, salts abundant everywhere), of the *capita mortua* that suck from the air vitriol, salts, etc.

xxx) If such flammable exhalations are gathered in a too northern space, they will create a flame and ignite by themselves. This proposal is proven by a large number of physical experiments known to everyone.

xxxi) As chemistry testifies, all the somewhat tenuous exhalations are gathered in less space because of the cold, some easily, others less easily. I do not speak here about this gathering of exhalations that condense them and bring them back to the form of a fluid, but about the gathering by which they only gather closer to each other alternatively.

xxxii) As the Sun moves back from the equator towards the south, the northern regions of the air end up becoming cold and they become colder the more northern they are and the more the Sun moves back. Moreover, the regions of the air of the same parallel or of the same climate are affected by an approximately identical coming of the cold. I say approximately because there are causes other than the Sun that compete with it, but the absence of the Sun is without a doubt the main cause of cold. This principle is manifest and needs no further proof.

xxxiii) I add to it a proposal that I dare not present as absolutely certain, but that I am compelled to present for the benefit of posterity. This proposal is as follows: the flammable exhalations, helping warmth, mix and merge with the watery steams, and are afterwards split up from them when cold comes, when the watery steams gather faster than the flammable exhalations. I now assert that when the Sun is in the boreal regions, or when warm winds blow, the air becomes warmer and the very tenuous watery steams are lifted into the air and mix with the flammable exhalations (see §XXXIII and §XXIX). When the Sun moves back towards the south, cold takes its place and is distributed evenly in the same climate (see §XXXII), and it gathers in small, very tenuous clouds the flammable exhalations which mix with the watery steams (see §XXI and §XIX). The flammable exhalations gradually leave these small clouds and gather on top of them (see §XXXIII). Once a sufficient amount of these exhalations has gathered up, it produces a flame and catches fire (see §XXX). A new gathering is formed by this ignition, it catches fire and the history repeats itself as many times as possible. I think that suits admirably to them what the very famous Professor Wolff said about them when he called matter of lightning the matter of the polar lights.

xxxiv) This being said, we easily understand how luminous matter is generated (see §XIX). Its positioning follows the evolution of the climate (see §XV) and there are the same number of repeated ignition and extinguishing of the small luminous clouds (see §V, no. 3 and 4). It is also possible that repeated bolts that occur during the summertime have the same origin.

xxxv) As the watery steams are gradually left behind by the flammable exhalations, they become heavier, stabilize and become visible. Since this is set and rely on what was said previously (see §XVII), we understand how the phenomena described in §IV, no. 1, §V, no. 3 and 4 and §VI, no. 2 work.

xxxvi) Thus, I was able to explain almost every phenomenon that usually occurs during the polar lights. This matter is too vast for me to exhaust it completely, so I was forced to say many things briefly in order to be concise. If the occasion and the interest of the matter permit, I will one day be able to say more about the distance of the luminous matter from the Earth. I can indeed determine it geometrically, given the height of the arc, the horizontal width of the legs and if my previously mentioned hypothesis is accepted (see §XV). In fact, if q is the curve of the sky's elevation, m is the curve of the arc at its maximal height, g is half the width of the legs, a is half the diameter of the Earth and *b* is the curve of the angle equal to the maximal height of the arc and to the height of the equator taken at the same time, the distance from the matter at the top of the arc to the observer will be of 2 amqqgg/r(rrbb-ggqq. I will explain the discovery of this law and its application to examples another time.

xxxvii) Finally, I enjoy adding that the polar lights were also known at times older than ours: there were men who reported a lot of prodigies, also listed comets and looked for their causes. Aristotle was among those who studied comets: he described them very clearly in the first book of the *Meteorology* (Chapters 4 and 5): he compares the light, which I call vertical, to a flame that appears in a field because of the burning thatch, and he calls the trembling flames goats, but calls the other type of arc *chasma*, using a term that was then used and taken from the interior dark space of the arc that precisely represents a chasm. Our beams are girders to him, and he says that they go up from the *chasma*. Pliny also describes the phenomena of the polar lights in the second book of his *Natural History* (Chapters 26 and 27). Those who considered this meteor an extraordinary and astonishing phenomenon are very numerous. I will add two old testimonies that my dear Bayer provided to me. This first is in the Chronicle of Edessa, slip 407 (which is in Giuseppe Simone Assemani's *Oriental Library*, Volume 1), with these words: "during the year 813 of the Seleucid era[9], a large amount of fire was seen in the northern part of the sky, the fire burnt all night." The other testimony is from Robert the Monk in the fifth books of his *Historia Hierosolymitana* in 1097, who says: "during this night, a comet was shining among other stars of the sky and was making rays of its own light and between the north and the south a redness of fire was shining in the sky." I add to it the testimony of Marcello Squarcialupi who, in an essay on comets, gave quite a good representation of their lights with the following description:

> The sky, during year 1575, during the harvest month, burnt all night, or seemed to burn. Whereas the moon stayed hidden, a great light appears towards Silesia and Sarmatia, just like at aurora. There was in the horizon an expanse that rose in a form of an arc, oblong; and from this luminous trail, coming from its highest end, great rays, sometimes here, sometimes there, were crossing it straightly. These rays stayed motionless for a long time in the shape of a cone or they disappeared and new ones suddenly appeared. And then, as I observed all this with amazement, gigantic flames were shining by coming out of the intervals between the rays, trembling and soon going towards the top of the sky, one after another, in a very quick race. In this place, some stayed for a while, but other soon vanished. I for my part saw, a bit before the middle of the night, some rays that were hasting from the two ends of the light towards the middle of the space [the zenith] and very violently rushed against each other in an opposed and adverse

---

9 AD 501/502.

way, like two armies of enemies. Then there were a shock and the mingling of the rays that were bending themselves, urging the adverse rays, sometimes taking flight, standing up and twisting themselves with an incredible variety of colors, purple, violet, yellow and red. During this celestial fight and struggle, as the oriental part bent, the occidental rays seemed to be stronger, more powerful and winner. Those who like to believe themselves said that there was to be soon a dreadful war between us and the Mohammedans etc.

I think that we have to relate other older and newer stories here that evoke armies, weapons, fights, etc. that have been seen in the sky.

References

Acta literaria sveciæ (1720–1724). *Acta Literaria Sveciæ, Upsaliae Publicata. Volumen Primum, Continesn, Annos 1720. 1721. 1722. 1723. & 1724*. Regole italiane di catalogazione: esempi [Online]. Available at: https://norme.iccu.sbn.it/index.php?title= Reicat:Acta_literaria_Sveciae.

An Historical Miscellany (1794). The Aurora Borealis, or Northern Lights. *An Historical Miscellany of Curiosities and Rarities in Nature and Art*, Volume the Third. Champante and Whitrow, London, 40–49.

Anonymous [probably Gerhard Friedrich Müller] (1753a). Lettre d'un Officier de la Marine russe à un seigneur de la cour concernant la carte des nouvelles découvertes au Nord de la Mer du Sud, & le Mémoire qui y sert d'explication publié par Mr De l'Isle. A Paris en 1752. Translated from the original Russian. *Nouvelle Bibliothèque Germanique, ou Histoire Littéraire de l'Allemagne, de la Suisse, et des Pays du Nord, par Mr Samuel Formey, Juillet, Août et Septembre 1753*. Pierre Mortier, Amsterdam, 46–87.

Anonymous (1753b). L'Épilogueur moderne a jugé nécessaire de publier ce qui suit touchant une carte de Mr Delisle. *L'Épilogueur moderne. Historique, Galant et Moral, Pour servir de suite au Patriote, et autres Précédents, Tome IX*. Isaac Buyn, Amsterdam, 160–161.

Appleby, J.H. (2001). Mapping Russia: Farquharson, Delisle and the royal society. *Notes and Records of the Royal Society of London*, 55(2), 191–204.

Arago, F. (1855). Chapitre VIII : Quels ont été les premiers observateurs des taches solaires ? *Astronomie Populaire*. Gide et J. Baudry, Paris, and T.O. Weigel, Leipzig.

Aristotle (2009). *Meteorology*. Translated by E.W. Webster [Online]. Available at: http://classics.mit.edu//Aristotle/meteorology.html.

Atlas Russien (1745). *Atlas russien : contenant une carte générale et dix-neuf cartes particulières de tout l'Empire de Russie et des pays limitrophes*. Typis Academiae Imperialis Scientarium, Petropoli.

Baldini, H. (2007). Dizionario Biografico degli Italiani, Volume 68 [Online]. Available at: https://www.treccani.it/enciclopedia/eustachio-manfredi_%28Dizionario-Biografico%29/.

Bartholmess, C. (1850). *Histoire Philosophique de l'Académie de Prusse*. Librairie de Marc Ducloux, Paris.

Bayuk, D. (2018). Father Antoine Gaubil, S.J. (1689–1759) and his election to the Saint Petersburg Academy of Sciences. *Conference on History of Mathematical Sciences: Portugal and East Asia V*, 331–350.

Becquerel, A.C. (1847). *Éléments de Physique terrestre et de Météorologie*. Firmin Didot Frères, Paris.

Benitez, M. (1986). Trois lettres de J.-N. Delisle au comte de Plélo. L'exploration de la Sibérie. *Dix-huitième siècle*, 18, Littératures françaises, 191–200.

Bernardi, G. (2016). *The Unforgotten Sisters, Female Astronomers and Scientists Before Caroline Herschel*. Springer International Publishing, Cham.

Bernier, F. (1684). *Abrégé de la philosophie de Gassendi*, volume 5. Anisson, Posuel & Rigaud, Lyon.

Bernoulli, J. (1701). Nouveau phosphore ; Lettre de M. Bernoulli, Professeur à Groningue, touchant son nouveau Phosphore. *MARS*[1] (1743), 1–9, 137–148.

Bernoulli, D.J. (1752). Nouveaux Principes de Méchanique et de Physique, tendans à expliquer la Nature & les Propriétés de l'Aiman. Pour concourir au prix de 1746. *Recueil des pièces qui ont remporté des prix de l'Académie royale des sciences. Pièces de 1744 & 1746*. Gabriel Martin, J.B. Coignard, Hippolyte-Louis Guerin, Jombert, Paris, 117–144.

Bertholon, P. (1787). *De l'Électricité des Météores*. Bernuset, Lyon.

Bertrand, J. (1869). *L'Académie des Sciences et les académiciens de 1666 à 1793*. J. Hetzel, Paris.

Bigourdan, G. (1895). *Inventaire général et sommaire des manuscrits de la Bibliothèque de l'Observatoire de Paris*. Observatoire de Paris, Paris.

Bigourdan, M.G. (1917). Lettres de Léonard Euler, en partie inédites. *Bulletin Astronomique, Série I*, 34, 258–319.

Bilfinger, G.B. (1725). *Dilucidationes Philosophicæ De Deo, Anima, et Mundo*. Tübingen.

Bilfinger, G.B. (2021). Georg Bernhard Bilfinger, Württembergischer Philosoph, Baumeister, Mathematiker und Theologe [Online]. Available at: https://de.wikipedia.org/wiki/Georg_Bernhard_Bilfinger.

Biot, J.-B. (1820). Considérations sur la nature et les causes de l'Aurore boréale. *Journal des Savants*, 342–354.

Birkeland, K. (1908). *The Norwegian Aurora Polaris Expedition, 1902–1903, Volume 1*. Longmans, London.

---

1 Abbreviation of *Mémoires de l'Académie Royale des Sciences*.

Bonno, G. (1948). La culture et la civilisation britanniques devant l'opinion française de la paix d'Utrecht aux Lettres philosophiques (1713–1734). *Transactions of the American Philosophical Society Held at Philadelphia for Promoting Useful Knowledge, New Series – Volume 38, Part 1*. The American Philosophical Society, Philadelphia.

Boschiero, L. (2004). The young and the restless: Scientific institutions in late 17th-century and early 18th-century Italy. *Working Papers, Columbia University* [Online]. Available at: https://academiccommons.columbia.edu/doi/10.7916/D8V98FDP.

Bradley, J. (1728). A letter from the Reverend Mr. James Bradley Savilian Professor of Astronomy at Oxford, and F.R.S. to Dr. Edmond Halley Astronom. Reg. &c. giving an account of a new discovered motion of the fix'd stars. *Phil. Trans. R. Soc. Lond.*[2], 35(406), 637–661.

Bradley, R.E. (2007). *Leonhard Euler: Life, Work and Legacy*. Elsevier, Amsterdam.

Brekke, A. (2018). Pioneers of electric currents in geospace. In *Electric Currents in Geospace and Beyond*, Keiling, A., Marghitu, O., Wheatland, M. (eds). American Geophysical Union, New York.

Brekke, A. and Egeland, A. (1983). *The Northern Light. From Mythology to Space Research*. Springer Verlag, Berlin.

Briggs Jr., J.M. (1967). Aurora and enlightenment eighteenth-century explanations of the aurora borealis. *Isis*, 58(4), 491–503.

Bruneau, O. and Passeron, I. (2015). Des lions et des étoiles : Dortous de Mairan, un physicien distingué. *Revue d'Histoire des Sciences*, 68(2), 259–279.

Brunet, P. (1970). *L'introduction des théories de Newton en France au xviiie siècle avant 1738*. Slatkine Reprints, Geneva.

Calendriers Saga (2022). Les calendriers/La réforme grégorienne : un jour ou l'autre [Online]. Available at: https://icalendrier.fr/calendriers-saga/etudes-thematiques/reforme-gregorienne.

Calinger, R. (1996). Leonhard Euler: The first St. Petersburg years (1727–1741). *Historia Mathematica*, 23, 121–166.

Canton, J. (1753). Electrical experiments, with an attempt to account for their several phænomena; Together with some observations of thunder-clouds. *Phil. Trans. R. Soc. Lond.*, 48, 350–358.

Cassini, J.-D. (1693). Découverte de la lumière céleste qui paraît dans le zodiaque. In *Recueil d'Observations faites en plusieurs voyages par ordre de Sa Majesté pour perfectionner l'astronomie et la géographie, avec divers Traités Astronomiques*. Imprimerie Royale, Paris.

Cassini, J.-D. (1708). Réflexions sur la comète qui a paru vers la fin de l'année 1707. *MARS* (1730), 89–101.

---

2 Abbreviation for *Philosophical Transactions of the Royal Society of London*.

Cassini, J. (1718). De la grandeur et de la figure de la Terre. *Suite des mémoires de l'Académie Royale des Sciences* (1720).

Cassini, J. (1733). Réflexions sur la hauteur du baromètre observée sur diverses montagnes. *MARS* (1735), 40–48.

Cassini, J. (1735). De la révolution du soleil et des planètes autour de leur axe. *MARS* (1738), 453–454.

Cassini, J. (1737). De la Comète qui a paru aux mois de Février de Mars et d'Avril de cette année 1737. *MARS* (1740), 170–182.

Castelnau, J. (1858). *Mémoire historique et biographique sur l'ancienne société royale des sciences de Montpellier*. Boehm, Montpellier.

Cavazza, M. (2002). The institute of science of Bologna and the Royal Society in the eighteenth century. *Notes and Records of the Royal Society of London*, 56(1), 3–25.

Cavendish, H. (1790). On the height of the luminous arch which was seen on Feb. 23, 1784. *Phil. Trans. R. Soc. Lond.*, 80, 101–105.

Celsius, A. (1735). Observations of the aurora borealis made in England by Andr. Celsius, F.R.S., Secr. R.S. of Upsal in Sweden. *Phil. Trans. R. Soc. Lond.*, 39(441), 241–244.

Chabin, M.-A. (1983). Les français et la Russie dans la première moitié du XVIIIème siècle. La famille Delisle et les milieux savants. Thesis for the archivist-paleogeographer diploma, École des Chartes, Paris.

Chabin, M.-A. (1996). Moscovie ou Russie ? Regard de Joseph-Nicolas Delisle et des savants français sur les États de Pierre le Grand. *Dix-huitième siècle*, L'Orient, 28, 43–56.

Chassefière, E. (2021a). *Physics of the Terrestrial Environment, Subtle Matter and Height of the Atmosphere: Conceptions of the Atmosphere and the Nature of Air in the Age of Enlightenment*. ISTE, London, and John Wiley & Sons, New York.

Chassefière, E. (2021b). Conceptions de l'atmosphère et nature de l'air au XVIIIe siècle : l'héritage cartésien. *Revue d'histoire des sciences*, 74(2), 407–439.

Chassefière, E. (2021c). Aurora borealis systems in the German-Russian world in the first half of the eighteenth century: The cases of Friedrich Christoph Mayer and Leonhard Euler. *Annals of Science*, 78(2), 162–196, DOI: 10.1080/00033790.2021.1891284.

Chassefière, E. (2021d). Obstacles encountered by four major European astronomical observatories belonging to academies in the eighteenth century. *Journal for the History of Astronomy*, 52(4), 414–444 [in press].

Chassefière, E. (2022a). L'aurore boréale enjeu du mécanisme cartésien et la dispute entre Paris et Montpellier, le choix français. *Almagest*, 13(1), 58–78.

Chassefière, E. (2022b). Le choix de Joseph-Nicolas Delisle face aux cartésiens de l'Académie des Sciences (1715–1725). *Dix-huitième siècle*, 54, 409–426 [forthcoming].

de Chaufepié, J.-G. (1753). *Nouveau Dictionnaire Historique et Critique pour servir de Supplément ou de Continuation au Dictionnaire Historique et Critique de Mr Pierre Bayle*. Z. Chatelain, Amsterdam, Pierre de Hondt, The Hague.

Clairaut.com (2022). Chronologie de la vie de Clairaut (1713–1765) [Online]. Available at: http://www.clairaut.com/.

Collection académique (1772). *Collection Académique, composée de l'histoire & des mémoires des plus célèbres académies & sociétés littéraires de l'Europe, Tome onzième de la partie étrangère, contenant les mémoires de l'académie des Sciences de Stockholm*. Panckoucke, Paris.

Commentarii de Bononiensi (1732). Commentarii de Bononiensi Scientarium et Artium Instituto Atque Academia. Natural History Museum Library, London [Online]. Available at: https://archive.org/details/commentariidebon1731unse/page/n5/mode/2up.

Commentarii de Petropolitanae (1728). Commentarii Academiae Scientarium Imperialis Petropolitanae. Biodiversity Heritage Library [Online]. Available at: https://www.biodiversitylibrary.org/item/113726#page/9/mode/1up.

Cornes, R. (2008). The barometer measurements of the royal society of London, 1774–1842. *Weather*, 63(8), 230–235.

Cornes, R. (2010). Early meteorological data from London and Paris: Extending the North Atlantic oscillation series. Thesis, University of East Anglia.

Correspondance de Joseph-Nicolas Delisle (2021). Correspondance de Joseph-Nicolas Delisle : inventaire détaillé, 1709–1767, B1/1-8, E1/13, B2/5. Bibliothèque numérique de l'Observatoire de Paris [Online]. Available at: https://bibnum.obspm.fr/.

Cotte, L. (1774). *Traité de météorologie*. Imprimerie Royale, Paris.

Crépel, P. and Schmit, C. (2017). *Autour de Descartes et Newton ; Le paysage scientifique lyonnais dabs le premier XVIIIe siècle*. Hermann, Paris.

D'Alembert, J.L.R. (1821). Éloge de La Motte. *Œuvres complètes tome III*. Belin, Paris.

D'Alembert, J.L.R. (1893). *Discours préliminaire de l'Encyclopédie*. Librairie de la Bibliothèque Nationale, Paris.

Dalton, J. (1834). *Meteorological Observations and Essays*, 2nd edition. Harrison and Crosfield, Manchester.

Dawson, N.-M. (2000). *L'Atelier Delisle, L'Amérique du Nord sur la table à dessin*. Éditions du Septentrion, Sillery, Quebec.

Debarbat, S. (1986). Newton, Halley et l'Observatoire de Paris. *Revue d'histoire des sciences*, 39(2), 127–154.

Debarbat, S. (1990). L'arc géodésique le plus long : Delisle, les Struve et l'Observatoire de Pulkovo. In *Inertial Coordinate System on the Sky*, Lieske, J.H. and Abalakin, V.K. (eds). International Astronomical Union/Union Astronomique Internationale. Springer, Dordrecht [Online]. Available at: https://doi.org/10.1007/978-94-009-0613-6_5, 25–28.

Delisle, J.-N. (1720). Nouvelles réflexions sur la figure de la Terre, par Joseph-Nicolas De l'Isle, communiquées à M. l'Abbé Bignon en avril 1720. Avec un petit essai fait dans le même temps sur la distance de la tour de Poquansi à Chartres, A7/7 (4). *Inventaire des archives de Joseph-Nicolas Delisle*, Observatoire de Paris.

Delisle, J.-N. (1728). *Discours lu dans l'Assemblée Publique de l'Académie des Sciences, le 2 mars 1728, par Mr De l'Isle, avec la réponse de Mr Bernoulli*. Imprimerie de l'Académie des Sciences, Saint-Petersburg.

Delisle, J.-N. (1734–1738). Lettres au comte de Maurepas et à Mr Bernoulli avec leurs réponses au Sujet de la mesure du parallèle pour déterminer la figure de la terre suivant ma méthode, 1734, 1735, 1737, 1738, A7/7 (7). In *Inventaire des archives de Joseph-Nicolas Delisle*, Observatoire de Paris.

Delisle, J.-N. (1735). Extract of a letter from Mr. Jos. Nic. De l'Isle, F.R.S. to John Machin, Secr. R.S. & Pr. Astr. Gresh. Dated Petersburg, 6/17 Feb. 1732/3, containing several literary communications concerning the construction of a quicksilver thermometer, and his observations on the eclipses of Jupiter's satellites, Annis 1731, and 1732. Translated from the French by Phil. Henry Zollman. F.R.S. *Phil. Trans. R. Soc. Lond.*, 39(441), 221–229.

Delisle, J.-N. (1737). Projet de Mesure de la Terre en Russie, lu dans l'Assemblée de l'Académie des Sciences de St Petersbourg, le 21 janvier 1737, à St Petersbourg, de l'Imprimerie de l'Académie des Sciences, 1737. *Bibliothèque Germanique, ou Histoire Littéraire de l'Allemagne, de la Suisse, et des Pays du Nord, Année MDCCXXXVII, Tome XXXIX.* Pierre Humbert, Amsterdam.

Delisle, J.-N. (1738). *Mémoires pour servir à l'histoire et au progrès de l'astronomie, de la géographie et de la physique*. Imprimerie de l'Académie des Sciences, St. Petersburg.

Delisle, J.-N. (1749). Observations du thermomètre, faites pendant les grands froids de la Sibérie. *MARS* (1753), 1–14.

Delisle, J.-N. (1753). *Nouvelles cartes des découvertes de l'Amiral de Fonte*. Paris.

Denina, A.C.G.M. (1798). De l'influence qu'a eue l'Académie de Berlin sur d'autres grands établissements de la même nature. *Mémoires de l'Académie Royale des Sciences et Belles Lettres Depuis l'Avènement de Frédéric Guillaume II au Trône*. George Decker, Berlin, 562–573.

Descartes, R. (1824). *Les météores, Œuvres de Descartes, Tome cinquième*. F.G. Levrault, Paris and Strasbourg.

Descartes, R. (1965). *Discourse on Method, Optics, Geometry, and Meteorology*. Translated by Paul J. Olscamp. The Bobbs-Merrill Company, Inc., Indianapolis.

Dictionnaire de physique (1793). *Dictionnaire de physique*. Monge, G., Cassini, J.-D., Bertholon, P.-N. (eds). Hôtel de Thou, Paris.

Ditisheim, P. (1925). L'Observatoire Royal de Greenwich. *Société Astronomique de France*, décembre 1925. John G. Wolbach Library, Harvard-Smithsonian Center for Astrophysics. Provided by the NASA Astrophysics Data System.

Dragoni, G. (1993). *Marsigli, Benedict XIV and the Bolognese Institute of Science. Renaissance and Revolution: Humanists, Scholars, Craftsmen and Natural Philosophers in Early Modern Europe.* Field, J.V., Cambridge University Press.

Drake, S. and O'Malley, C.D. (1960). *The Controversy on the Comets of 1618.* University of Pennsylvania Press, Philadelphia.

Dufay, C.F.C. (1730). Suite des observations sur l'aimant. *MARS* (1732), 142–157.

Dulac, G. (2008). Deux réseaux au service de l'académie des sciences de Saint-Pétersbourg: Autour de Ribeiro Sanches et de Johann-Albrecht Euler. *Dix-huitième siècle*, 40, 193–210.

Eather, R.H. (1980). *Majestic Lights: The Aurora in Science, History, and the Arts.* American Geophysical Union.

École, J. (1964). Cosmologie wolffienne et dynamique leibnizienne : essai sur les rapports de Wolff avec Leibniz. *Les études philosophiques*, janvier-mars 1964, Nouvelle Série, 19e Année, No. 1, Varia Germanica (janvier-mars 1964), 3–9.

Edleston, M.A.J. (1850). *Correspondence of Sir Isaac Newton and Professor Cotes, Including Letters from Other Eminent Men.* John W. Parker, John Deighton, London, Cambridge.

Ekman, M. (2016). The man behind "Degrees Celsius". A Pioneer in Investigating the Earth and its Change. The Summer Institute for Historical Geophysics. Åland Islands, Sweden.

Elena, A. (1991). "In lode della filosofessa di Bologna": An Introduction to Laura Bassi. *Isis*, 82(3), 510–518.

Encyclopédie de Diderot et d'Alembert (1751–1772). *Encyclopédie ou Dictionnaire raisonné des Sciences, des Arts et des Métiers, par une Société de Gens de Lettres, 28 t.* Diderot, D. and Le Rond D'Alembert, J. (eds). Briasson, David, Le Breton, Durand, Paris, puis Faulche et compagnie, Neufchastel.

Esteve, M.R.M. (2018). The circulation of scientific knowledge in Euler's first stage at Saint Petersburg Academy of Sciences. In *The Scientific Dialogue Linking America, Asia and Europe Between the 12th and the 20th Century*, F. D'Angelo (ed.). Associazione culturale Viaggiatori, Naples.

Euler, L. (1746). Recherches Physiques sur la cause de la queue des comètes, de la lumière boréale, et de la lumière zodiacale. *Histoire de l'Académie Royale des Sciences et des Belles-Lettres de Berlin*, 117–140.

Euler, L. (1802). *Letters of Euler on Different Subjects in Natural Philosophy. Addressed to a German Princess*, 2nd edition, Vol. 1. Murray and Highley, J. Cuthell, Vernor and Hood, Longman and Rees, Wynn and Scholey, G. Cawthorn, J. Harding and J. Mawman, London.

Findlen, P. (1993). Science as a career in enlightenment Italy: The strategies of Laura Bassi. *Isis*, 84(3), 441–469.

Findlen, P. (2016). The Pope and the Englishwoman: Benedict XIV, Jane Squire, the Bologna Academy, and the Problem of Longitude. *Benedict XIV and the Enlightenment: Art, Science, and Spirituality*, Messbarger, R., Johns, C.M.S., Gavitt, P. (eds). University of Toronto Press.

Fontenelle, B. (1699). Préface. *HARS*[3] (1732), i–xix.

Fontenelle, B. (1715). Sur l'éclipse solaire du III mai. *HARS* (1741), 47–58.

Fontenelle, B. (1716a). De par le roi. *HARS* (1718), 1–5.

Fontenelle, B. (1716b). Sur une lumière septentrionale. *HARS* (1718), 6–7.

Fontenelle, B. (1717). Sur une lumière horizontale. *HARS* (1719), 3–5.

Fontenelle, B. (1723). Sur les ombres des corps. *HARS* (1753), 90–101.

Fontenelle, B. (1726). Sur la lumière septentrionale. *HARS* (1753), 3–7.

Fontenelle, B. (1727). Éloge de M. Neuton. *HARS* (1729), 151–172.

Fontenelle, B. (1730). Sur la lumière septentrionale, et sur une autre lumière. *HARS* (1732), 6–9.

Fontenelle, B. (1732). Sur un Système de l'Aurore Boréale. *HARS* (1735), 1–21.

Fontenelle, B. (1733a). Sur l'Aurore Boréale. *HARS* (1735), 23–24.

Fontenelle, B. (1733b). Sur une nouvelle méthode pour les longitudes. *HARS* (1735), 79–81.

Fontenelle, B. (1735). Nouvelle idée [de A. Celsius] sur la mesure de la lumière. *HARS* (1735), 5–8.

Fontenelle, B. (1766a). *Œuvres de Monsieur de Fontenelle, des Académies, Française, des Sciences, des Belles-Lettres, de Londres, de Nancy, de Berlin & de Rome, Nouvelle édition, Tome onzième.* Regnard, Paris.

Fontenelle, B. (1766b). *Éloges des Académiciens de l'Académie Royale des Sciences, Morts depuis l'an 1699*, Nouvelle édition, Tome second. Libraires Associés, Paris.

Formey, S. (1750). SUÈDE, Stockholm. In *Nouvelle Bibliothèque Germanique, ou Histoire Littéraire de l'Allemagne, de la Suisse, et des Pays du Nord, par Mr Samuel Formey, Janvier, Février et Mars 1750, Tome sixième, Première partie*. Pierre Mortier, Amsterdam.

Formey, S. (1755). Mémoire sur la Vie & les Ouvrages de Mr Krafft. In *Nouvelle Bibliothèque Germanique, ou Histoire Littéraire de l'Allemagne, de la Suisse, & des Pays du Nord, Janvier, Février et mars 1755, Tome seizième, Première partie*. J. Schreuder, & Pierre Mortier le Jeune, Amsterdam.

Forster, G. (1777). Chap. IV. Run from the Cape to the Antarctic Circle; first season spent in high Southern Latitudes. – Arrival on the Coast of New Zealand. In *Voyage Round the World*, White, B., Robson, J., Elmsly, P., Robinson, G. (eds). London.

---

3 Abbreviation for *Histoire de l'Académie Royale des Sciences*.

Fouchy, J.-P.G. (1753). Éloge de M. Sloane. *HARS* (1757), 305–320.

Fouchy, J.-P.G. (1768). Éloge de M. de l'Isle. *HARS* (1770), 167–183.

Frängsmyr, T. (1975). Christian Wolff's mathematical method and its impact on the eighteenth century. *Journal of the History of Ideas*, 36(4), 653–668.

Frängsmyr, T. (1990). The mathematical philosophy. In *The Quantifying Spirit in the 18th Century*, Frängsmyr, T., Heilbron, J.L., Reider, R.E. (eds). University of California Press, Berkeley, Los Angeles, Oxford.

Franklin, B. (1779). Suppositions and conjectures towards forming and hypothesis, for the explanation of the Aurora Borealis. *Political Miscellaneous and Philosophical Pieces*. J. Johnson, London.

Fressoz, J.-B. and Locher, F. (2015). L'agir humain sur le climat et la naissance de la climatologie historique, XVIIe-XVIIIe siècles. *Revue d'histoire moderne & contemporaine*, 62(1), 48–78.

Frize, M. (2013). *Laura Bassi and Science in 18th Century Europe, The Extraordinary Life of Italy's Pioneering Female Professor*. Collection Points Sciences, Le Seuil, Paris.

Gapaillard, J. (1993). *Et pourtant elle tourne : le mouvement de la Terre*. Le Seuil, Paris.

Garcelon, P. (2017). Joseph Nicolas Delisle (1688–1768). *Astronomie – Grands Astronomes – Le grand siècle, site de vulgarisation sur le thème de l'astronomie* [Online]. Available at: https://pg-astro.fr/.

Glas, O. (1877). *Essai sur la Société Royale des Sciences d'Upsal et ses rapports avec l'Université d'Upsal*. Berling, Uppsala.

Godin, L. (1726). Sur le Météore qui a paru le 19 octobre de cette année. *HARS* (1753), 287–302.

Graham, G. (1724a). An account of observations made of the variation of the horizontal needle at London, in the latter Part of the Year 1722, and beginning of 1723. *Phil. Trans. R. Soc. Lond.*, 33(383), 96–107.

Graham, G. (1724b). Observations of the dipping needle, made at London, in the beginning of the year 1723. *Phil. Trans. R. Soc. Lond.*, 33(389), 332–339.

Graham, G., Halley, E., Celsius, A., Bevis, J., Milner, J. (1738). A collection of the observations made on the eclipse of the moon, on March 15, 1735–1736. Which were communicated to the Royal Society. *Phil. Trans. R. Soc. Lond.*, 40(445), 14–18.

Gros, M. (2010). European voyages around the Lapland Expedition (1736–1737). The circulation of science and technology. *Proceedings of the 4th International Conference of the European Society for the History of Science*, Barcelona, 18–20 November 2010, 322–326.

Guerlac, H. (1981). *Newton on the Continent*. Cornell University Press, Ithaca.

Hahn, R. (1975). Scientific research as an occupation in eighteenth-century Paris. *Minerva*, 13, 501–513.

Hall, M.B. (1982). The Royal Society and Italy 1667–1795. *Notes and Records of the Royal Society of London*, 37(1), 63–81.

Halley, E. (1687). A discourse concerning gravity, and its properties, wherein the descent of heavy bodies, and the motion of projects is briefly, but fully handled: Together with the solution of a problem of great use in gunnery. *Phil. Trans. R. Soc. Lond.*, 16(179), 3–21.

Halley, E. (1692). An account of the cause of the change of the variation of the magnetic needle. With an hypothesis of the structure of the internal parts of the earth: As it was proposed to the Royal Society in one of their late meetings. *Phil. Trans. R. Soc. Lond.*, 17(195), 563–578.

Halley, E. (1705). Astronomiæ cometicæ synopsis. *Phil. Trans. R. Soc. Lond.*, 24(297), 1882–1899.

Halley, E. (1717). An Account of the late Surprizing Appearance of the Lights seen in the Air, on the sixth of March last, with an Attempt to explain the Principal Phænomena thereof. *Phil. Trans. R. Soc. Lond.*, 29(347), 406–428.

Hansteen, C. (1827). On the Polar Lights, or Aurora Borealis and Australis. In *The Philosophical Magazine, or Annals of Chemistry, Mathematics, Astronomy, Natural History, and General Sciences, Vol. II, July–December 1827*. Richard Taylor, London.

Hartsoeker, N. (1707). De la nature et des propriétés de l'aimant. *Cours de physique, Livre troisième*, Chapter V, 191–226.

Hine, E.M. (1995). Dortous de Mairan and eighteenth century "System Theory". *Gesnerus*, 52, 54–65.

Hiorter, O.P. (1747). Om magnet-nålens åtskillige ändringar, som af framledne professoren herr And. Celsius blifvit i akt tagne och sedan vidare observerade, samt nu framgifne. *Kongl. Svenska Vetenskaps Academiens Handlingar*, 8, 27–43.

Histoire de l'Académie Royale de Berlin (1752). *Histoire de l'Académie Royale des Sciences et des Belles Lettres, depuis son origine jusqu'à présent*. Haude et Spener, Libraires de la Cour & de l'Académie Royale, Berlin.

Histoire de l'Académie Royale des Sciences (1699). *HARS* (1699). BnF Gallica [Online]. Available at: https://gallica.bnf.fr/ark:/12148/bpt6k35013#.

Histoire de la Société Royale des Sciences de Montpellier (1766). Où l'on expose ce qui s'est passé de plus considérable dans cette compagnie depuis son établissement jusqu'en 1717 ; Lettres-Patentes du Roi données au mois de Février 1706, portant établissement d'une Société Royale des Sciences à Montpellier; Statuts de la Société Royale des Sciences établie à Montpellier. *Histoire de la Société Royale des Sciences, établie à Montpellier, avec les mémoires de mathématique et de physique*. Benoit Duplain, Lyon, 1–41.

Histoire de la Société Royale des Sciences de Montpellier (1778). Sur deux Aurores Boréales observées en 1726 & 1730. *Histoire de la Société Royale des Sciences, établie à Montpellier, avec les mémoires de mathématique et de physique*. Jean-Martel Aîné, Montpellier, 4–23.

History of the Leopoldina (2021). "Never Idle" – The establishment of the Leopoldina [Online]. Available at: https://www.leopoldina.org/en/about-us/about-the-leopoldina/history/the-history-of-the-leopoldina/.

History of the Royal Observatory (2021). *History of the Royal Observatory. From Greenwich to Herstmonceux*. The Observatory Science Centre, Herstmonceux [Online]. Available at: https://www.the-observatory.org/rgohistory.

de Humboldt, A. (1855). *Cosmos. Essai d'une description physique du monde*. Gide et J. Baudry, Paris.

Huygens, C. (1690). *Discours de la cause de la pesanteur*. In *Traité de la lumière*, vander Aa, P., Leide.

Huygens, C. (1937). Le magnétisme (1680). *Œuvres complètes, Tome dix-neuvième, Mécanique théorique et physique, De 1666 à 1695*. Martinus Nijhoff, The Hague.

Isnard, A. (1915). Joseph-Nicolas Delisle, sa biographie et sa collection de cartes géographiques à la Bibliothèque Nationale. *Bulletin de la Section de Géographie*. Imprimerie Nationale, Paris.

Jaquel, R. (1976). L'astronome français Joseph-Nicolas Delisle (1688–1768) et Christfried Kirch (1694–1740) directeur de l'Observatoire de Berlin (1716, 1740). *Actes du 97ème congrès national des sociétés savantes, Nantes, 1972, Section des sciences*. Bibliothèque Nationale, Paris.

Jodra, S. (2004). La Russie au XVIIIe siècle. *Imago Mundi, Encyclopédie gratuite en ligne* [Online]. Available at: http://www.cosmovisions.com/ChronoRussie18.htm.

Joly, B. (2012). Étienne François Geoffroy, entre la Royal Society et l'Académie royale des sciences : ni Newton, ni Descartes, Methodos. *Savoirs et textes*, 12 [Online]. Available at: https://doi.org/10.4000/methodos.2855.

Journal des Savants (1748). Nouvelles littéraires, Russie, de S. Petersbourg. *Journal des Savants pour l'année M. DCC. XLVIII. Janvier*. Gabriel-François Quillau, Paris, 500–507.

Jurin, J. (1723). Invitatio ad Observationes Meteorologicas communi consilio instituendas. A Jacobo Jurin, M.D. Soc. Reg. Secr. & Colleg. Med. Lond. Socio. *Phil. Trans. R. Soc. Lond.*, 32(379), 422–427.

Klein, O. (2001). Un voyage scientifique au XVIIIème siècle : le voyage dans le nord de la Russie de Louis Delisle de la Croyère (1727–1730). Mémoire de Maîtrise d'Histoire, sous la direction de Mme le Professeur Marie-Noëlle Bourguet, Université Paris 7 – Denis Diderot, Paris.

Kœnigsfeld, T. (1768). Extrait d'un Voyage fait en 1740 à Beresow en Sibérie, aux dépens de la Cour Impériale, par M. Delisle, Doyen de l'Académie Royale des Sciences, alors Professeur d'Astronomie à l'Académie de Petersbourg, pour y observer le passage de Mercure sur le disque du Soleil, & du Journal de M. Koenigsfeld, qui l'accompagnait. In *Continuation de l'Histoire Générale des Voyages, Tome soixante-douzième*. Rozet, Paris.

Krafft, G.W. (1737). Observationum meteorologicarum ab anno 1726 usque in finem anni 1736 fractarum, comparatio. Praelecto prima. *Commentarii Academiae scientiarum imperialis Petropolitanae* (1744), 316–343.

Kragh, H. (2012). Newtonianism in the Scandinavian Countries, 1690–1790. *RePoSS: Research Publications on Science Studies 19*. Centre for Science Studies, University of Aarhus [Online]. Available at: http://www.css.au.dk/reposs.

de La Condamine, C.M. (1735). Manière de déterminer astronomiquement la différence en longitude de deux lieux peu éloignés l'un de l'autre. *HARS* (1738), 1–11.

de La Hire, P. (1713). Sur la hauteur de l'atmosphère. *MARS* (1739), 53–64.

Laboulais, I. (2006). Les systèmes : un enjeu épistémologique de la géographie des lumières. *Revue d'Histoire des Sciences*, 59, 97–125.

Lalande, J. (1790). *Voyage en Italie*, Tome second. Geneva.

Lalande, J. (1801). *Histoire céleste française, contenant les observations faites par plusieurs astronomes français*. Imprimerie de la République, Paris.

Lamoine, G. (1993). L'Europe de l'esprit ou la Royal Society de Londres. *Dix-huitième siècle*, 25, 167–198.

Lauridsen, P. (2014). *Vitus Bering: The Discoverer of Bering Strait*, S.C. Griggs & Company, Chicago [Online]. Available at: https://www.gutenberg.org/files/46032/46032-h/46032-h.htm.

Le Gars, S. (2015). Dortous de Mairan et la théorie des aurores polaires : trajectoire et circulation d'une idée, de 1733 à 1933. *Revue d'histoire des sciences*, 68, 311–333.

Le Monnier, P.C. (1741). *Histoire Céleste ou Recueil de toutes les Observations Astronomiques faites par Ordre du Roi*. Briasson, Paris.

Léger, L. (1919). L'académie des sciences de Petrograd du XIIIe au XXe siècle. *Journal des Savants (juillet-août 1919)*, 203–210.

Legrand, J.-P. and Le Goff, M. (1987). Louis Morin et les observations météorologiques sous Louis XIV. *La Vie des Sciences, Comptes rendus, série générale*, 4(3), 251–281.

Lémonon Waxin, I. (2019). La Savante des Lumières françaises, histoire d'une persona : pratiques, représentations, espaces et réseaux. PhD Thesis, École des Hautes Études en Sciences Sociales.

Leonov, V. (2005). *Libraries in Russia*. De Gruyter Saur, Munich.

L'Esprit des Journaux (1783). Vie de Samuel Klingenstierna, précepteur du roi de Suède, secrétaire d'état, &c extraite de la Bibliothèque Suédoise. *L'Esprit des Journaux, françois et étrangers, par une société de gens-de-lettres, décembre 1783, Tome XII, douzième année*. Valade, Paris, Jean-Jacques Tutot, Liège, 235–243.

de Limiers, H.P. (1723). *Histoire de l'Académie Appelée l'Institut des Sciences et des Arts, Établi à Boulogne [Bologne] en 1712*. Amsterdam.

Lisac, I., Martinovic, I., Vujnovic, V. (2011). Ruder Boskovic's insights on polar light, Published in the Treatise De aurora boreali in 1738. *The International Scientific Symposium, Philosophy of Ruder Josip Boskovic*, Faculty of Philosophy of the Society of Jesus, Zagreb, November 4.

Mailly, É. (1867). *Essai sur les Institutions Scientifiques de la Grande-Bretagne et de l'Irlande*. M. Hayez, Brussels.

de Mairan, J.-J.D. (1733). *Traité physique et historique de l'aurore boréale* (1731). Imprimerie Royale, Paris.

de Mairan, J.-J.D. (1738). Journal d'Observations des Aurores Boréales, Qui ont été vues à Paris ou aux environs, à Utrecht, & à Petersbourg, dans le cours de l'année 1734, Avec quelques Observations de la Lumière Zodiacale. *Suite des Mémoires de Mathématiques et de Physique de l'Académie Royale des Sciences de l'année M. DCCXXXIV*. Pierre Mortier, Amsterdam, 769–800.

de Mairan, J.-J.D. (1742). Éloge de M. l'abbé de Molières. *MARS* (1745), 195–205.

de Mairan, J.-J.D. (1747). *Éloges des Académiciens de l'Académie Royale des Sciences, morts dans les années 1741, 1742, & 1743*. Durand, Paris.

de Mairan, J.-J.D. (1749). *Dissertation sur la Glace, ou Explication Physique de la formation de la Glace, et de ses divers phénomènes*. Imprimerie Royale, Paris.

de Mairan, J.-J.D. (1754). *Traité physique et historique de l'aurore boréale*, 2nd edition. Imprimerie Royale, Paris.

de Mairan, J.-J.D. (1860). Lettres inédites [unpublished letters] de Mairan à Bouillet. *Bulletin de la Société Archéologique de Béziers (Hérault)*, volume 2. Imprimerie de Mme Veuve Millet, Béziers.

Maraldi, J.-P. (1716). Observations d'une lumière septentrionale. *MARS* (1718), 91–107.

Maraldi, J.-P. (1721). Observations de deux météores. *MARS* (1723), 231–245.

Maraldi, J.-P. (1723). Diverses expériences d'optique. *MARS* (1753), 111–142.

Maraldi, J.-P. (1726). Description de l'Aurore Boréale du 26 septembre, & de celle du 19 octobre. Observées au Château de Breuillepont, Village entre Pacy & Ivry, Diocèse d'Evreux. *MARS* (1753), 198–215.

Marchand, J. (1929). Le départ en mission de l'astronome J.-N. Delisle pour la Russie (1721–1726). *Revue d'Histoire Diplomatique*. Berger-Levrault, Paris, 373–396.

de Maupertuis, P.L.M. (1731). Problème astronomique. *HARS* (1764), 464–465.

de Maupertuis, P.L.M. (1738). *Relation du voyage fait par ordre du roi au cercle polaire, pour déterminer la figure de la Terre*. Paris.

Maury, L.-F.A. (1864). *L'ancienne Académie des Sciences*. Didier & Cie, Paris.

Mayer, F.C. (1726). De Luce Boreali. *Commentarii Academiae scientiarum imperialis Petropolitanae* (1728), 351–367.

Mayer, F.C. (1729). De Luce Boreali. *Commentarii Academiae Scientiarum Imperialis Petropolitanae* (1735), 121–129.

Mazauric, S. (2007). *Fontenelle : et l'invention de l'histoire des sciences à l'aube des Lumières.* Fayard, Paris.

Mervaud, M. (2009). Voltaire et l'Académie des sciences de Saint-Pétersbourg. *Revue des études slaves,* 80(4), 459–471.

Messbarger, R. (2002). *The Century of Women: Representations of Women in Eighteenth-Century Italian Public Discourse.* University of Toronto Press.

Miscellanea berolinensia (1710). Miscellanea berolinensia ad incrementum scientiarum. Hathi Trust Digital Library [Online]. Available at: https://babel.hathitrust.org/cgi/pt?id=mdp.39015039502011&view=1up&seq=1.

Mommertz, M. (2005). The invisible economy of science, a new approach to the history of gender and astronomy at the eighteenth-century Berlin Academy of Science. In *Men, Women and the Birthing of Modern Science,* Zinsser, J. (ed.). Northern Illinois University Press.

van Musschenbroek, P. (1731). Ephemerides meteorologicæ, barometricæ, thermometricæ, epidemicæ, magneticæ, ultrajectinæ, conscrita à Petro Van Muschenbroek, L.A.M. Med. & Phil. D. Phil. & Mathes Profess. in Acad. Ultraj. Anno 1729. *Phil. Trans. R. Soc. Lond.,* 37(425), 357–384.

van Musschenbroek, P. (1739). *Des météores ignés. Essai de Physique.* Samuel Luchtman, Leiden.

van Musschenbroek, P. (1741). Observations météorologiques, barométriques, thermométriques, épidémiques, & magnétiques, faites à Utrecht en 1730 & 1731. *Transactions Philosophiques de la Société Royale de Londres, Année M. DCC. XXXI.,* translated by M. de Brémond. Piget, Paris, 239–260.

Neveu, S. (2014). L'a priori, l'a posteriori, le pur et le non pur chez Christian Wolff et ses maîtres. Philosophy PhD Thesis, Université du Luxembourg.

Nevskaja, N.I. (1973). Joseph-Nicolas Delisle (1688–1768). *Revue d'histoire des sciences,* 26(4), 289–313.

Newton, I. (1722). *Traité d'Optique sur les réflexions, réfractions, inflexions, et les couleurs de la lumière, seconde édition française.* Montalant, Paris.

Newton, I. (1759). *Principes mathématiques de la philosophie naturelle, par Feue Madame la Marquise du Chastellet, Tome second.* Desaint & Saillant, Lambert, Paris.

Newton, I. (1999). *The Principia.* Translated by I.B. Cohen and A. Whitman. University of California Press, Berkeley.

Note bibliographique, A7/7 (6). *Inventaire des archives de Joseph-Nicolas Delisle,* Observatoire de Paris.

Omont, M.H. (1917). Lettres de J.-N. Delisle au Comte de Maurepas et à l'Abbé Bignon sur ses travaux géographiques en Russie (1726–1730). *Bulletin de la Section de Géographie*, Tome XXXII. Imprimerie Nationale, Paris.

Passeron, I. (2013). L'Académie des sciences et l'Observatoire de Paris sont-ils parisiens ? In *T. Belleguic et L. Turcot. Histoire de Paris. De l'âge classique à la modernité (XVIIe–XVIIIe siècles) Tome I*. Editions Hermann, Paris.

Philosophical Transactions (1665). *Phil. Trans. R. Soc. Lond., 1(1)*. The Royal Society Publishing [Online]. Available at: https://royalsocietypublishing-org.insu.bib.cnrs.fr/toc/rstl/1665/1/1.

Picard, J. (1736). Voyage d'Uranibourg. In *Ouvrages de mathématiques de M. Picard*, Mortier, P. (ed.). Amsterdam.

Pigeon, M. (2014). Review of the book: Frize, M. (2013). *Laura Bassi and Science in 18th Century Europe: The Extraordinary Life of Italy's Pioneering Female Professor.* Springer-Verlag, Berlin and Heidelberg. *Recherches féministes. Femmes extrêmes*, 27(1), 273–278.

de Plantade, F. (1706). *Discours prononcé à la première assemblée publique de la Société Royale des Sciences*. Jean Martel, Montpellier.

Poleni, G. (1731). Viri Celeberrimi Johannis Marchionis Poleni, R.S.S. ad virum Doctissimum Jacobum Jurinum, M.D.R.S.S. Epistola, qua continetur Summarium Observationum Meteorologicarum per Sexennium Patavij habitarum. *Phil. Trans. R. Soc. Lond.*, 37(421), 201–216.

Poleni, G. (1741). De M. le Marquis Jean Poleni de la Société Royale, à M. Jacques Jurin, Docteur en Médecine, de la Société Royale, contenant le précis (a) des Observations Météorologiques, qu'il a faites pendant six ans à Padoue. *Transactions Philosophiques de la Société Royale de Londres*, M. DDC. XXXI. Piget, Paris, 259–280.

Poleni, G. (1747). An Account of the red Lights seen Dec. 5, 1737, as observed at Padua. *The Philosophical Transactions (From the Year 1732, to the Year 1744) Abridged*, 529–532.

Portraits de médecins (2017). Anna Morandini Manzolini, 1716–1774, Médecin et anatomiste italienne, Professeur à l'université de Bologne. *Portraits de médecins* [Online]. Available at: https://www.medarus.org/Medecins/MedecinsTextes/morandini_anna.html.

Rabouin, D. (2009). *Mathesis Universalis : l'idée de "mathématique universelle" d'Aristote à Descartes*. Presses Universitaires de France, Paris.

de Ratte, E.-H. (1743). *Éloge de Mr de Plantade*. Assemblée Publique de la Société Royale des Sciences, tenue dans la grande salle de l'Hôtel de Ville de Montpellier, le 21 novembre 1743. Imprimerie Jean Martel, Montpellier.

Rey, A.-L. (2004). La dynamique de Leibniz : un autre visage de la science. *Les nouvelles d'Archimède*, Ahmed Djebbar, 36.

Rey, A.-L. (2013). Le leibnizo-newtonianisme : la construction d'une philosophie naturelle complexe dans la première moitié du 18e siècle. La méthode d'Émilie du Châtelet entre hypothèses et expériences. *Dix-huitième siècle*, 1(45), 115–129.

Rjeoutski, V. (2007). *La langue française en Russie au siècle des Lumières. Éléments pour une histoire sociale*, halshs-00273216.

Robertson, E. and O'Connor (2012). J. Eustachio Manfredi. *MacTutor History of Mathematics Archive* [Online]. Available at: https://mathshistory.st-andrews.ac.uk/Biographies/Manfredi/.

Schaffer, S. (2014). *La fabrique des sciences modernes*. Le Seuil, Paris.

Schiebinger, L. (1987). Maria winkelmann at the berlin academy: A turning point for women in Science. *Isis*, 78(2), 174–200.

Schlegel, K. and Silvermann, S. (2011). Johann Christian Heuson, a little-known auroral scholar of the early 18th century. *History of Geo- and Space Sciences*, 2, 89–95.

Schmit, C. (2015). Les dynamiques de Jean-Jacques Dortous de Mairan. *Revue d'Histoire des Sciences*, 68(2), 281–309.

Schmit, C. (2020). *La philosophie naturelle de Nicolas Malebranche au XVIIIe siècle. Inertie, causalité et théorie des petits tourbillons*. Classiques Garnier, Paris.

Schröder, W. (2005). Changes in the interpretation of aurora and the example of the night of march 17, 1716. *Acta Geodaetica et Geophysica Hungarica*, 40(1), 105–112.

Scienza a Due Voci (2004–2010). Manfredi Maddalena. *Scienza a Due Voci. Le donne nella scienza italiana dal Settecento al Novecento* [Online]. Available at: http://scienzaa2voci.unibo.it/biografie/1212-manfredi-maddalena.

Seguin, M.S. (2012). Fontenelle et l'Histoire de l'Académie Royale des Sciences. *Dix-huitième siècle*, 44, 365–379.

Shafranovskij, K.I. (1967). Les salles de l'Académie des sciences de Saint-Pétersbourg – en 1741. *Cahiers du monde russe et soviétique*, 8(4), 604–615.

Shank, J.B. (2008). *The Newton Wars and the Beginning of the French Enlightenment*. University of Chicago Press, Chicago.

Sigrist, R. (2013). Les communautés savantes européennes à la fin du siècle des Lumières. *M@ppemonde*, 110(2013.2) [Online]. Available at: http://mappemonde-archive.mgm.fr/num38/articles/art13204.html.

Sigrist, R. and Moutchnik, A. (2015). Entre Ciel et Terre : les fonctions de l'astronomie sans la Russie du 18ème siècle. *Almagest, International Journal for the History of Scientific Ideas*, 6/2, 85–124.

Siscoe, G.L. (1986). An historical footnote on the origin of "Aurora Borealis". *History of Geophysics*, 2, 11–14.

Stan, M. (2012). Newton and Wolff: The Leibnizian reaction to the Principia. 1716–1763. *The Southern Journal of Philosophy*, 50(3), 459–481.

Stempels, H.C. (2011). Anders Celsius' contributions to meridian arc measurements and the establishment of an astronomical observatory in Uppsala. *Baltic Astronomy*, 20, 179–185.

Sten, J.C.-E. (2014). *A Comet of the Enlightment, Anders Johann Lexel's Life and Discoveries*. Springer/Birkhaüser, Cham.

Struve, O. (1845). Esquisse historique de l'Observatoire de l'Académie des Sciences de Saint-Pétersbourg et des travaux astronomiques exécutés en Russie pour l'avancement de la Géographie du pays. *Description de l'Observatoire Astronomique Central de Poulkova*, Imprimerie de l'Académie Impériale des Sciences, St. Petersburg, 6–23.

Struve, O. (1847). Sur les manuscrits de Joseph de l'Isle conservés à l'Observatoire de Paris. *Recueil des Actes de la séance publique de l'Académie Impériale des Sciences de Saint-Pétersbourg tenue le 11 janvier 1847*, Leopold Voss, Leipzig.

van Swinden, J.H. (1780). Recherches sur les aiguilles aimantées, et sur leurs variations régulières, qui ont partagé le prix pour l'année 1777. *Mémoires de Mathématiques et de Physique*, Tome VIII, Moutard, Panckoucke, Paris.

van Swinden, J.H. (1785). Dissertation sur les mouvements irréguliers de l'aiguille aimantée. *Analogie de l'électricité et du magnétisme ou Recueil de Mémoires*. Veuve Duchesne, Paris.

Taton, R. (1957). *La science antique et médéviale, des origines à 1450*. Presses Universitaires de France, Paris.

The Description of an Aurora Borealis (1728). The Description of an Aurora Borealis mention'd in the foregoing Letter. *Phil. Trans. R. Soc. Lond.*, 35(399), 304–305.

The Prince of Cassano (1747). An Account of the red Lights seen Dec. 5, 1737, as observed at Naples. *The Philosophical Transactions (From the Year 1732, to the Year 1744) Abridged, and Disposed under General Heads, the Latin papers being translated in English, by John Martyn, F. R. S., Vol. VIII, Part II*. W. Innys, C. Itch, T. Astley, in Pater Noster Row, T. Woodward, C. Davis in Holbourn, and R. Manby and H. S. Cox on Ludgate-Hill, London, 527–529.

The Royal Observatory Greenwich (2021). The Royal Observatory Greenwich… where east meets west. Graham Dolan [Online]. Available at: http://www.royalobservatorygreenwich.org/articles.php?article=1053.

Transactions philosophiques (1760). *Transactions philosophiques de la Société Royale de Londres*, Année M. DCC. XLI. Briasson, David l'aîné, Le Breton. Durand, Paris.

Truesdell, C. (1984). *An Idiot's Fugitive Essays on Science, Methods, Criticism, Training, Circumstances*. Springer Verlag, Berlin.

Turner, A.J. (2002). The observatory and the quadrant in eighteenth-century Europe. *Journal for the History of Astronomy*, xxxiii, 373–385.

Vanzo, A. (2015). Christian Wolff and experimental philosophy. *Oxford Studies in Early Modern Philosophy*, 7, 225–255.

des Vignoles, A. (1721). Éloge de Madame Kirch à l'occasion de laquelle on parle de quelques autres Femmes & d'un Paysan Astronome. *Journal littéraire d'Allemagne, de Suisse et du Nord, Tome Troisième*. Pierre Humbert & Fils, Amsterdam, 157–186.

des Vignoles, A. (1741). Éloge de M. Kirch le fils, Astronome de Berlin. *Journal littéraire d'Allemagne, de Suisse et du Nord.* Isaac Beauregard, The Hague, 300–351.

Viik, T. (2012). Anders Celsius – mees, kellelt saime temperatuuriskaala. *Geodeet*, 42(2), 66–75, trad. de Suzanne Héral [Online]. Available at: http://viik.planet.ee/Celsius_FRENCH.pdf.

Wargentin, P.W. (1752). A Letter from the Secretary of the Royal Academy of Sciences in Sweden, to Cromwell Mortimer, M.D. and R.S. Sec. concerning the variation of the magnetic needle. *Phil. Trans. R. Soc. Lond.*, 47, 126–131.

Weidler, J.F. (1747). Auroræ Boreales observed at Wittenberg in 1733, in 1734. *The Philosophical Transactions (From the Year 1732, to the Year 1744) Abridged*, 550–551.

Weld, C.R. (1848). *A History of the Royal Society with Memoirs of the Presidents.* John W. Parker, London.

Werrett, S. (2010). The schumacher affair: Reconfiguring academic expertise across dynasties in eighteenth-century russia. *Osiris*, 25, 1, Expertise and the Early Modern State, 104–126.

Wolf, C. (1902). *Histoire de l'Observatoire de Paris de sa fondation à 1793.* Gauthier-Villars, Paris.

Wolf, A. (1938). *A History of Science Technology and Philosophy in the 18th Century.* George Allen & Unwin Ltd, London.

Wolff, C. (1716). *Gedanken über das ungewöhnliche Phænomenon, Lectione Publica.* Auf der Universität zu Halle erhöffnet, Halle.

Yevlashin, L.S., Starkov, G.V., Chernous, S.A. (1986). M.V. Lomonosov and the Study of Polar Aurorae. *Geomagnetism and Aeronomy*, 26(6), 749–752 [Online]. Available at: https://arxiv.org/ftp/arxiv/papers/1709/1709.08847.pdf.

von Zach, F.X. (1806). *Monatliche Correspondenz zur Beförderung der Erd- un Himmels-Kunde.* Verlage der Beckerschen Buchhandlung, Gotha.

von Zach, F.X. (1819). *Correspondance Astronomique, Géographique, Hydrographique et Statistique du Baron de Zach, Second Volume.* Bonaudo, Gênes.

von Zach, F.X. (1822). *Correspondance Astronomique, Géographique, Hydrographique et Statistique du Baron de Zach, Seventh Volume.* Bonaudo, Gênes.

Zanotti, E. (1747). An Account of the red Lights seen Dec. 5, 1737, as observed at the Observatory of the Institute of Bononia. *The Philosophical Transactions (From the Year 1732, to the Year 1744) Abridged*, 532–536.

Index

A, B

aberration of stars, 70, 154, 183
Académie de Béziers, 38, 49
Académie de Bordeaux, 40
Académie Royale des Sciences de Paris, 17, 23, 26, 27, 29, 30, 32, 37, 41, 200, 232, 255, 259, 266
  Histoire de, 1, 34–36, 40, 54, 146, 200, 210
  Mémoires de mathématique et de physique de, 1, 17, 35, 88, 90, 109, 169, 184, 203
Academy of Sciences of the Institute of Bologna, 25, 75, 231, 237, 247, 248, 253, 255, 262, 263, 266, 270, 276, 280, 282, 283, 287
Accademia degli Inquieti, 231, 247, 249, 259, 262, 280
Accademia dei Ricovrati, 283
Accademia del Cimento, 259, 280
Accademia delle Traccia, 259
Accademia Naturae Curiosorum, 260, 280
Acta literaria sveciæ, 181, 183, 215, 217
active forces, 39, 155, 157
Adelbulner, 177
Adodurov, 168
Æpinus, 168
Agensi, 253, 282, 286

Algarotti, 154, 249, 253, 267
Amontons, 201, 202, 204, 206
Areskin, 121
Aristotle, 138, 147
Arnelius, 10
Arnold, 238–240
astrology, 64, 241, 242, 255
astronomical household, 236, 237, 239, 245, 247, 249, 259, 275, 281
Astronomical Observatory of Bologna, 188, 235, 237, 247, 269, 270
Astronomical Observatory of Uppsala, 172, 178, 184, 191, 194, 220, 222
astronomy, 20, 35, 38, 50, 55, 57, 64, 66, 68, 77, 83, 85, 86, 89, 93, 96, 98, 100, 106, 111, 112, 116, 117, 120, 122, 126, 144, 147, 154, 164, 168, 169, 172, 173, 175, 177, 181, 183, 190, 214, 222, 225, 228, 231, 235, 236, 238, 240, 243, 245, 247, 249, 250, 255, 256, 262, 266, 269, 271, 280, 281
Atlas Russicus, 103, 112
atmosphere (*see also* solar), 4, 6, 7, 73, 115, 129, 139, 144, 148, 187
  height of the, 10, 21, 128, 144, 148, 174, 205
  lunar, 63, 72, 73
attraction at a distance, 36, 37, 157, 158, 175

aurora borealis, 1, 3–5, 7, 11, 12, 14, 19, 41, 45, 46, 52, 55, 57, 60, 91, 112, 115, 117, 124, 125, 129, 131, 132, 134, 135, 138, 144, 145, 171, 172, 176, 177, 179, 183–185, 188, 190, 192–194, 196, 198, 206, 210, 233, 235–237
 system, 17, 30, 39, 40, 115, 143, 145, 148, 184
Auzout, 26
Bacon, 35, 270
barometer, 10, 192, 203, 206, 207, 266
Bassi, 253, 255, 282, 283, 288
Beccari, 188, 267
Becquerel, 198
Benedict XIV, 253, 254, 267, 280, 283
Benzelius, 179, 182, 214, 216, 218, 219
Bering, 94, 193
Bernoulli D., 116, 132, 134, 153, 154, 158, 160
Bernoulli Je., 39, 48, 49, 159, 174
 & Ja., 27, 50, 272
Berzelius, 207
Bestoujev, 106
Bianchini, 178, 252
Bignon, 15, 17, 23, 27, 29, 32, 33, 35, 41, 52, 55, 63, 64, 79, 80, 89, 96, 255, 257, 258
Bilfinger, 49, 131, 133, 154, 158, 174
Biot, 145
Blumentrost, 64, 66, 82, 91–93, 123
Bode, 246
boiling point, 202, 207
Bon, 20
Bonde, 175, 217
Boscovich, 140, 188
Bouguer, 50, 159
Bouillet, 6, 37, 49
Bouillier, 31
Boyle, 43, 200
Bradley, 45, 70, 154, 179, 191
Braun, 168
British Library, 50
Bruce, 122

Buache, 95, 257
Burman, 173, 175, 182, 214, 215, 220, 221

C, D

Caldani, 280
calendar, 122, 132, 182, 201, 221, 239, 240, 242, 244, 245, 255, 259, 261, 272, 274–276, 278, 281
Camposanpiero, 284
Cantemir, 123
Carafa, 282
Cardinal de Fleury, 31, 38
Cartesianism, 6, 37, 97, 175
Cartesians, 37, 42, 48, 49, 53, 62, 71, 76, 89, 146, 154, 158, 159, 172, 176, 212
cartography, 90, 91, 93, 99, 111, 112, 118, 159, 192
Cassano, 187, 190
Cassini J., 11, 49, 63, 64, 66, 70, 72, 78, 82, 86, 87, 90, 159, 172, 180, 220, 228, 258
Cassini J.-D., 6, 20, 22, 27, 29, 38, 46, 48, 63, 249
 & J., 88
Cassini de Thury, 84
Castel, 30, 32, 36, 48–50
Castelnau, 17, 21
celestial mechanics, 49, 62, 68, 70, 132
Celsius A., 49, 56, 100, 101, 146, 171, 176, 179, 184, 189, 190, 192, 194, 195, 198, 202, 206, 207, 211, 215, 218, 220, 231, 232, 280
Celsius N., 173
censorship, 89
Châtelet, 176, 282
Chazelles, 65
Chouvalov, 105
Clairaut, 160
Clapiès, 20
climate, 10, 22, 23, 134, 226
clouds, 135, 147, 148, 199

Colbert C.-J., 23
Colbert J.-B., 26, 27
Collège Royal, 55, 65, 66, 70, 89, 94, 110, 112, 255, 257, 259
Colombière, 47
comet, 46, 48, 70, 71, 109, 115, 144, 182, 183, 222, 223, 241, 245, 269, 274
Commentarii Academiae scientiarum imperialis Petropolitanae, 60, 112, 132, 135, 140, 150, 158, 167, 168
Instituti Bononiensis, 267, 269
Conduitt, 36
Connaissance des temps, 255
Copernic, 152
Corazzi, 265
Cotes, 35, 47
Cotte, 57, 128, 171, 198, 199, 209, 210, 237
Couplet, 67
D'Alembert, 34, 160
Dacier, 284
Darbisse, 257–259
Delisle A., 236, 256–258, 282
Delisle C., 255
Delisle G., 62, 64, 91, 95, 192, 255, 258, 259
Delisle J.-N., 49, 50, 55, 72, 87, 92, 97, 115, 122, 131, 146, 159, 161, 164, 167, 171, 178, 180, 184, 186, 194, 202, 206, 209, 222, 228, 232, 242, 244, 255, 269, 278
Delisle S.-C., 255
Desaguliers, 50
Descartes, 1, 3, 9, 21, 26, 36, 37, 39, 48, 53, 128, 147, 175, 212, 248, 270
diffraction, 63, 71, 73, 75, 109, 269
Dimberg, 175
Doppelmayr, 56, 177, 192
Duhamel, 26

E, F

Earth, 3, 6, 9, 12, 21, 46, 49, 63, 87, 139, 143, 144, 151, 153, 154, 158, 174, 177, 187, 198
  shape of the, 29, 50, 62, 68, 70, 79, 80, 82, 89, 91, 97, 101, 109, 111, 158, 159, 172, 175, 178, 180, 182, 258
earthquakes, 3
eclipse, 20, 22, 49, 56, 57, 60, 62, 63, 69, 71, 75, 79, 90, 99, 101, 109, 176, 221, 222, 224, 266, 269
Eimmart, 66
Ekström, 183, 184, 191, 207, 223, 227
electric matter, 1, 3, 14, 19, 21, 52, 138, 190
electricity, 157, 171, 198, 199, 283
Ellicot, 179
Elvius, 173, 175, 214
empirism, 42, 128
ephemerides, 27, 232, 239, 243, 248, 250, 274
Erskine, 91
Esling, 177
espionage, 101, 104, 111, 112, 119, 167
ether, 3, 158
Euler, 98, 104, 108, 115, 132, 134, 143, 146, 148, 154, 159, 167, 169, 279
  sytem, 116
exhalations, 2, 3, 9, 11, 40, 52, 73, 115, 138
experimental
  philosophy, 41, 277, 278
  physics, 120, 146, 164, 175, 264, 266, 267
Farquharson, 91, 99, 104, 106, 122
Feuillée, 65, 72
Flamsteed, 43, 45, 50, 66, 184, 223
flying fires, 3, 45, 138
Folkes, 51, 177
Fonte, 95
Fontenelle, 1, 9, 16, 25, 29, 30, 32, 37, 38, 48, 73, 76, 82, 139, 154, 210, 251, 265, 283
Formey, 31
Forster, 44, 200

G, H

Galileo, 267, 270, 282
Gallie, 75, 89
Gascoigne, 66
Gassendi, 9, 26, 126
Gauteron, 21, 22, 25, 33
geodesy, 70, 123
Geoffroy C., 49
Geoffroy C.J., 62
Geoffroy É.F., 62
geography, 57, 94, 96, 108, 109, 118, 120, 123, 144, 159, 164, 168, 181, 216, 221, 256, 266, 276, 281
Gilbert, 3
Glas, 211, 219
Gmelin, 98, 107, 164
Godin, 10, 12, 50, 67, 172
Goldbach, 134, 150
Gouye, 29
Graham, 179, 190, 194, 195, 267, 270
Grammaticus, 70
Gravesande, 32, 47, 267
Grimaldi, 73, 75
Grishow, 112, 168
Guglielmini, 27, 248, 272, 280
Hadley, 179
Halley, 1, 2, 14, 21, 35, 43, 45, 46, 48, 49, 50, 52, 56, 70, 91, 116, 138, 174, 179, 190, 191, 198, 228
 system, 11, 19, 51, 54, 149
Hartsoeker, 3, 27, 50, 272
Heinsius, 104, 112, 147
heliocentrism, 116, 153, 173, 253, 267
Hermann, 132, 134
Heuson, 138
Hevelius, 46, 234, 239, 243
Hiorter, 56, 171, 181, 192, 194, 195, 198, 207, 222, 223, 226
Histoire Céleste/Celestial History, 55, 57, 65, 83, 86–88, 93, 112
Hobbes, 26
Hoffmann, 243, 272
Hooke, 66, 200
Horn, 216
Horrebow, 206
Hubin, 202
humidity, 63, 128, 200, 201
Huygens, 3, 26, 158
hygrometry, 199

I, J

ice, 40, 41
Imperial Academy of Sciences of St. Petersburg, 49, 60, 66, 71, 107, 115, 116, 119, 125, 131, 147, 149, 158, 168, 175, 202, 279
Imperial Observatory of St. Petersburg, 92, 100, 148, 222
infinitesimal calculus, 29
inflammation, 1, 3, 40, 138
Jablonski, 273, 274, 276
Jombert, 70
Journal des Savants, 47, 50, 164, 215, 257
Jupiter's satellites, 57, 60, 70, 79, 90, 112, 152, 222, 224, 269
Jurin, 199, 200

K, L

Keill, 35, 47, 50
Kepler, 49, 62, 115, 144
Kirch Christfried, 56, 66, 100, 176, 186, 205, 231, 240, 243, 244, 274, 278
Kirch, Christine, 231, 238, 240, 244, 276
Kirch G., 9, 233, 238, 241, 243, 272
Kirch M., 231, 240
Kirch sisters, 244, 245, 254
Kirilov, 110
Klingenstierna, 174, 220
Knutzen, 175
Kœnigsfeld, 101
Koopmann, 239
Korff, 162
Kourganov, 168

Krafft, 59, 98, 108, 125, 131, 133, 140, 144, 147
Krasilnikov, 99
Krosick, 243
L'Épilogueur, 95
La Chétardie, 107
La Condamine, 50, 80, 159
La Croyère, 94, 95, 134, 193
La Hire, 7, 27, 29, 55, 64, 65, 70, 73, 85, 88, 200, 202
La Lande, 67
Lacaille, 226
Lalande, 88, 112, 264
Lambertini, 253, 267, 282, 288
Laplace, 45
Le Cat, 31
Le Roy, 60
Leibniz, 116, 119, 124, 155, 233, 261, 270–277, 282
Lémery, 10
Lemonnier, 3, 84, 86, 179, 183, 184, 206
Leopoldina, 261
Leroy, 159
Lestocq, 106
Lieutaud, 255
Linnaeus, 177, 218
Locke, 128, 200
Lomonosov, 107, 161, 162, 164, 168
Louville, 9, 49, 55, 62, 63, 67, 69, 71, 73
Louvois, 27

M, N

Maffei, 280
magnet, 1, 3, 157, 194
magnetism, 3, 45, 52, 148, 157, 160, 171, 190, 192, 194, 198, 199
Mairan, 1, 5, 11, 25, 37, 42, 52, 57, 77, 111, 116, 134, 140, 142, 144, 177, 184, 187, 190, 197, 198, 235, 237
  system, 6, 11, 19, 53, 54, 146
Malebranche, 42, 47
Malézieu, 72

Malvasia, 260
Manfredi E., 177, 185, 231, 235, 247, 248, 259, 262, 264, 266, 267, 280
Manfredi G., 247, 248, 249
Manfredi, H., 249
Manfredi M., 232, 247–249
Manfredi sisters, 235, 236, 247, 252, 254, 281, 283, 288
Manfredi T., 232, 247–249
Manzolini, 282, 283, 288
Maraldi J.P., 4, 8, 22, 40, 63, 72, 77, 84, 96, 139, 147, 178, 200, 234
  & J.-D., 88
Maraldi J.-D., 200
Margraff, 65
Marinoni, 105, 205, 269
Mariotte, 26
Marsigli, 25, 248, 253, 262, 265, 266, 276, 280
Martello, 248
mathesis, 147
Maupertuis, 50, 81, 101, 140, 159, 172, 174, 179, 191, 277, 278, 280
Maurepas, 18, 31, 82, 87, 95, 256
Maury, 27, 29
Mayer, 59, 115, 125, 131, 132, 135, 139, 142, 144, 145, 147, 190
  system, 132, 147
mechanics, 35, 48, 54, 68, 70, 89, 132, 167, 176, 184, 214, 220, 222, 283
mechanistic, 37, 54, 76, 78, 97
Mémoires de Trévoux, 33, 47, 50, 82, 258
Mémoires littéraires de la Grande Bretagne, 50
Menchikov, 122
Mercure de France, 50, 257
Mercury, 21, 22, 38, 46, 57, 58, 67, 101, 109, 183, 202, 203, 224, 227, 239, 240, 270
meridian, 20, 22, 64, 74, 78–80, 82, 91, 92, 177, 193, 249, 264
Mersenne, 26

Messier, 81, 112, 209
meteorological observations, 20, 38, 128, 131, 184, 193, 199, 203, 205, 210, 221
meteorology, 38, 57, 128, 168, 199, 200, 236, 241
Miscellanea Berolinensia, 202, 241, 272
monads, 156, 158
Montanari, 260
Montmor L.H., 26
Montmort P.R., 49, 62
moon, 57, 63, 68, 70, 71, 75, 79, 90, 109, 132, 150, 152, 159, 160, 179, 184, 222, 224, 269, 270
Morgagni, 262, 267
Morin, 199, 209
Mortimer, 51, 179, 196
Müller, 95, 98, 107, 164
Muratori, 248, 267
Musschenbroek, 3, 138, 186, 193, 267
Nartov, 104, 162
navigation, 94, 118, 120, 122, 182, 196
needle, 57, 171, 184, 191, 192, 194, 196, 199
  declination of the, 52, 191
  inclination of the, 192, 194
network, 55, 65, 108, 110, 111, 168, 171, 237
Newton, 27, 35, 39, 42, 45, 48, 50, 53, 62, 68, 69, 74, 75, 79, 91, 160, 175, 253
Newtonianism, 29, 36, 47, 53, 97, 146, 154, 156, 175, 253, 267
Newtonians, 42, 53, 54, 62, 89, 100, 116, 117, 146, 148, 155, 172, 178
Nouvelle Bibliothèque Germanique, 95

O, P

Observatoire Royal de Paris, 22, 46, 62, 64, 66, 72, 89, 97, 172, 180, 220
Oldenburg, 43
optics, 35, 39, 62, 69, 77, 138, 175, 253
Orlov, 169
Orsi, 248, 267
Ostermann, 104
parallax, 6, 46, 132, 140, 153, 154, 270
Pascal, 26
passive force, 155
Pemberton, 50
Pembroke, 35, 50
Perrault, 26
Philosophical Transactions, 2, 35, 43, 46, 50, 60, 179, 199, 202
Picard, 22, 27, 46, 55, 79, 88, 173, 210
Pignatelli, 282, 288
Pillonière, 36
Piscopia, 284
planets, 39, 48, 56, 57, 59, 63, 70, 79, 100, 132, 152, 154, 157, 240, 242, 243, 262
Plantade, 1, 6, 13, 19, 20, 37, 52, 72, 190
Plélo, 99
Poleni, 177, 186, 187, 190, 200, 206, 210, 280
Polhammar, 214
Pontchartrain, 23, 27
Popov, 99, 168
Pound, 70
pressure, 21, 128, 206, 209
Printzen, 272, 273
Privat de Molières, 49, 50
Ptolemy, 46

Q, R

Querini, 280
rain, 27, 199, 209, 210
Ramus, 234
Rast, 70
Razoumovski, 106, 164
Réaumur, 32, 39, 40, 64, 77, 202
Régis, 21
Rheyle, 22
Roberval, 26
Rœmer, 27, 66, 234, 272
Rossi, 286

Royal Academy of Sciences of Prussia, 108, 115, 116, 119, 186, 218, 231, 233, 238, 241, 260, 270, 274, 277, 280, 281, 288
Royal Academy of Sweden, 172, 174, 180, 196, 207, 217, 224–226
Royal Observatory of Berlin, 56, 235, 242, 243, 278
Royal Observatory of Greenwich, 27, 43, 50, 91
Royal Society of London, 2, 27, 35, 42, 43, 45, 46, 49, 50, 60, 62, 70, 91, 119, 177, 196, 200, 221, 236, 259, 267
Royal Society of Uppsala, 172, 174, 179, 181, 211, 218, 223
Russia
  map of, 101, 103, 110, 111, 159, 256

S, T

Sanches, 106, 107
Sandri, 262
Saporta, 22
Sarrochi, 282
Saturn, 70, 152, 241
Saulmon, 49
Saurin, 29, 47, 158
Schmettau, 276, 280
Schubert, 169
Schumacher, 62, 92, 97, 100, 105, 116, 121, 161, 167
Scudery, 284
Sédileau, 55, 200
shooting stars, 3
Sisson, 267, 270
Sloane, 43, 49, 50, 179
snow, 200, 242
Société Royale des Sciences de Montpellier, 1, 13, 17, 20, 21, 24, 27, 33, 38, 52, 53
solar
  atmosphere, 6, 73, 144
  matter, 1, 6, 7, 139, 144, 190, 197

rays, 78, 145
Souciet, 82, 258
Spole, 173, 175
Stancari, 75, 247, 248, 262, 269
statistical, 134, 177, 185, 214
Störmer, 223, 226
subtle air, 1, 7, 144
subtle matter, 16, 37, 48, 54
sun, 6, 7, 20, 22, 46, 48, 57, 62, 64, 65, 67, 68, 71, 73, 101, 109, 132, 144, 152, 154, 160, 174, 176, 202, 221, 224, 239, 241, 255, 269
sunspots, 7, 20, 39, 274
Svedberg, 214, 219
system, 1, 6, 15–17, 41, 147
systematic spirit, 17
tail, 115, 144
Teinturier, 63
temperature, 101, 171, 183, 184, 199, 200, 202, 203, 242
Teplov, 164
Thalès, 3
thermometer, 38, 57, 101, 112, 171, 174, 184, 202, 203, 205, 207, 209, 221, 266
Tompion, 190
transit, 21, 46, 57, 71, 101, 109, 227, 270
Tschirnhaus, 27
Tycho Brahe, 46, 55, 65, 152, 173

U, V

Uppsala University, 172–174, 178, 182, 214, 216, 221
Vakhtang, 123
Vallerius, 175
Vallisneri, 284, 286
vapors, 1, 3, 9, 10, 12, 73, 143
Varignon, 29, 272
Vaugondy, 95
Venus, 46, 57, 63, 71, 72, 109, 224, 241, 270
Verzaglia, 248
Viereck, 276

Vignoles, 231, 233, 235, 241
Vignon, 94
Villemot, 47
Volpi, 284, 285
Vorontsov, 97, 105, 108, 113
vortices, 36, 37, 39, 42, 48, 49, 54, 90, 157, 158, 160, 175

W, Z

Wagner, 56, 66, 243, 245, 278
Wargentin, 56, 104, 108, 172, 183, 196, 208, 221, 222, 226
Weidler, 176, 189, 190
Weitbrecht, 98
Whiteside, 3
wind, 129, 199, 210, 242
Winkelmann, 126, 233, 234, 236, 238, 239, 241, 243, 274, 275
Wolff, 115, 121, 124, 126, 128, 143, 147, 148, 155, 174, 221
Wolffianism, 175
women
  exceptional, 247, 282, 286
  scientists, 253, 255, 286, 287
Wurzelbau, 56, 66, 83, 177, 239
Zach, 232
Zanotti E., 177, 183, 235, 267, 269, 270
Zanotti F.-M., 188, 249, 267, 280
Zanotti G., 249
zodiacal light, 6, 115, 145